住房和城乡建设部“十四五”规划教材
高等学校土木工程专业国际化人才培养全英文系列教材
河海大学留学生英文教材建设项目

Design of Concrete and Masonry Structure

混凝土与砌体结构设计

张　华　周广东　主编
周继凯　主审

中国建筑工业出版社
CHINA ARCHITECTURE & BUILDING PRESS

图书在版编目(CIP)数据

混凝土与砌体结构设计 = Design of concrete and masonry structure ：英文 / 张华，周广东主编. — 北京：中国建筑工业出版社，2022.12

住房和城乡建设部“十四五”规划教材 高等学校土木工程专业国际化人才培养全英文系列教材 河海大学留学生英文教材建设项目

ISBN 978-7-112-28114-5

Ⅰ. ①混… Ⅱ. ①张… ②周… Ⅲ. ①混凝土结构-高等学校-教材-英文②砌块结构-高等学校-教材-英文 Ⅳ. ①TU37②TU36

中国版本图书馆 CIP 数据核字(2022)第 202141 号

责任编辑：仕 帅 吉万旺
文字编辑：卜 煜
责任校对：李辰馨

住房和城乡建设部“十四五”规划教材
高等学校土木工程专业国际化人才培养全英文系列教材
河海大学留学生英文教材建设项目
Design of Concrete and Masonry Structure
混凝土与砌体结构设计
张 华 周广东 主编
周继凯 主审
*
中国建筑工业出版社出版、发行(北京海淀三里河路9号)
各地新华书店、建筑书店经销
北京科地亚盟排版公司制版
天津翔远印刷有限公司印刷
*
开本：787 毫米×1092 毫米 1/16 印张：13¼ 插页：1 字数：388 千字
2022 年 12 月第一版 2022 年 12 月第一次印刷
定价：**48.00** 元（赠教师课件）
ISBN 978-7-112-28114-5
(40192)

Brief Introduction

This book is written based on the training objectives of civil engineering major, the latest design codes and a large number of relevant textbooks and literature. The book is a textbook of design of concrete and masonry structure for civil engineering major in colleges and universities. The book contains four chapters, namely introduction, single-storey industrial building, multi-storey frame structure and masonry structure, respectively. In order to facilitate teachers' teaching and students' learning, each chapter sets up knowledge points, learning objectives and exercises, and adds engineering design examples.

This book is written for undergraduate students of civil engineering major in colleges and universities. The book has clear logic, fluent language, rich engineering cases and it is easy to read and understand. The book is not only used as a textbook of civil engineering and other related majors in ordinary colleges and universities, but also used as a reference for teachers and students of higher vocational colleges, adult colleges and universities, as well as engineering and technical personnel.

本书根据土木工程专业的培养目标，结合最新规范，以及参考大量相关教材和文献编写。本书为高等院校土木工程专业混凝土与砌体结构设计课程教材。全书共包含四章，分别为绪论、单层厂房结构、多层框架结构和砌体结构。为了便于教师教学和学生学习，每章设置知识要点、学习目标和思考练习，并增加设计实例和工程案例。

本书面向高等院校土木工程专业本科生编写，逻辑清晰，语言流畅，工程案例丰富，易读易懂。不仅可以作为普通高等院校土木工程专业及其他相关专业的教材和参考教材，也可供高职、高专、成人高校师生使用以及工程技术人员参考。

为了更好地支持教学，我社向采用本书作为教材的教师提供课件，有需要者可与出版社联系，邮箱 jckj@cabp. com. cn，电话（010）58337285。

出版说明

党和国家高度重视教材建设。2016 年，中办国办印发了《关于加强和改进新形势下大中小学教材建设的意见》，提出要健全国家教材制度。2019 年 12 月，教育部牵头制定了《普通高等学校教材管理办法》和《职业院校教材管理办法》，旨在全面加强党的领导，切实提高教材建设的科学化水平，打造精品教材。住房和城乡建设部历来重视土建类学科专业教材建设，从“九五”开始组织部级规划教材立项工作，经过近 30 年的不断建设，规划教材提升了住房和城乡建设行业教材质量和认可度，出版了一系列精品教材，有效促进了行业部门引导专业教育，推动了行业高质量发展。

为进一步加强高等教育、职业教育住房和城乡建设领域学科专业教材建设工作，提高住房和城乡建设行业人才培养质量，2020 年 12 月，住房和城乡建设部办公厅印发《关于申报高等教育职业教育住房和城乡建设领域学科专业“十四五”规划教材的通知》（建办人函〔2020〕656 号），开展了住房和城乡建设部“十四五”规划教材选题的申报工作。经过专家评审和部人事司审核，512 项选题列入住房和城乡建设领域学科专业“十四五”规划教材（简称规划教材）。2021 年 9 月，住房和城乡建设部印发了《高等教育职业教育住房和城乡建设领域学科专业“十四五”规划教材选题的通知》（建人函〔2021〕36 号）。为做好“十四五”规划教材的编写、审核、出版等工作，《通知》要求：（1）规划教材的编著者应依据《住房和城乡建设领域学科专业“十四五”规划教材申请书》（简称《申请书》）中的立项目标、申报依据、工作安排及进度，按时编写出高质量的教材；（2）规划教材编著者所在单位应履行《申请书》中的学校保证计划实施的主要条件，支持编著者按计划完成书稿编写工作；（3）高等学校土建类专业课程教材与教学资源专家委员会、全国住房和城乡建设职业教育教学指导委员会、住房和城乡建设部中等职业教育专业指导委员会应做好规划教材的指导、协调和审稿等工作，保证编写质量；（4）规划教材出版单位应积极配合，做好编辑、出版、发行等工作；（5）规划教材封面和书脊应标注“住房和城乡建设部‘十四五’规划教材”字样和统一标识；（6）规划教材应在“十四五”期间完成出版，逾期不能完成的，不再作为《住房和城乡建设领域学科专业“十四五”规划教材》。

住房和城乡建设领域学科专业“十四五”规划教材的特点：一是重点以修订教育部、住房和城乡建设部“十二五”“十三五”规划教材为主；二是严格按照专业标准规范要求编写，体现新发展理念；三是系列教材具有明显特点，满足不同层次和类型的学校专业教学要求；四是配备了数字资源，适应现代化教学的要求。规划教材的出版凝聚了作者、主审及编辑的心血，得到了有关院校、出版单位的大力支持，教材建设管理过程有严格保障。希望广大院校及各专业师生在选用、使用过程中，对规划教材的编写、出版质量进行反馈，以促进规划教材建设质量不断提高。

住房和城乡建设部“十四五”规划教材办公室

2021 年 11 月

Preface

In recent years, a large number of infrastructures that have attracted worldwide attention have been built and put into use in China. At the same time, China has also undertaken a large number of overseas construction projects, which has become one of the core industries of China's export and foreign exchange earnings. However, due to insufficient understanding of China's standard system and infrastructure level, most countries along the Belt and Road are unfamiliar with Chinese design norms and construction standards of concrete and masonry structures. Therefore, it is urgent to compile English textbooks for the course "Concrete and Masonry Structural Design" based on Chinese standards, and at the same time integrate Chinese culture and the achievements of Chinese infrastructure, so that international students in China can deeply understand China's structural specification system and infrastructure level and promote Chinese infrastructure technology around the world. This book is organized and compiled to meet the requirements of civil engineering teaching and personnel training under the new situation.

This book is written on the basis of ***Code for Design of Concrete Structures*** (GB 50010—2010) and ***Code for Design of Masonry Structures*** (GB 50003—2011). The book contains four chapters, namely introduction, single-storey industrial building, multi-storey frame structure and masonry structure, respectively. This book is written by Hohai University. Chapter 1 and Chapter 4 are written by Zhou Guangdong; Chapter 2 and Chapter 3 are written by Zhang Hua. In addition, Bai Lingyu, Yang Lei, Pan Luoyu and Jin Chuanjun also participate in the drawing and layout of the book. The book is co-revised and finalized by Zhang Hua and Zhou Guangdong. The authors would like to express their sincere gratitude for the reference to some textbooks, monographs and professional literature in the process of writing this book. The authors hope this book can provide help for readers to study and work, and the readers can criticize and correct the mistakes and inappropriate points in the book.

Authors

June 2022

前言

近年来，一大批举世瞩目的基础设施在我国建成并投入使用。同时，中国还承担了一大批海外建设项目，这已经成为我国出口创汇的核心产业之一。然而，绝大多数“一带一路”沿线国家仍然对中国规范体系和基建水平认识不足，对中国混凝土结构设计规范和施工标准不熟悉。因此，亟待以中国规范为准则，编写《混凝土与砌体结构设计》课程的英文教材，同时融入中国文化和中国基建的成就，让来华留学生深入了解中国的结构规范体系和基建水平，助推中国基建技术在全世界推广。本书正是为了适应新形势下土木工程专业教学和人才培养的要求而组织编写的。

本书以《混凝土结构设计规范》GB 50010—2010 和《砌体结构设计规范》GB 50003—2011 为基础编写。全书共包含四章，分别为绪论、单层厂房结构、多层框架结构和砌体结构。本书由河海大学编写，第 1 章、第 4 章由周广东编写；第 2 章、第 3 章由张华编写。此外，白凌宇、杨磊、潘罗宇、靳传军参与了本书画图和排版工作。全书由张华、周广东修改定稿。本书编写过程中参考了部分教材、专著和专业文献，在此表示诚挚的感谢。希望本书能为读者学习和工作提供帮助，书中错误和不当之处敬请读者批评指正。

作　者

2022 年 6 月

Contents

Chapter 1
Introduction

Prologue

Main points

1. Basic knowledge of structures.
2. The process of structural design.
3. Development of buildings in China.

Learning requirements

1. Know the definition and component of structures.
2. Know the task in each step of structural design.
3. Master the selection and layout of concrete structures with different purposes.
4. Know the development of concrete structures.

1.1 Overview of structure

A structure is defined as a system of interconnected members assembled in a stable configuration and used to support a load or combination of loads under the equilibrium of various external forces and internal reactions. This includes buildings, but the term "structure" can also be used to refer to any body of connected parts that are designed to bear loads, even if it is not intended to be occupied by people. Engineers sometimes refer to these as "non-building" structures. Typical examples include aqueducts, viaducts, bridges, canals, cooling towers, chimneys, dams, railways, roads, retaining walls and tunnels. The structure can be of many geometric shapes, patterns, sizes, materials and configurations. Only buildings are discussed in this book. It should be noted that this book does not strictly distinguish between the building and the structure, both of which refer to a structure with a roof and walls standing more or less permanently in one place, such as a house or factory.

According to the material of the structure, structures can be classified into reinforced concrete structures, steel structures, timber structures, masonry structures, composite material structures, etc. Specifically, the reinforced concrete structure refers to these structures whose main members are made of concrete and steel bars. Other types of structures, such as steel structures and masonry structures, can be similarly defined. This book only presents reinforced concrete structures and masonry structures.

A structure would bear various loads throughout its lifetime. In Chinese design codes, loads that act on structures are divided into three broad categories: dead loads, live loads and accidental loads. Dead loads, which are sometimes called permanent loads, are those constant in magnitude and fixed in location. Usually the major part of the dead load is the weight of the structure itself. This can be calculated with good accuracy from the design configuration, structural dimensions and structural density. Live loads are loads that can change in magnitude and position. They include occupancy loads, warehouse materials, rain loads, snow loads, wind loads, thermal loads, construction loads, overhead service cranes, equipment operating loads and many others. Their magnitude and distribution at any given time are uncertain, and even their maximum intensities throughout the lifetime of the structure are not known with precision. The minimum live loads for which the floors and roof of a building should be designed are usually specified in the ***Load Code for the Design of Building Structures*** (GB 50009—2012). Accidental loads are those that may not occur but have a very large amplitude once they occur, like seismic loads, blast loads and tsunami loads.

Seismic load is a kind of typical accidental load. Many areas of the world are in earthquake territory, and in those areas, it is necessary to consider seismic forces in design for all types of structures. Through the centuries, there have been catastrophic failures of buildings during earthquakes. It has been estimated that as many as 50,000 people lost their lives in the 1988 earthquake in Armenia. The 1989 Loma Prieta and 1994 Northridge earthquakes in California caused many billions of dollars of property damage as well as a considerable loss of life. The 2008 earthquake in Sichuan Province, China, caused 69,000 fatalities and another 18,000 missing. Recent earthquakes have clearly shown that these buildings not been designed for earthquake forces are seriously destroyed. Most structures can be economically designed and constructed to withstand the forces caused during most earthquakes. The cost of providing seismic resistance to existing structures (called retrofitting), however, can be extremely high.

To serve its purpose, a structure must satisfy three basic requirements: safety, serviceability and durability. Safety requires that the strength of the structure be adequate for all loads that may foreseeably act on it. Serviceability requires that deflections be adequately small; that cracks, if any, be kept to tolerable limits; that vibrations be minimized; etc. Durability requires that the steel will not be seriously corroded and the concrete and other materials will not be seriously carbonized and corroded within their lifetime.

There are two other considerations that a sensible designer ought to bear in mind: economy and aesthetics. One can always design a massive structure, which has more-than-adequate safety, serviceability and durability, but the ensuing cost of the structure may be exorbitant, and the end product, far from aesthetic. It is indeed a challenge, and a responsibility, for the structural designer to design a structure that is not only appropriate for the architecture, but also strikes the right balance between safety and economy.

1.2 Structural system

Any structure is made up of structural components (load-carrying, such as beams and columns) and non-structural components (such as partitions, false ceilings and doors). The structural components, put together, constitute the "structural system". Its function is to resist effectively the action of dead, live and accidental loads, and to transmit the resulting forces to the supporting ground, without significantly disturbing the geometry, integrity and serviceability of the structure.

Specifically, these load-carrying components include beams, slabs, roofs, columns, walls, stairs, foundations and other components. Other components are those not commonly used in ordinary structures, such as rods, arches and shells. According to the position of components, they can be divided into components of superstructure (like beams, slabs, roofs, columns, walls and stairs) and components of substructure (like foundations). According to the direction of components, they can be divided into vertical components (like columns, walls and foundations), horizontal components (like beams, roofs and slabs) and oblique components (like stairs).

For convenience, a structural system could be separated into three load-resisting subsystems, viz. the horizontal load-resisting subsystem (floor subsystem), the vertical load-resisting subsystem (framing subsystem) and the footing, although, in effect, these three subsystems are complementary and interactive, as shown in Fig. 1-1. The horizontal load-resisting subsystem consists of those components bearing vertical load effects, such as beams, slabs, roof, stairs, etc.; while the vertical load-resisting subsystem is comprised of those components bearing horizontal load effects, such as columns, walls, etc. The footing, which is usually placed below the surface of the ground, is those structural components used to support columns and walls and transmit their loads to the underlying soil or rock.

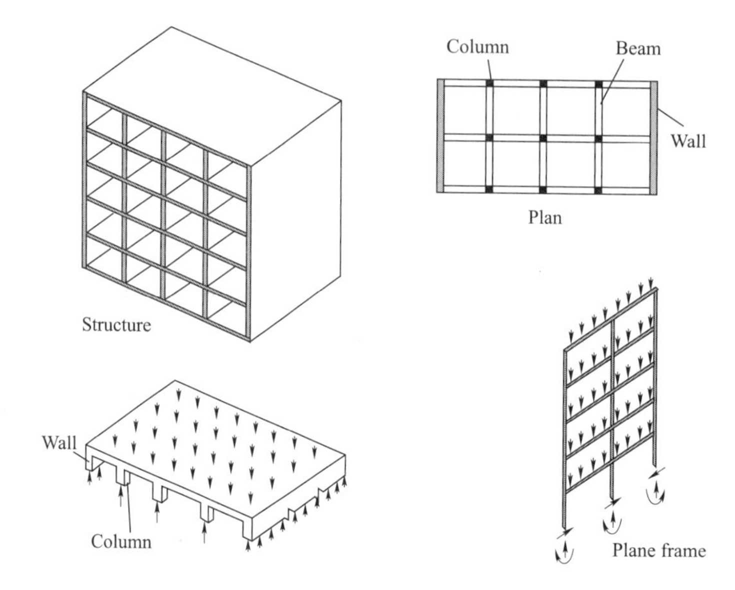

(a) Vertical load transmission

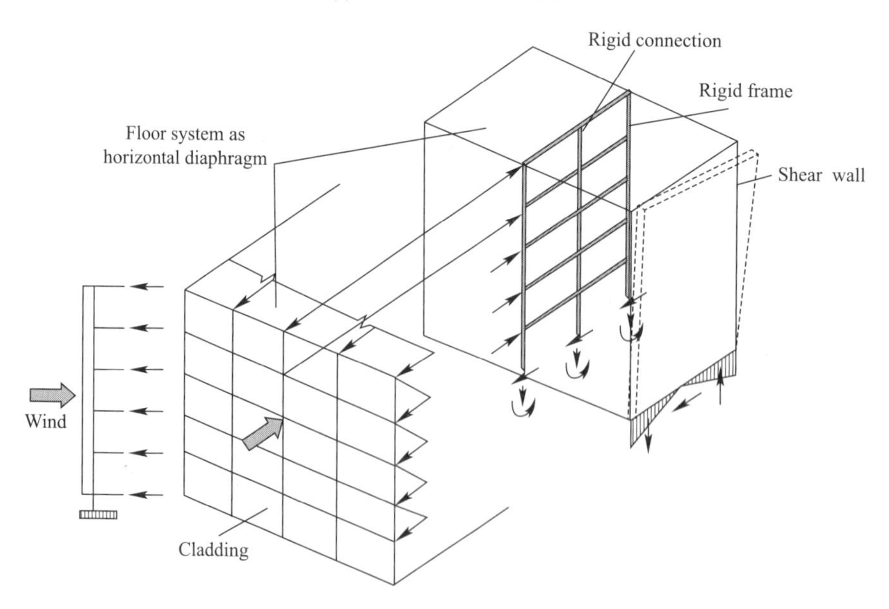

(b) Lateral load transmission rigid connection

Fig. 1-1 Load transmission mechanisms

1.2.1 Floor subsystem

The (horizontal) floor subsystem resists the gravity loads (dead loads and live loads) acting on it and transmits these to the vertical framing system. In this process, the floor subsystem is subjected primarily to flexure and transverse shear, whereas the vertical frame elements are generally subjected to axial compression, often coupled with flexure and shear (Fig. 1-1a). In cast-in-situ reinforced concrete construction, the floor subsystem is usually classified into four types, including the wall-supported slab subsystem, the beam-supported slab subsystem, the flat plate subsystem and the flat slab subsystem.

In the wall-supported slab subsystem, the floor slabs, generally 100—200 mm thick with spans ranging from 3 to 7.5 m, are supported on load-bearing walls (masonry), as shown in Fig. 1-2. This subsystem is mainly adopted in low-rise buildings. The slab panels are usually rectangular in shape.

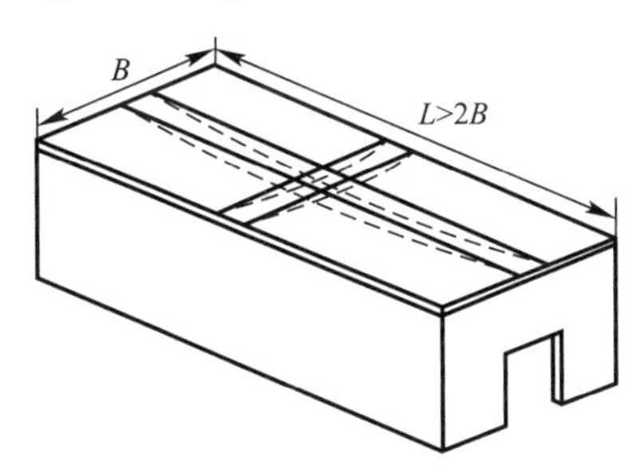

Fig. 1-2 Wall-supported slab subsystem

The beam-supported slab subsystem is similar to the wall-supported slab subsystem, except that the floor slabs are supported on beams (instead of walls). The beams are cast monolithically with the slabs in a grid pattern, as shown in Fig. 1-3, with spans ranging from 3 to 7.5 m. This subsystem is commonly adopted in high-rise building construction, and also in low-rise framed structures. The gravity loads acting on the slabs are transmitted to the columns through the network of beams.

In the flat plate subsystem, the floor slabs are supported directly on the columns, without the presence of stiffening beams, except at the periphery, as shown Fig. 1-4. It has a uniform thickness of about 125—250 mm for spans of 4.5—6 m. Its load carrying capacity is restricted by the limited shear strength and hogging moment capacity at the column supports. Because it is relatively thin and has a flat undersurface, it is called a flat plate, and certainly has much architectural appeal.

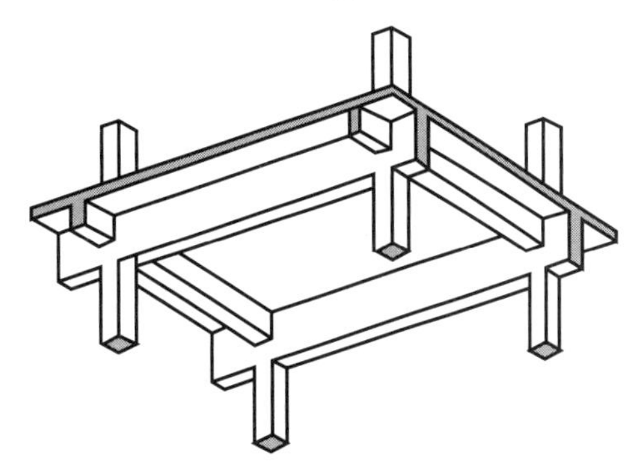

Fig. 1-3 Beam-supported slab subsystem

The flat slab subsystem is adopted in some office buildings, as shown in Fig. 1-5. The flat slabs are plates that are stiffened near the column supports by means of "drop panels" and/or "column capitals" (which are generally concealed under "drop ceilings"). Compared to the flat plate subsystem, the flat slab subsystem is suitable for higher loads and larger spans, because of its enhanced capacity in resisting shear and hogging moments near the supports. The slab thickness varies from 125 to 300 mm for spans of 4—9 m.

1.2.2 Framing subsystem

The vertical framing subsystem resists the gravity loads and lateral loads from the floor subsystem and transmits these effects to the foundation and ground below. The framing subsystem is made up of a three-dimensional framework of beams and columns. For convenience, we may divide the framsversal into separate plane frames in the transversal and longitudinal directions of the building. In cast-in-situ reinforced concrete structures, the vertical framing subsystem is usually classified into columns and walls.

Columns are skeletal structural elements, whose cross-sectional shapes may be rectangular, square, circular, L-shaped, etc. — often as specified by the architect. The size of the column section is dictated, from a structural viewpoint, by its height and the loads acting on it — which, in turn, depend on the type of floor system, spacing of columns, number of storeys, etc. The column is generally designed to resist axial compression combined with (biaxial) bending moments that are induced by "frame action" under gravity and lateral loads. These load effects are more pronounced in the lower storeys of tall buildings; hence, high strength concrete (up to 50 MPa) with high reinforcement area (up to 6% of the concrete area) is frequently adopted in such cases to minimize the column size.

Walls are vertical elements, made of masonry or reinforced concrete. They are called bearing walls if their main structural function is to support gravity loads, and are referred to shear walls if they are mainly required to resist lateral loads due to wind and earthquake. The thickness of reinforced concrete bearing walls varies from 125 to 200 mm; however, shear walls may be considerably thicker in the lower storeys of tall buildings. The walls around the lift cores of a building often serve as shear walls.

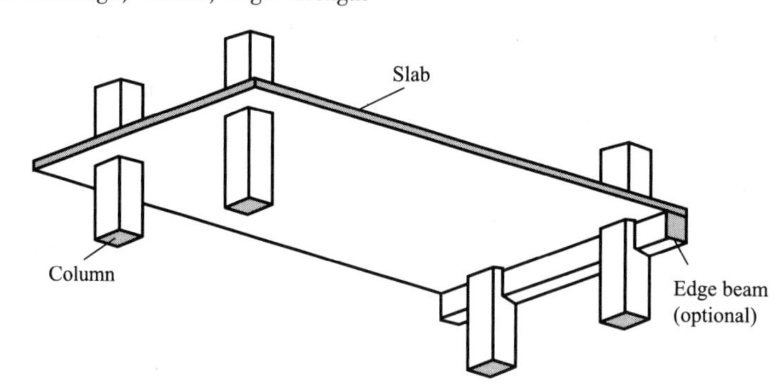

Fig. 1-4 Flat plate subsystem

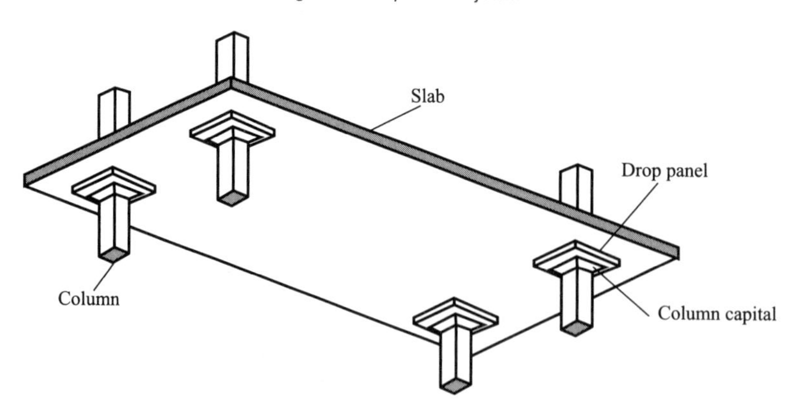

Fig. 1-5 Flat slab subsystem

1.2.3 Footings

The purpose of the footing is to effectively support the superstructure by transmitting the applied load effects (reactions in the form of vertical and horizontal forces and moments) to the soil below, without exceeding the "safe bearing capacity" of the soil. At the same time, the footing should ensure that the settlement of the structure is within tolerable limits, and as nearly uniform as possible. Common footings include isolated footings, combined footings and wall footings.

For ordinary structures located on reasonably firm soil, it usually suffices to provide a separate footing for every column. Such a footing is called an isolated footing. It is generally square or rectangular in plan; other shapes are resorted to under special circumstances. The footing basically comprises a thick slab which may be flat (of uniform thickness), stepped or sloped (on the upper surface), as shown in Fig. 1-6(a).

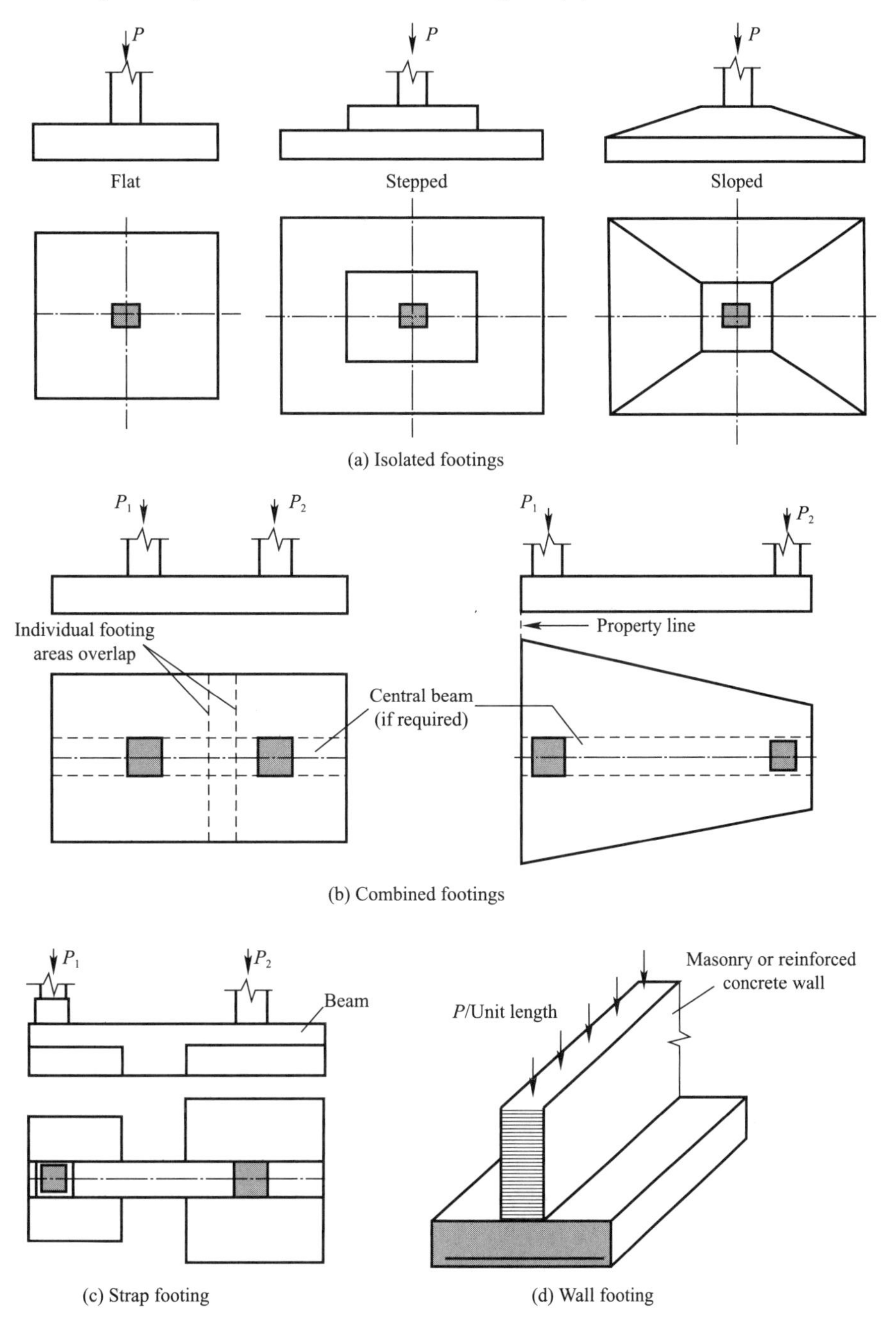

Fig. 1-6 Types of footings

In some cases, it may be inconvenient to provide separate isolated footings for columns (or walls) on account of inadequate areas available in plan. This may occur when two or more columns (or walls) are located close to each other and/or if they are relatively heavily loaded and/or rest on soil with low safe bearing capacity, resulting in an overlap of areas if isolated footings are attempted. In such cases, it is advantageous to provide a single combined footing.

For the columns, as shown in Fig. 1-6 (b), the term "combined footing" is often used when two columns are supported by a common footing, the term "continuous strip footing" is used if the columns (three or more in number) are aligned in one direction alone, and the term "raft foundation" ("mat foundation") is used when there is a grid of multiple columns. The combining of footings contributes to improved integral behavior of the structure. An alternative to the conventional combined footing is the strap footing, in which the columns are supported essentially on isolated footings, but interconnected with a beam, as shown in Fig. 1-6(c).

Reinforced concrete footings are required to support reinforced concrete walls, and are also sometimes employed to support load-bearing masonry walls. Wall footings distribute the load from the wall to a wider area, and are continuous throughout the length of the wall, as shown in Fig. 1-6(d).

1.2.4 Lateral load resisting subsystem

The horizontal and vertical subsystems of a structural system interact and jointly resist both gravity loads and lateral loads. Lateral load effects (due to wind and earthquake) predominate in tall buildings, and govern the selection of the structural system. Lateral load resisting subsystem of buildings generally consists of frames, shear walls and tubes.

Frames are generally composed of columns and beams, as shown in Fig. 1-1 and Fig. 1-7. Their ability to resist lateral loads is entirely due to the rigidities of the beam-column connections and the moment-resisting capacities of the individual members. They are often (albeit mistakenly) called "rigid frames", because the ends of the various members framing into a joint are "rigidly" connected in such a way as to ensure that they all undergo the same rotation under the action of loads. In the case of the "flat plate" or "flat slab" subsystem, a certain width of the slab, near the column and along the column line, takes the place of the beam in "frame action". Frames are used as the sole lateral load resisting subsystem in buildings with up to 15 to 20 storeys.

Shear walls are solid walls made of reinforced concrete or masonry, which usually extend over the full height of the building. They are commonly located at the lift/staircase core regions. Shear walls are also frequently placed along the transverse direction of a building, either as exterior (facade) walls or as interior walls. The walls are very stiff, having considerable depth in the direction of lateral loads; they resist loads by bending like vertical cantilevers, fixed at the base. The various walls and co-existing frames in a building are linked at the different floor levels by means of the floor subsystem, which distributes the lateral loads to these different systems appropriately. The interaction between the shear walls and the frames is structurally advantageous in that the walls restrain the frame deformations in the lower storeys, while the frames restrain the wall deformations in the upper storeys. Frame-shear wall systems are generally considered in buildings up to about 40 storeys.

Tubes are subsystems in which closely-spaced columns are located along the periphery of a building. Deep spandrel beams located on the exterior surface of the building interconnect these columns. The entire system behaves like a perforated box or framed tube with a high flexural rigidity against lateral loads. When the (outer) framed tube is combined with an "inner tube" (or a central shear core), the system is called a tube-in-tube. When the sectional plan of the building comprises several perforated tubular cells, the system is called a bundled tube or "multi-cell framed tube". Tubular systems are effective up to 80 storeys.

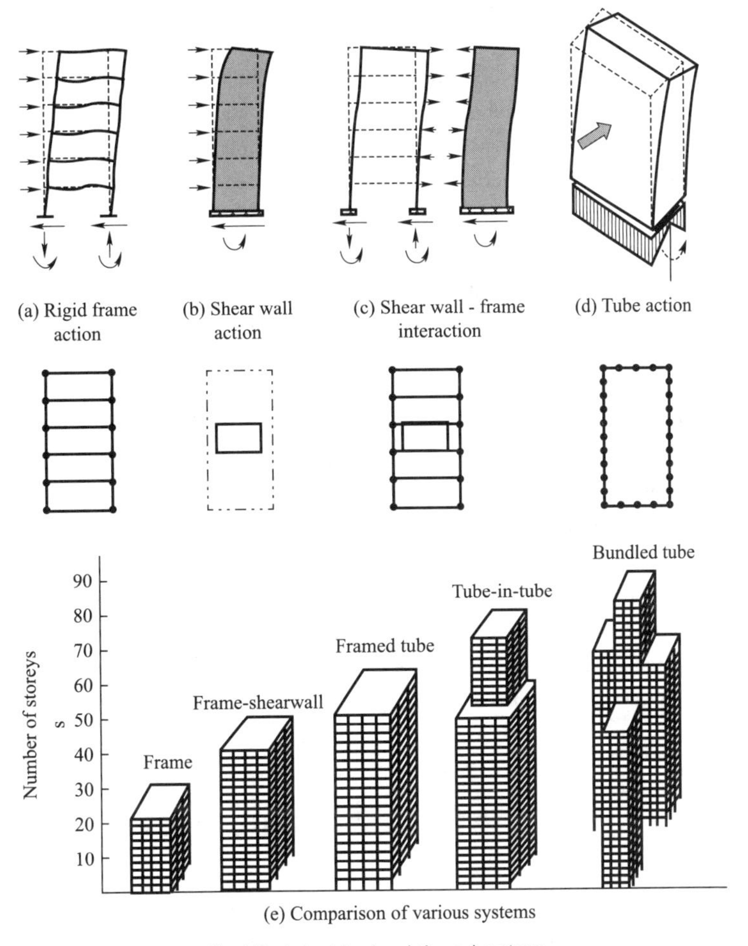

Fig. 1-7 Lateral load resisting subsystems

1.3 Reinforced concrete structures

Concrete is a stone-like material obtained by permitting a carefully proportioned mixture of cement, sand and gravel or other coarse aggregate, and water to harden in forms of the shape and dimensions of the desired structure. The bulk of the material consists of fine and coarse aggregate. Cement and water interact chemically to bind the aggregate particles into a solid mass. Additional water, over and above that needed for this chemical reaction, is necessary to give the mixture the workability that enables it to fill the forms and surround the embedded reinforcing steel prior to hardening.

Concrete may be remarkably strong in compression, but it is equally remarkably weak in tension. Its tensile "strength" is approximately one-tenth of its compressive "strength". Hence, the use of plain concrete as a structural material is limited to situations where significant tensile stresses and strains do not develop, as in hollow (or solid) block wall construction, small pedestals and "mass concrete" applications (in dams, etc.). Concrete would not have gained its present status as a principal building material, but for the invention of reinforced concrete, which is concrete with steel bars embedded in it. The idea of reinforcing concrete with steel has resulted in a new composite material, having the potential of resisting significant tensile stresses, which is hitherto impossible. Thus, the construction of load-bearing flexural members, such as beams and slabs, becomes viable with this new material. The steel bars (embedded in the tension zone of the concrete) compensate for the concrete's incapacity for tensile resistance, effectively taking up all the tension, without separating from the concrete. Currently, reinforced concrete is the most widely used building material in the world. Accordingly, the structure mainly made of reinforced concrete is called reinforced concrete structure.

Recent developments in concrete composites have resulted in several new products that aim to improve the tensile strength of concrete and impart ductility. As a result, the construction of structures with higher heights, larger spans, and longer lengths becomes possible. Fiber-reinforced concrete, which is concrete with the addition of fibers in convenient quantities (normally up to about 1% or 2% by volume) to conventional concrete, shows excellent resistance to cracking and impact. Concrete-filled steel tubes, which are composite members consisting of a steel tube infilled with concrete, have extremely high axial strength capacity. Steel-concrete composite elements use concrete's compressive strength alongside steel's resistance to tension, and when tied together this results in a highly efficient and lightweight unit that is commonly used for structures such as multi-storey buildings and bridges. Fiber-reinforced polymer reinforcements, which offer a number of advantages such as corrosion resistance, non-magnetic properties, high tensile strength, lightweight and ease of handling, are considered as a substitute for steel bars.

1.4 Masonry structures

Masonry is the oldest building material known to humans, masonry has been used in one form or another since the dawn of history. The term masonry refers generally to brick, tile, stone, concrete block, etc., or their combination, bonded with mortar. However, many different definitions of masonry are in vogue. The ***International Building Code*** (IBC 2009) defines masonry as "a built-up construction or combination of building units or materials of clay, shale, concrete, glass, gypsum, stone or other approved units bonded together with or without mortar or grout or other accepted methods of joining". ASTM E631—2015 defines masonry as "construction usually in mortar, of natural building stone or manufactured units such as brick, concrete block, adobe, glass, block tile, manufacture stone or gypsum block". The ***McGraw-Hill Dictionary of Scientific and Technical Terms*** defines masonry as "construction of stone or similar materials such as concrete or brick". A commonality in these various definitions is that masonry essentially is an assemblage of individual units which may be of the same or different kind, and which have been bonded together in some way to perform intended function.

Masonry structure refers to those buildings whose walls and columns are mainly made of natural or manufactured units and mortar. Brick, stone and concrete blocks are the most common materials used in masonry construction. Therefore, masonry structure can be simply divided into brick masonry structure, block masonry structure and stone masonry structure. The most important characteristic of masonry structure is its simplicity. Laying pieces of units on top of each other, either with or without cohesion via mortar, is a simple (though adequate) technique that has been successfully used ever since remote ages. The Sphinx, Coliseum (the Flavian Amphitheater in Rome), Parthenon, Roman aqueducts, the Great Wall in China, and many castles, cathedrals, temples, mosques, and dams and reservoirs all over the world stand as a testimony of enduring and aesthetic quality of masonry. Masonry structure continues to be used for many types of buildings, ranging from multi-storey high-rises to low-income apartment buildings.

A masonry structure also consists of the horizontal load-resisting subsystem, the vertical load-resisting subsystem and the footing. The difference is that the masonry structure may contain ring beams, structural reinforced concrete columns, cantilever beams and lintels. The ring beam is placed between the wall top and the slab bottom in the form of circle or polygon. Unlike normal beams that take vertical loads, ring beams normally take lateral loads. However, if the ring beam is not supported on a continuous wall but is supported on columns at intervals, the ring beams take both the horizontal as well as vertical loads. The structural reinforced concrete column sets in the wall of multi-storey masonry building in order to enhance the integrity and stability of the structure. Structural reinforced concrete columns are generally connected with the ring beams, forming a spatial frame that can resist bending and shear load. It is an effective measure to prevent the collapse of the masonry building under earthquake. The cantilever beam is a beam that is half cantilevered and half stretched into the masonry wall, which is used to support cantilevered corridors, balconies and other structures. A lintel is a horizontal beam supporting loads over an opening, which can be made of brick, steel or concrete.

1.5 Structural design

Design is the determination of the general shape and specific dimensions so that a structure will perform the function for which it is created and will safely withstand the influences that will act on it throughout its lifetime. These influences are primarily the loads and other forces to which it will be subjected, as well as other detrimental agents, such as temperature fluctuations and foundation settlements.

In China, structural design is governed by the probability-based limit states method. The method aims for a comprehensive and rational solution to the design problem, by considering safety at ultimate loads and serviceability at working loads. The probability-based limit states method uses a multiple safety factor format which attempts to provide adequate safety at ultimate loads as well as adequate serviceability at service loads, by considering all possible "limit states". There are two types of limit states: ultimate limit states and serviceability limit states. The ultimate limit states deal with strength, overturning, sliding, buckling, fatigue fracture, etc. Serviceability limit states deal with discomfort to occupancy and/or malfunction, caused by excessive deflection, crack-width, vibration, leakage, etc., and also loss of durability, etc. The selection of the various multiple safety factors is supposed to have a sound probabilistic basis, involving the separate consideration of different kinds of failure, types of materials and types of loads.

The objective of limit states design is to ensure that the probability of any limit state being reached is acceptably low. This is made possible by specifying appropriate multiple safety factors for each limit state. Of course, in order to be meaningful, the specified values of the safety factors should result (more-or-less) in a "target reliability". Evidently, this requires a proper reliability study to be done by the code-making authorities.

The multiple safety factor format is:

$$R_d = S_d \quad (1\text{-}1)$$

Where,

R_d——the design resistance computed using the reduced material strengths;

S_d—— the design load effect computed for the enhanced loads, involving separate partial load factors for dead load, live load and accidental load.

1.5.1 Process

The design of large-scale building engineering can be divided into three phases: the technical design phase, the in-depth design phase and the engineering drawing phase. For these simple or general structures, the in-depth design phase could be omitted.

1. Technical design phase

The main work of this phase is to determinate the structural scale and layout, functional requirements and the total investment. The finish of this work needs the cooperation of different disciplines. This phase is responsible for the preparation of structural design introduction, structural types, construction plan and structural layout. The structural design introduction includes design criteria, key points and specification.

2. In-depth design phase

The in-depth design phase is necessary for these very complicated structures. Its purpose is to solve some challenging problems that can not be solved in the technical design phase. It is the concretization of the design based on the preliminary design. If necessary, some complex problems can be comprehensively studied by researchers.

3. Engineering drawing phase

The engineering drawing phase is the most important design stage before the construction of the project. All possible technical problems in the construction should be solved in the phase. All construction drawings with details should be prepared for engineer in situ. In real practice, the design and the drawings may be modified to satisfy the construction requirement.

1.5.2 Codes

Structural design is mainly governed by four codes, the ***Load Code for the Design of Building Structures*** (GB 50009—2012), the ***Code for Design of Concrete Structures*** (GB 50010—2010), the ***Code for the Design of Masonry Structures*** (GB 50003—2011) and the ***Code for Seismic Design of Buildings*** (GB 50011—2010). These codes are generally updated every 10 years.

The ***Load Code for the Design of Building Structures***(GB 50009—2012) provides the most unfavorable loads, including dead load, live load on floors and roofs, crane load, snow load, wind load, thermal action, accidental load and their combinations.

The ***Code for Design of Concrete Structures***(GB 50010—2010) provides reliable calculation methods for resistance of reinforced concrete beams, slabs and columns under tensile, compressive, bending, shear and torsional load, as well as design principles for adequate serviceability and durability of reinforced concrete members.

The ***Code for the Design of Masonry Structures***(GB 50003—2011) provides reliable calculation methods for resistance of masonry (i.e., walls and columns) under tensile, compressive, bending, shear and torsional load, as well as design principles for adequate serviceability and durability of reinforced concrete members.

The ***Code for Seismic Design of Buildings*** (GB 50011—2010) provides methods to reduce the earthquake damage of buildings, avoid casualties and reduce economic losses under earthquakes.

1.6 Development of buildings in China

Since reform and opening up, along with the rapid development of national science and technology, China's building technology has undergone three stages (i.e., learning and following in the 1980s, tracking and improving in the 1990s, and innovating and transcending since the start of the 21st century), and completed the transition from "follower" to "competitor," and finally to "leader." Due to the great efforts of Chinese researchers and engineers, a great number of technical problems in materials, design, construction and management have been solved. High-performance materials, such as C50 and C60 concretes, Q500 steel, and 2000 MPa steel wire, have been invented. Rational design methods and effective vibration control technologies have been proposed. Various construction equipments and strategies have been developed to address super-high buildings. Advanced information technologies, including building information modeling technology, big data technology and automatic detection technology, have been offered for structural management. These achievements make it possible to construct complex buildings with the high performance and efficiency.

Before the 1980s, the development of building technology in China was slow. The number of high-rise buildings (an important symbol of building technology) was small. The tallest building in China in the 1950s was Beijing National Hotel (Fig. 1-8), which was built in 1959. It has 12 storeys and is 47.4 m high. It is a reinforced concrete frame structure. The tallest building in the 1960s was Guangzhou Hotel (built in 1968) with 27 storeys and a height of 87.6 m. It is a reinforced concrete frame-shear wall structure. In the 1970s, high-rise buildings had been preliminarily developed. The highest building in the 1970s is Guangzhou Baiyun Hotel (Fig. 1-9), which was built in 1976. It has 33 storeys and is 114.1 m high. It adopted reinforced concrete frame-shear wall structure.

Fig. 1-8 Beijing National Hotel

Fig. 1-9 Guangzhou Baiyun Hotel

From 1980s to 1990s, China's building technology entered a period of rapid development. A large number of high-rise buildings have been finished in Beijing, Shanghai, Guangzhou, Shenzhen and other big cities. The building height exceeded 200 m. The simple reinforced concrete frame structure and reinforced concrete frame-shear wall structure can not meet the architects' desire for building height. Structural systems with stronger lateral load-bearing capacity, such as reinforced concrete frame-tube structures, reinforced concrete tube-in-tube structures, concrete-filled steel tube columns and steel-concrete composite structures, have begun to be adopted to high-rise buildings. In 1985, Shenzhen International Trade Center Building (Fig. 1-10) was completed. It adopted a reinforced concrete tube-in-tube structure, with 50 floors and a height of 158.7 m. It was the highest building in China at that time. In 1987, Guangzhou International Building (Fig. 1-11) was finished, with a reinforced concrete tube-in-tube structure. It has 63 floors and the height exceeded 200 m for the first time.

Fig. 1-10 Shenzhen International Trade Center Building

In the 21st century, China's building technology is developing faster. At present, it has surpassed most countries, including the United States, Britain and Japan. High-rise buildings with a height of more than 300 m are all over the country. For example, Shanghai Tower (Fig. 1-12), Shanghai World Financial Center, Shenzhen Pingan Financial Center, Goldin Finance 117, Nanjing Zifeng Tower (Fig. 1-13), etc. China's high-rise buildings have become the world's high-rise buildings exhibition hall.

Shanghai Tower (上海中心大厦) is a 128-storey, 632-meter (2073 ft)-tall megatall skyscraper in Shanghai, as shown in Fig. 1-12. It is the world's second-tallest building by height to architectural top and it shares the record of having the world's highest observation deck within a building or structure at 562 m. The core of Shanghai center building is circular, and the combination of the mega frame, the core tube and the steel-concrete composite outrigger truss forms the system to resist the lateral load. The eddy-current tuned mass damper is adopted for the first time in the world to reduce the vibration induced by wind. The damper with a weight of 1000 t is independently developed by China and is located on the 125th floor. The damper can reduce the wind-induced vibration by more than 43%. Even in strong wind weather, more than 90% of the people in the building will feel very comfortable.

Fig. 1-11 Guangzhou International Building

Nanjing Zifeng Tower (南京紫峰大厦) is located in Nanjing, China, as shown in Fig. 1-13. It covers an area of 18,721 m^2, with a total construction area of 26,1075 m^2, of which the main building has 4 floors underground and 89 floors above the ground, with a height of 450 m. In 2011, Nanjing Zifeng Tower won the first prize of the 7th Excellent Architectural Structure Design Award of China Architectural Society. The foundation of Nanjing Zifeng Tower adopts pile raft foundation to meet the requirements of bearing capacity and settlement control, and employs reinforced concrete foundation slab with a thickness of 3.4 m. The plan of Nanjing Zifeng Tower is triangular. The main building is a steel frame with reinforced concrete core tubes, and the frame columns on the periphery are rigid structures. The main building of Nanjing Zifeng Tower has three steel truss floors on the 10th, 35th and 60th floors respectively, and the height of each truss floor is 8.4 m.

Fig. 1-12 Shanghai Tower

Fig. 1-13 Nanjing Zifeng Tower

Exercises

1.1 What is structure? How to classify structures?

1.2 What are the characteristic of the reinforced concrete structure and masonry structure?

1.3 List some world-famous structures and analyze their properties.

Chapter 2
Single-storey Industrial Building

Prologue

Main points

1. Structure type, structure composition and structure layout of single-storey industrial building.
2. Determination of calculation diagram of bent frame structure.
3. Internal force calculation method of bent frame under various loads.
4. Designing method and detailing of bent column (including corbel).
5. Designing method and structural detailing of independent foundation under column.

Learning requirements

1. Be familiar with structure type and layout of single-storey industrial building.
2. Master the internal force analysis method of bent frame structure.
3. Master the designing method and detailing of bent column (including corbel).
4. Master the designing method and detailing of independent foundation under column.

2.1 Structure composition and layout of single-storey industrial building

2.1.1 Structure types of single-storey industrial building

1. Properties of single-storey industrial building

Industrial building can be divided into three categories according to the number of floors, that is, single-storey industrial building, multi-storey industrial building and industrial building with mixed-storey. For some industrial building with large-scale mechanical equipment production and maintenance, single-storey industrial building is often used to meet the needs of large space and vibration.

The main properties of single-storey industrial building are as follows:

1) Single-storey industrial building with a large space has great adaptability to all types of industrial production, especially the production of heavy products. Therefore, single-storey industrial building is widely used in the steel-making, rolling, casting, forging, metalworking, assembly and other large plants of metallurgical or mechanical production.

2) The columns of single-storey industrial building are the main load-bearing members, which bear the loads and transmit the loads to the foundation.

3) Horizontal transportation tools, such as cranes, can be set in theindustrial building to facilitate the production and transportation of heavy products. Therefore, the dynamic load should be considered in structural design of the industrial building.

4) Natural lighting and natural ventilation devices can beequipped on the roof.

2. Classification of single-storey industrial building

1) According to structure material

Single-storey industrial building can be divided into hybrid structure, concrete structure and steel structure according to the material of load-bearing members.

(1) Hybrid structure

Brick masonry is used as the material of the column and foundation, and timber, reinforced concrete or light steel is generally used for the roof truss (or girder) for single-storey industrial building with hybrid structure. The single-storey industrial building with hybrid structure is suitable for a small workshop with crane tonnage ≤5 t, span ≤15 m, elevation of column top ≤8 m and no special production requirements.

(2) Steel structure

The single-storey industrial building with steel structure adopts steel column and steel roof truss, which is suitable for large industrial building with crane tonnage ≥ 250 t, span ≥ 36 m and special production requirements (such as industrial building with 10 t or larger than 10 t forging hammer, high temperature industrial building, etc.).

(3) Concrete structure

Reinforced concrete column, reinforced concrete or steelroof truss (or girder) are generally used for single-storey industrial building with concrete structure, which is suitable for medium-sized industrial building with crane tonnage ≤ 250 t, span 18—30 m.

Except for some special cases, prefabricated reinforced concrete structure is generally used for single-storey industrial building. The main members commonly used in the prefabricated concrete industrial building are processed and produced in the industrial building, and then transported to the construction site for assembly to form the overall structure of the industrial building. The main members with a unified specification and dimensions are all standardized customized production so as to greatly shorten the design time and construction time.

2) According to structure types

Structure types of single-storey industrial building include bent frame structure and rigid frame structure.

(1) Bent frame structure

The load-bearing structure members of bent frame structure mainly consist of roof truss (or girder), column and foundation. The connection of column and roof truss (or girder) is hinged, and the connection of column and foundation is fixed.

Bent frame structure is generally composed of prefabricated reinforced concrete roof truss (or girder), crane beam, column and foundation. During the process of construction, the column is generally hoisted and inserted into the cup of prefabricated foundation, and then filled with fine aggregate concrete to ensure the fixed connection between the column and the foundation.

According to the production process and use requirements, the bent frame structure can be made into single-span and multi-span bent frame structure with equal height as shown in Fig. 2-1(a), unequal height bent frame structure as shown in Fig. 2-1(b) and zigzag-type bent frame structure as shown in Fig. 2-1(c). Zigzag-type bent frame structure is often used in large-area operation of textile factories, which is conducive to one-way lighting.

(2) Rigid frame structure

Different from the hinged connection between the roof truss (or girder) and the column of the bent frame structure, this kind of industrial building structure is called the rigid frame structure when the roof truss (or girder) and the column are rigidly connected as a whole.

The rigid frame structure can be divided into three types, i.e., three hinged rigid frame, two hinged rigid frame and non-hinged rigid frame.

If the top connections of girder and column and the connections of column and foundation are fixed, this kind of rigid frame structure is called non-hinged rigid frame as shown in Fig. 2-2(a). If the top connections of girder and column are made into rigid connection and the connection of column and foundation are hinged, it is called two hinged rigid frame as shown in Fig. 2-2(b). If the top joint of rigid frame and the connection of column and foundation are hinged, this kind of rigid frame structure is called three hinged rigid frame as shown in Fig. 2-2(c). In the actual project, we often design the girder or column into variable cross-sections along the axis direction to conform to the change of the bending moment of the member, which can greatly save the material consumption and reduce the self-weight.

In order to facilitate the construction, the two hinged rigid frame and non-hinged rigid frame are usually made into three sections as shown in Fig. 2-2(a), and the splicing joint is set at the place where the bending moment is zero or very small.

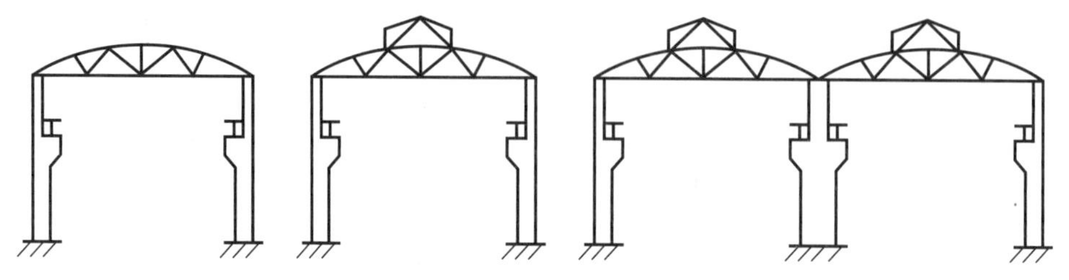

(a) Equal height bent frame structure

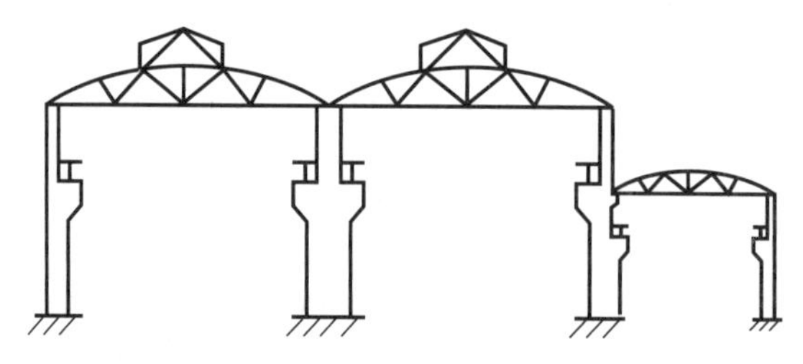

(b) Unequal height bent frame structure

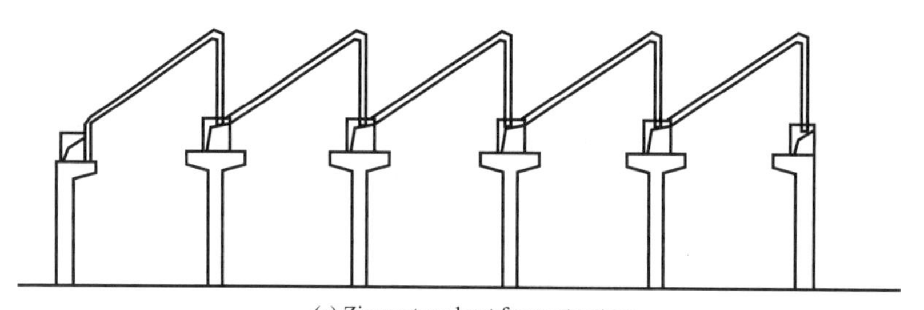

(c) Zigzag-type bent frame structure

Fig. 2-1 Types of bent frame structure

Portal rigid frame structure is generally suitable for small and medium-sized single-storey industrial building without crane or crane tonnage ≤10 t, span ≤18 m and eave height ≤10 m.

This chapter mainly describes the design of prefabricated reinforced concrete bent frame structure of single-storey industrial building.

2.1.2 Structure composition of single-storey industrial building

The single-storey industrial building structure usually consists of the following structural members as shown in Fig. 2-3.

1. Roof system

The roof structure is composed of roof slab (including gutter slab), roof truss or girder (including roof bracing), and sometimes there are skylight truss and bracket truss. Roof structure is divided into non-purlin roof system and purlin roof system. When the large roof slabs are directly supported (ensures three-point welding) on the roof truss (or girder), it is called non-purlin roof system and its roof stiffness is large. When the small roof slabs (or tiles) are supported on the purlins and the purlins are supported on the roof truss, it is usually called purlin roof system. The roof truss (or girder) bears the dead weight of roof structure and roof live loads (including snow load and other loads, such as ash load, suspended load, etc.) and transmits these loads to the column, so the roof truss (or girder) is called roof load-bearing structure. The skylight truss for ventilation and lighting is also a kind of roof load-bearing structure.

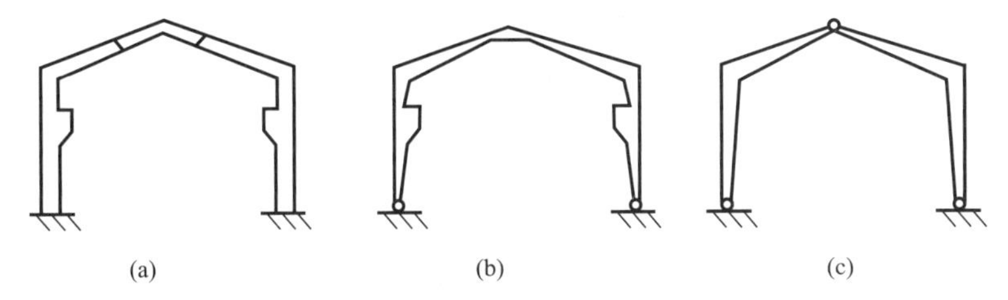

(a) (b) (c)

Fig. 2-2 Types of rigid frame structure

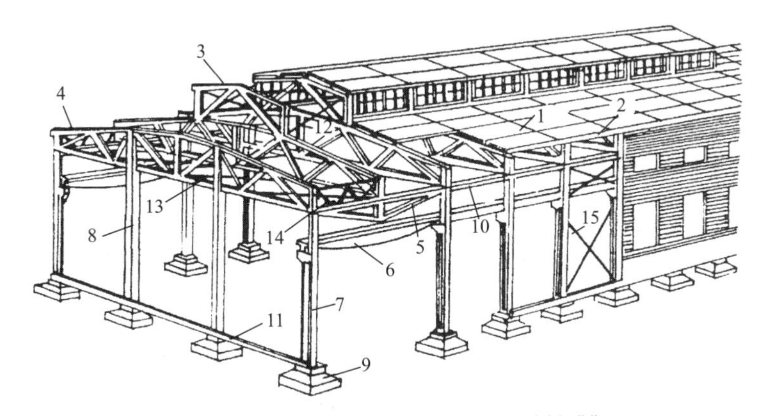

Fig. 2-3 Composition of a single-storey industrial building

1-Roof slab; 2-Gutter slab; 3-Skylight truss; 4-Roof truss; 5-Bracket truss; 6-Crane beam; 7-Bent frame column; 8-Wind-resisting column; 9-Foundation; 10-Coupling beam; 11-Foundation beam; 12-Vertical bracing between skylight trusses; 13-Transversal bracing of roof truss; 14-Vertical bracing between roof trusses; 15-Column bracing

2. Transversal bent frame structure

The transversal bent frame structure is a planar skeleton composed of roof truss (or girder), transversal bent columns and foundation, which is the basic load-bearing structure of the industrial building.

The vertical loads and transversal horizontal loads of the industrial building structure are mainly transmitted to the foundation through the transversal planar bent frame structure. The vertical loads include dead weight, roof live load, snow load, ash load and crane vertical load. Transversal horizontal loads include transversal wind load, transversal braking force of crane and transversal horizontal earthquake action.

3. Longitudinal bent frame structure

The longitudinal bent frame structure is composed of longitudinal columns, coupling beams, crane beams, column bracing and foundations as shown in Fig. 2-4.

The longitudinal bent frame structure can ensure the longitudinal stability and rigidity of the industrial building, also bear the longitudinal wind load, longitudinal horizontal braking force of crane, longitudinal earthquake action and temperature stress, and transmit these loads to the foundation.

4. Maintenance structure

The maintenance structure includes longitudinal wall, transversal wall, coupling beam, wind-resisting column and foundation beam, etc. In addition to the maintenance function, the maintenance structure also bears the self-weight of the wall and members, the wind load acting on the wall and transfers them to the load-bearing structure.

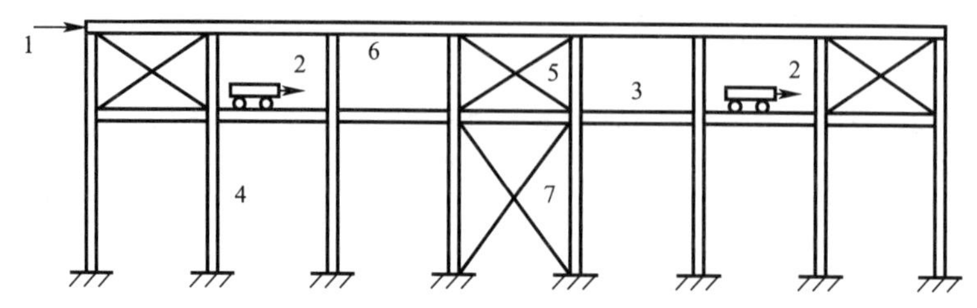

Fig. 2-4 Longitudinal bent frame structure
1-Wind load; 2-Braking force of crane; 3-Crane beam; 4-Column;
5-Upper column bracing; 6-Coupling beam; 7-Lower column bracing

5. Bracing

The bracing includes roof bracing and column bracing. The functions of bracing are to strengthen the spatial rigidity of the industrial building structure, ensure the stability and rigidity of the structure during the installation and using stage, and transfer the wind load, crane horizontal load or horizontal earthquake action to the corresponding load-bearing members.

6. Crane beam

The crane beam is generally fabricated and simply supported on the bracket of column. The crane beam bears the vertical load, the crane transversal load or the crane longitudinal load, and transmits the loads to the horizontal or longitudinal planar bent frame structure respectively.

2.1.3 Loading transfer path of single-storey industrial building

1. Load types

According to the direction of the loads, loads of single-storey industrial building are divided into vertical loads and horizontal loads. The vertical loads include dead load, live load on the roof, vertical load of crane, etc. The horizontal loads include transversal horizontal loads (transversal wind load, crane transversal horizontal load) and longitudinal horizontal loads (longitudinal wind load, crane longitudinal horizontal load, temperature stress).

2. Loading transfer path

The vertical load and transversal horizontal load are mainly transmitted to the foundation through the transversal bent frame structure; the longitudinal horizontal loads are mainly transmitted to the foundation through the longitudinal bent frame structure. The loading transfer path is shown in the Fig. 2-5.

2.1.4 Structure layout of single-storey industrial building

1. Layout of the column grid and positioning axis

1) Layout of the column grid

The grid formed by the arrangement of load-bearing columns in the plane is called column grid, and the spacing of grid is called column grid size. The spacing between neighbor transversal positioning axes along the longitudinal direction is called column spacing; the spacing along the transversal direction is called span as shown in Fig. 2-6.

According to ***Coordination Standard for Building Modulus of Industrial Building*** (GB/T 50006—2010), the main dimensions and elevation of single-storey industrial building shall conform to the building modulus. The building modulus shall be expressed in "M" (100 mm) as the basic unit. When the span is less than 18 m, 30M series (i.e., times of 3 m) shall be adopted, such as 9 m, 12 m, 15 m and 18 m. When the span is more than 18 m, 60M series (i.e., times of 6 m) shall be adopted, such as 24 m, 30 m, 36 m and so on.

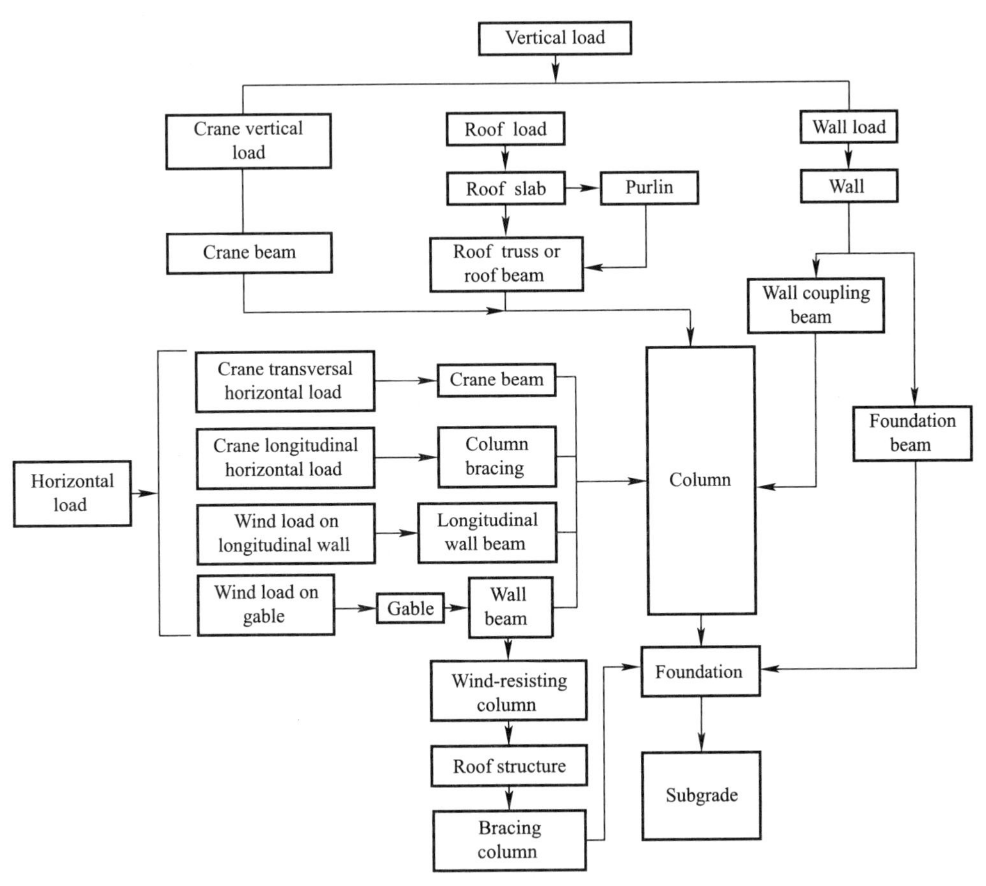

Fig. 2-5 Loading transfer path of the single-storey industrial building

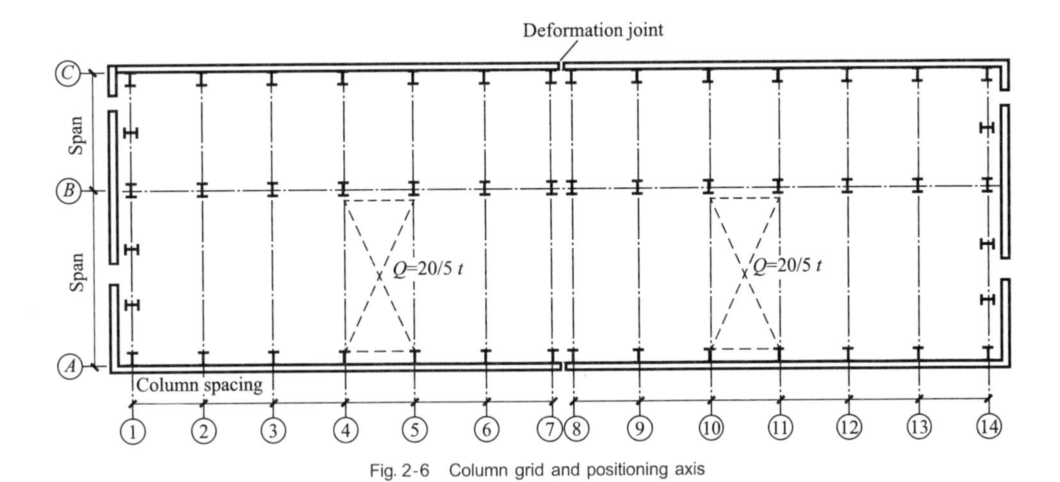

Fig. 2-6 Column grid and positioning axis

The column spacing is generally 6 m, sometimes 9 m and 12 m, which meets the requirements of 30M series (times of 3 m).

2) Transversal positioning axis

In Fig. 2-6, except for the end transversal positioning axis, other transversal positioning axes coincide with the center lines of columns. However, the end transversal positioning axis does not pass through the center line of the side column, but the center line of the end column moves 600 mm inward from the end transversal positioning axis.

3) Longitudinal positioning axis

The longitudinal positioning axes at the end of the industrial building generally pass through the outer edge of the bent column (that is, the inner edge of the longitudinal wall), such as axis *A* and axis *C* in the Fig. 2-6.

2. Deformation joint

Deformation joint includes expansion joint, settlement joint and earthquake proof joint.

1) Expansion joint

When the length of the industrial building is too large ($\geqslant$ 100 m indoor, > 70 m in open air), the expansion of the building caused by the temperature change is large. But if the deformation is constrained, it will produce excessive temperature stress, and the concrete will produce cracks due to the excessive main tensile stress. In that case, expansion joints shall be set to divide the industrial building into several temperature sections. The expansion joint will be set from the top of foundation, completely divides the superstructure into two temperature sections, and the joint gap is generally 20—40 mm. We generally use two rows of columns and two roof trusses arranged in parallel on both sides of the expansion joint, and use the connected foundation as shown in Fig. 2-7. The column center lines on both sides of the expansion joint move 600 mm from the transversal positioning axis.

2) Settlement joint

Bent frame structure is not sensitive to uneven settlement of foundation, so settlement joint is generally not set in a single-storey industrial building. However, in the case of the following conditions, such as large height difference between adjacent structure parts, great difference in lifting capacity of adjacent cranes, great change in subgrade layer and long construction time difference between parts, settlement joint shall be set to avoid uneven settlement of foundation.

The settlement joint will be set from the roof to the foundation, so as to prevent the whole building from being damaged in case of different settlements on both sides of the joint. The joint width is generally greater than 50 mm.

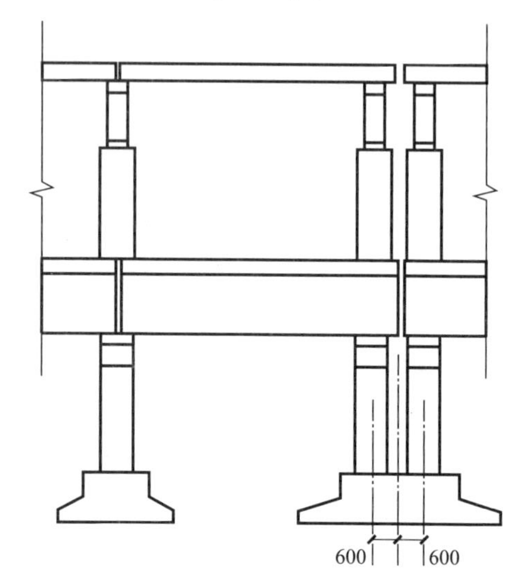

Fig. 2-7 Setting method of expansion joint

3) Earthquake proof joint

When the structural plane and elevation are too complicate or the height and rigidity of different parts are greatly different, the structure will be greatly twisted under the earthquake action. In order to reduce the earthquake damage of the industrial building, earthquake proof joint shall be set to separate the two adjacent parts.

The earthquake proof joint will be set to break the superstructure from the top of foundation. The width of the joint is 50—90 mm, which meets the corresponding requirements of seismic code.

3. Bracing layout

The bracing of a single-storey industrial building consists of roof bracing and column bracing. The function of the bracing is to strengthen the spatial stiffness of the industrial building, ensure the stability and normal operation of structural members, and transfer some horizontal loads (such as longitudinal wind load, crane longitudinal horizontal load and horizontal earthquake action) to the principal load-bearing elements. In addition, in the construction and installation stage, some temporary bracings should be set according to the specific situation to ensure the stability of structural members.

In the prefabricated reinforced concrete single-storey industrial building structure, although the brace is not the main load-bearing member, it is an important part of connecting various main structural members and connecting them into a whole. Engineering practice shows that if the bracing is not properly arranged, it will not only affect the normal use of the industrial building, but also may cause engineering accidents, so we should pay enough attention to it.

1) Layout of roof bracing

The roof bracing includes upper transversal horizontal bracing, lower transversal horizontal bracing, lower longitudinal horizontal bracing, vertical bracing and longitudinal tie rod.

The horizontal bracing in the upper and lower chord planes of the roof is generally in the cross form, and the division of the bracing sections should be adapted to the roof truss sections. The angle of the cross bar is generally 30°—60°, as shown in Fig. 2-8. The dotted lines in Fig. 2-8 represent the upper and lower chords of the roof truss.

Types of vertical bracing for roof truss (or girder) are as shown in Fig. 2-9.

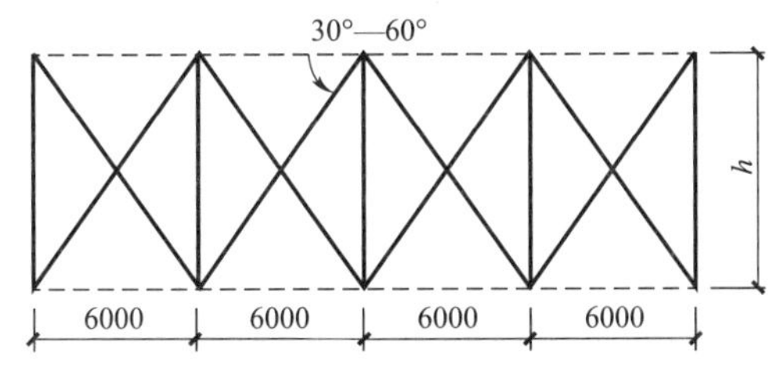

Fig. 2-8 Horizontal bracing of roof truss

(1) Layout of upper transversal horizontal bracing

Upper transversal horizontal bracing is generally arranged between the two adjacent roof truss chord members at both ends of the expansion joint section as shown in Fig. 2-8. When a large roof slab is adopted and the connection is reliable, the upper transversal horizontal bracing of the top chords may not be provided. The function of the upper transversal horizontal bracing is to enhance the overall rigidity of the roof system and effectively transmit the longitudinal wind load of the wind-resisting column to the bent column.

① When the roof is purlin roof system, or the roof is non-purlin roof system while the connection quality between roof slab and roof truss (or girder) can't be guaranteed, and the wind-resisting column is connected with the top chord of roof truss, upper transversal horizontal bracing should be arranged in the first column spacing or the second column spacing at the end of each expansion joint section.

② When the industrial building is equipped with skylight and the skylight passes through the second column spacing at the end of the industrial building or through the expansion joint, the top chord horizontal brace shall be set in the skylight range in the first column spacing or the second column spacing, and one to three longitudinal horizontal tie bars under compression shall be set in the skylight range.

(2) Layout of lower transversal horizontal bracing

Lower transversal horizontal bracing is generally arranged between the two adjacent roof truss chord members at both ends of the expansion joint section as shown in Fig. 2-8. The function of the lower transversal horizontal bracing is to enhance the overall rigidity of the roof, and effectively transfer the longitudinal wind load of the wind-resisting column to the bent column.

When there is a hanging crane at the lower chord of the roof truss, or there is a large vibration in the industrial building, or the wind load of the gable is transmitted to the lower chord of the roof truss through the wind-resisting column, the lower horizontal bracing of the roof should be set in the first column spacing or the second column spacing at both ends of each expansion joint section.

(3) Layout of lower longitudinal horizontal bracing

Generally, lower longitudinal horizontal

bracing can be omitted. Lower longitudinal horizontal bracing is only set when the industrial building has a large lifting bridge crane, wall crane or forging hammer and other vibration equipment, and the height or the span is large.

The function of lower longitudinal horizontal bracing is to enhance the space rigidity of the industrial building. Fig. 2-10 shows that the longitudinal horizontal bracing and the transversal horizontal bracing of the lower chord member form a rigid frame, which enhances the space rigidity between the bents.

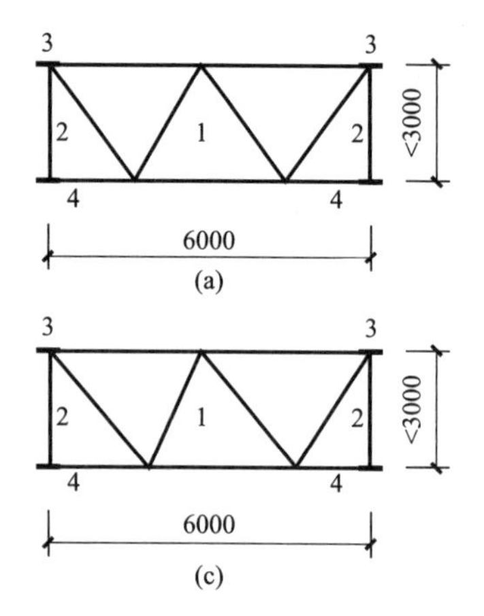

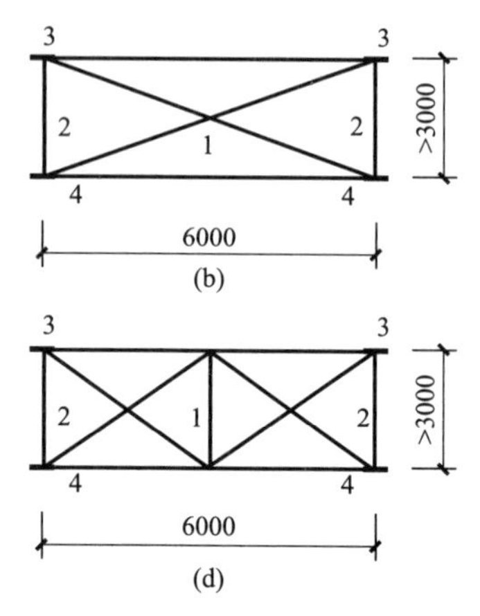

Fig. 2-9 Vertical bracing types of roof truss
1-Vertical brace of roof truss; 2-Vertical chord of roof truss;
3-Top chord of roof truss; 4-Lower chord of roof truss

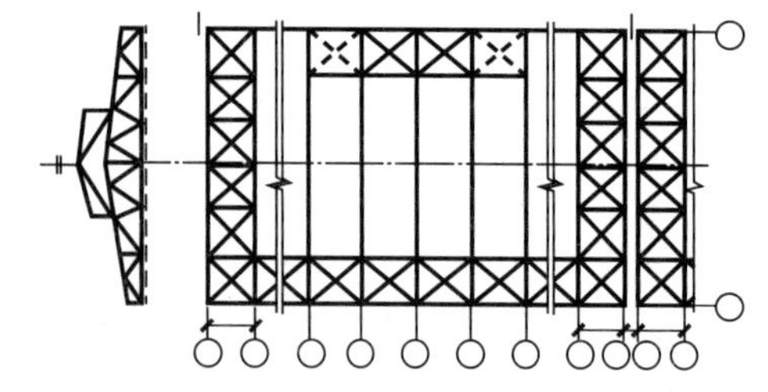

Fig. 2-10 Longitudinal horizontal bracing and transversal horizontal bracing of the lower chord

(4) Layout of vertical bracing of roof truss

The vertical bracing is set to ensure the out-of-plane stability of the roof system and the safety of structure during the roof truss is installed. The in-plane horizontal load of the top chord member of the roof truss is transferred to the in-plane lower chord member of the roof truss, which is used in cooperation with the transversal horizontal bracing. When the lower transversal horizontal bracing is set, the vertical bracing shall be arranged within the same column spacing with the lower horizontal bracing of the roof truss.

For the trapezoidal roof truss, in order to make the longitudinal horizontal force transmit reliably from the roof to the column top, and to ensure the out-of-plane stability of the roof truss during construction, a vertical brace should be set at both ends of the roof truss.

The vertical bracing should be set in the first column spacing or the second column spacing at the end of each expansion joint section. When the length of the expansion joint section of the industrial building is greater than 90 m, a vertical bracing should be added with the column spacing.

(5) Layout of longitudinal horizontal tie rod

The function of the tie rod is to act as the lateral bracing point of the upper and lower chords of the roof truss. In the top chord plane of roof truss, the ridge of large roof slab can play the role of rigid tie bar. When purlin is used, purlin can also play the role of tie bar, but the stability and load-bearing capacity of purlin should be checked. Before the roof structure is installed and the roof slab is in place, tie bars are set at both ends of the roof truss to ensure that the top chord of the roof truss has a good out-of-plane stiffness. When there is a skylight, it is important to set the tie rod at the roof ridge for stability of roof truss because there is no roof slabs or purlins in the skylight range.

2) Layout of column bracing

The column bracing includes the upper column bracing and the lower column bracing, as shown in Fig. 2-11. The function of the column bracing is to enhance the longitudinal stiffness and stability of the industrial building and transmit the longitudinal wind load and the longitudinal braking force of the crane to the foundation. The upper column bracing mainly bears the wind load, while the lower column bracing bears not only the wind load, but also the longitudinal horizontal load of the crane.

The steel structure is generally used for the column bracing, and the bearing capacity and stability of the members shall comply with the relevant code for design of steel structures. When theindustrial building is equipped with medium or light working crane, reinforced concrete structure can also be used for column bracing.

As shown in Fig. 2-11, the upper column bracing is usually made of tensioned crossing round steel. For the industrial building with crane, it is easy to relax if the upper column bracing adopts round steel bracing, so it is better to use the common single angle steel crossing bracing. The lower column bracing generally adopts crossing bracing, which is composed of single angle steel or two-angle steel; it can also be lattice crossing bracing composed of two-channel steel, H-section steel or steel pipe, and the cross-section size is determined by the in-plane load borne by the bracings.

The column bracing shall be provided in case of any of the following conditions.

① When the crane with A6—A8 or A1—A5 has a lifting capacity of 10 t or more;

② When the span of the industrial building is 18 m or above 18 m, or the height of the column is greater than 8 m;

③ When the total number of longitudinal columns in each row is less than 7;

④ When there is 3 t and more than 3 t hanging crane.

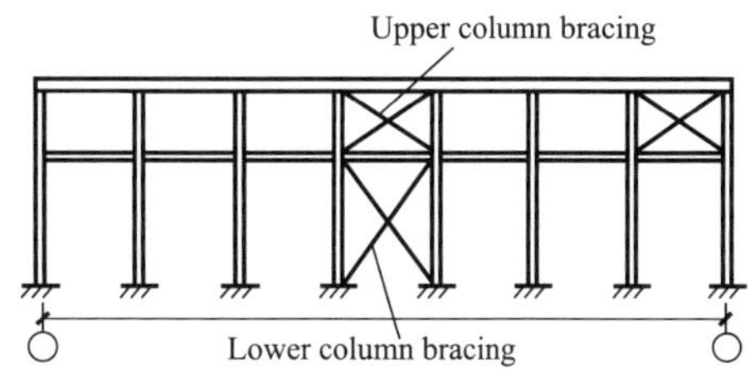

Fig. 2-11 Column bracing

The column bracing should be arranged in the center or near the center of the expansion joint section. The upper column bracing can also be set within the first column spacing at both ends of the industrial building.

4. Layout of maintenance structure

Maintenance structure of the industrial building includes roof slab, wall, wind-resisting column, ring beam, coupling beam, lintel and foundation beam, etc.

1) Wind-resisting column

The wind-resisting column is arranged on the gable. The wind-resisting column spacing is determined according to the wind load, and 6 m and 4.5 m spacing are used generally.

The top of the wind-resisting column is connected with the roof truss to transfer horizontal wind loads. The wind-resisting column does not bear the vertical loads of the roof truss. Therefore, the connection between the wind-resisting column and the roof truss must meet two requirements. Firstly, the wind-resisting column must be reliably connected with the roof truss in the horizontal direction so as to effectively transmit wind loads. Secondly, a certain relative displacement is allowed in the vertical direction to prevent columns from being damaged or occurring cracks when the settlement is uneven. Fig. 2-12 shows the wind-resisting column and the connection detailing.

The upper column of wind-resisting column is rectangular section in general, and its section size should not be less than 350 mm×300 mm. The lower column should be I-shaped or rectangular cross-section. When the column is high, double-leg column may also be used.

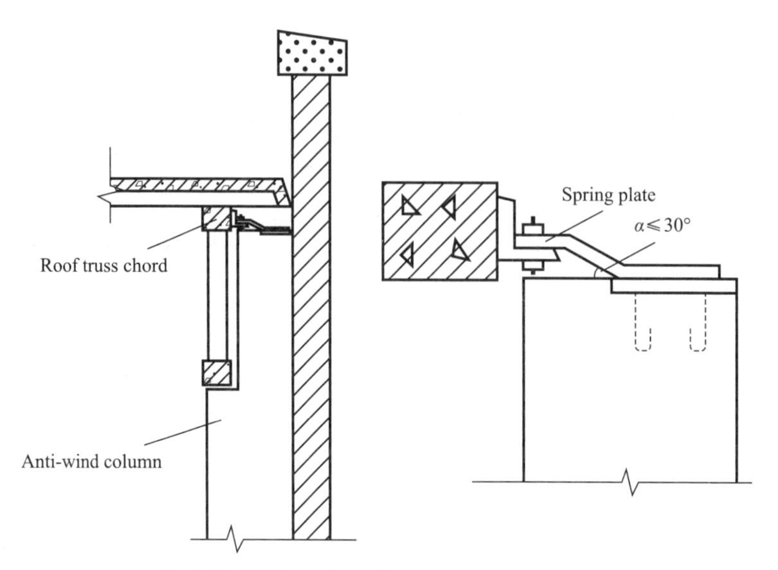

Fig. 2-12 Wind-resisting column and connection detailing

2) Ring beam, coupling beam, lintel and foundation beam

When brick masonry is used as the enclosure structure of the industrial building, ring beam or coupling beam, lintel and foundation beam should be set.

(1) Ring beam

The ring beam is located in the wall. The wall is hooped together with bent columns and wind-resisting columns by the ring beam to enhance the overall rigidity of the industrial building and prevent the building from being adversely affected by uneven settlement of the foundation or large vibration loads.

The arrangement of the ring beam is related to the height of the wall, the rigidity of the industrial building and the foundation condition. For the general single-storey industrial building, the following principles can be referred to.

① For the industrial building without bridge crane, when the wall thickness is ⩽ 240 mm and the eave elevation is 5—8 m, a ring beam should be arranged near the eave. When the height of eave is more than 8 m, one or more ring beams should be added;

② For the industrial building with bridge crane or large vibration equipment, in addition to the arrangement of ring beam at the eave or window top, it is advisable to add a ring beam at the elevation of crane beam or other appropriate positions. When the height of external wall is greater than 15 m, the ring beam should be added.

The ring beam should be set continuously on the same horizontal plane and form a closed beam. When the ring beam is cut off by the door or window, additional ring beam with the same section should be added above the door or window. The lap length of additional ring beam and ring beam shall not be less than twice of the vertical distance, and shall not be less than 1 m as shown in Fig. 2-13. When the wall thickness is not less than 240 mm, the width of the ring beam should not be less than 2/3 of the wall thickness. The height of ring beam shall be multiple of the brick thickness, and shall not be less than 120 mm. The longitudinal reinforcement bar of the ring beam should not be less than 4Φ10, and the stirrup spacing should not be greater than 300 mm. When the ring beam is

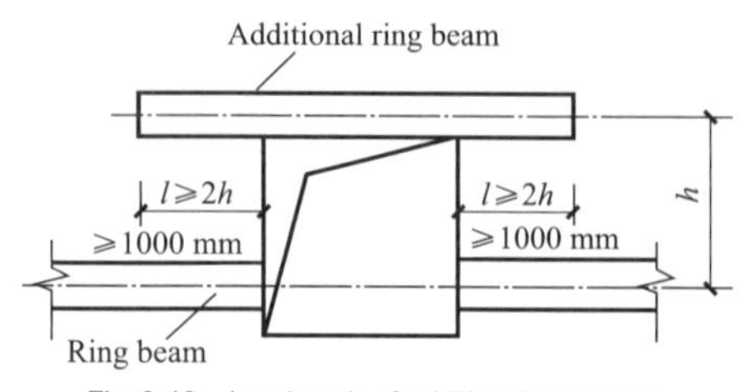

Fig. 2-13 Lap length of additional ring beam

also used as lintel, the amount of reinforcement bar shall be increased by lintel calculation. Ring beam can be cast-in-situ or prefabricated. The strength grade of concrete should not be lower than C20.

(2) Coupling beam

The function of the coupling beam is to connect the longitudinal columns to enhance the longitudinal stiffness of theindustrial building and transfer the wind load to the longitudinal column. In addition, the coupling beam also bears the weight of the upper wall. The coupling beam is usually prefabricated, and its two ends are placed on the corbel of the column, and can be connected by bolt or welding connection.

(3) Lintel

The role of the lintel is to support the weight of the wall on the door and window. The ring beam, coupling beam and lintel should be combined as whole as possible so as to save materials and simplify construction.

(4) Foundation beam

The foundation beam supports the weight of the wall, resting on the foundation and transmitting the wall weight to the top of the foundation.

Generally, the foundation beam and column can't be connected together in order to prevent uneven settlement. In order to prevent the crack of the foundation beam when the soil freezes and expands, a gap of 50—150 mm between subgrade and foundation beam shall be set. The foundation beam is directly placed on the cup mouth of the column foundation, or placed on the concrete cushion block above the foundation when the foundation is deeply embedded as shown in Fig. 2-14.

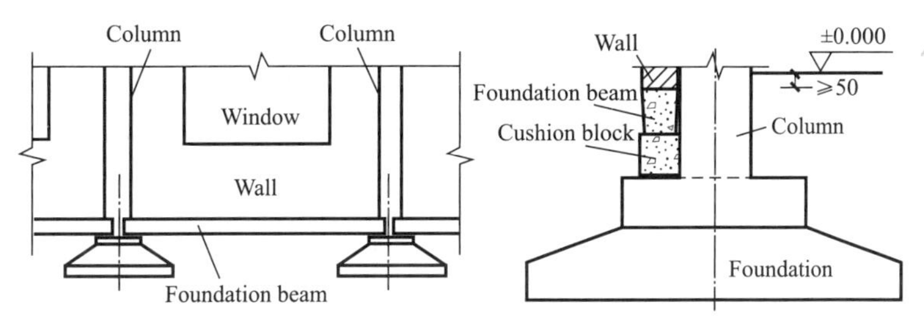

Fig. 2-14 Foundation beam

2.2 Calculation of bent frame structure

The single-storey industrial building structure is a space structure. In actual engineering, the space structure is simplified to a plane structure without considering the influence of adjacent bents in order to simplify the calculation. It is calculated as the transversal planar bent in the transversal direction (i.e., span direction) and the longitudinal planar bent in the longitudinal direction (i.e., column spacing direction). The transversal planar bent and the longitudinal planar bent do not affect each other and work independently. The transversal planar bent is the main bearing structure, so bent frame structure calculation in this section is the transversal planar bent.

The design of a single-storey industrial building includes structure type selection and layout, determination of structural calculation diagram, calculation of structural load, internal force analysis, internal force combination, calculation of bent column reinforcement (including hoisting calculation in construction calculation) and design of the foundation. The calculation of bent frame structure provides data for the design of the bent column and foundation.

2.2.1 Calculation diagram

1. Selection of calculation unit

If each column spacing is equal in the industrial building, a typical calculation unit can be obtained from the center line of the adjacent column spacing, as shown in oblique line of Fig. 2-15(a). The selection of a typical transversal planar bent in this part is shown in Fig. 2-15(b). In addition to the moving loads such as the crane load, the area of calculation unit is the load-bearing range of planar bent frame structure.

2. Assumptions for bent calculation

To simplify the calculation, assumptions are given as following.

(1) The ends of the columns are fixed to the top surface of the foundation. The top ends of the column are hinged to the roof truss (or girder).

(2) The roof truss (or girder) has no axial deformation.

The column and the foundation are poured together by fine aggregate concrete to form a whole, so we can consider the connection is fixed. However, in some cases, for example, when the subgrade is poor, the deformation is large or there is large ground load, the effect of foundation displacement and rotation on the internal force and deformation of the bent should be considered.

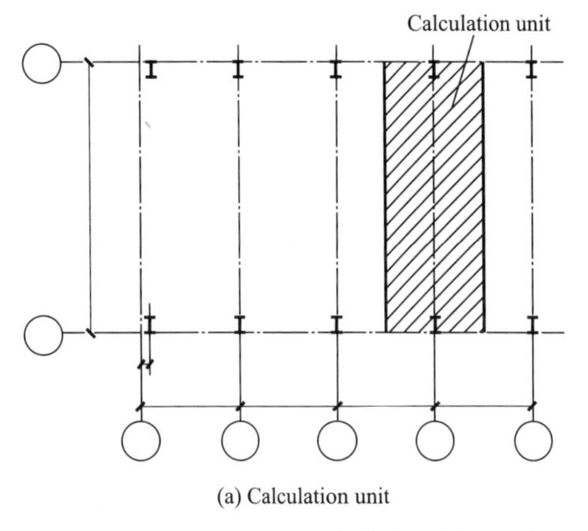

(a) Calculation unit

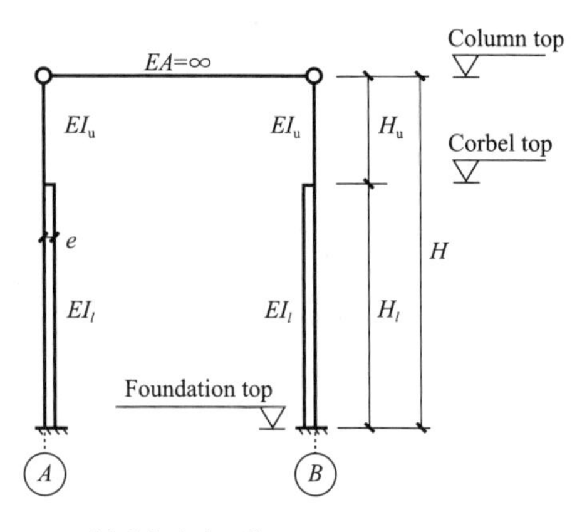

(b) Calculation diagram

Fig. 2-15 Calculation unit and calculation diagram of planar bent

3. Calculation diagram of planar bent

Based on the above, the calculation diagram is shown in Fig. 2-15(b). In the calculation diagram, the calculation axis of the column takes the center line of the upper and lower column sections, and roof truss (or girder) is expressed by a rigid rod without axial deformation.

Total height of column: H = elevation of column top + absolute value of the elevation of the foundation top.

The height of upper column: H_u = elevation of column top−elevation of corbel top.

The flexural rigidity of upper and lower column section are E_cI_u and E_cI_l, respectively, which are determined by concrete strength grade and cross-section shape and size of column. Here, I_u and I_l are the inertia moment of upper column and lower column, respectively.

2.2.2 Load calculation

The loads acting on the horizontal bent frame structure include dead load, live load on roof, snow load, ash load, crane load, wind load, etc. Except for crane load, other loads are calculated from the range of calculation unit. Fig. 2-16 shows the loads on the bent.

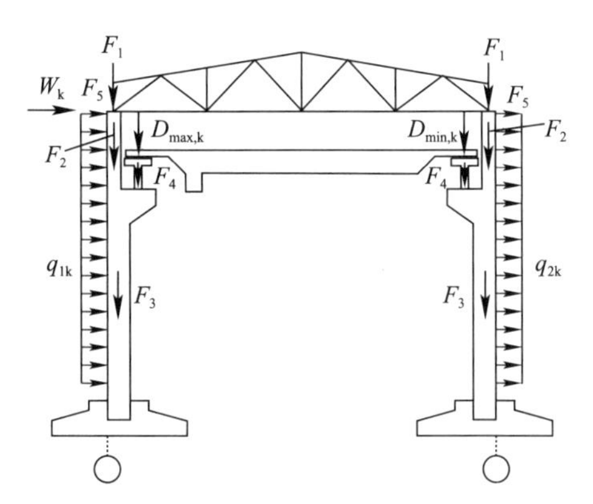

Fig. 2-16 Schematic diagram of actual loads on the bent

1. Permanent load

Permanent load includes the weight of roof F_1, weight of upper column F_2, weight of lower column F_3, weight of crane beams and track parts F_4 and weight of maintenance structure F_5.

1) Dead loads on roof (F_1)

Dead loads on roof (F_1) include weight of roof structure (such as thermal-insulating layer, waterproof layer, isolation layer, etc.), roof slab, gutter slab, roof truss, skylight truss, roof bracing and other dead loads, which can be calculated according to the detail drawing of roof construction, standard drawing of roof members and load code. The dead loads on roof are transmitted to the column top in the form of vertical concentrated force F_1 by the end of the roof truss (or girder). The action point of the dead loads on roof is at the intersection of the geometric center lines of the top and lower chords of the roof truss.

According to the position of action point, F_1 can be converted into the vertical force and the eccentric moment for the centroid position of the cross-section, as shown in Fig. 2-17 (a), where, $F_1=\overline{F_1}=\overline{F_1'}$; $M_1=F_1\times e_1$; $M_1'=F_1\times e_0$; $\overline{F_1}$, $\overline{F_1'}$ are axial pressures acting on upper column and lower column, respectively. The bent generates axial force under the action of the vertical force, and it is not necessary to analyze the internal force of the bent under the axial force, but the internal force of the bent needs to be analyzed under the action of the eccentric moment.

The calculation diagram under the action of loads on roof F_1 is shown in Fig. 2-17(b). The same analysis method is used for other vertical loads.

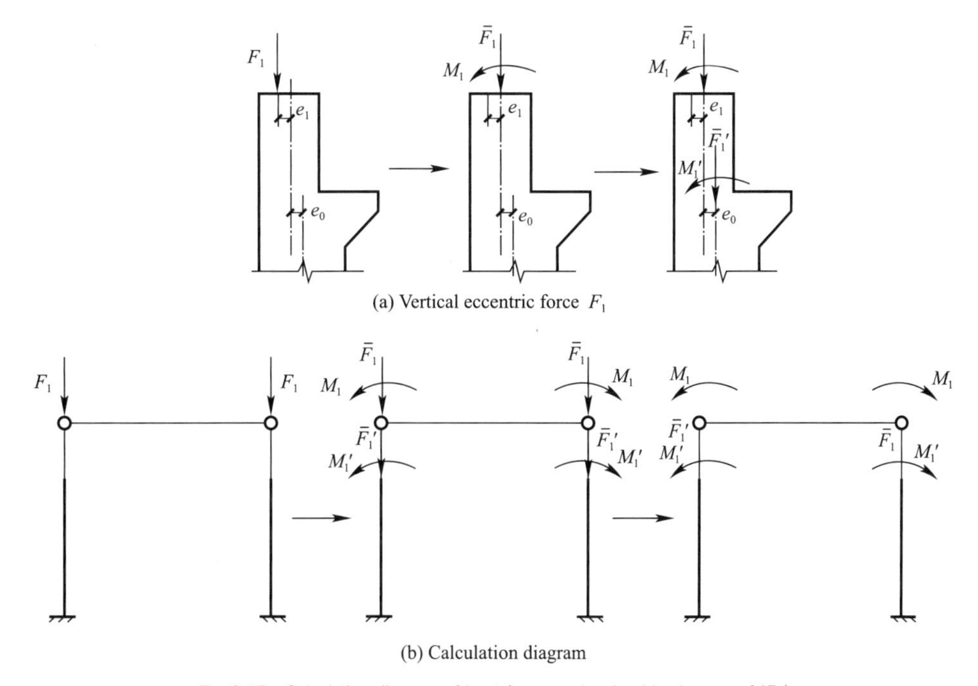

Fig. 2-17 Calculation diagram of bent frame under dead loads on roof (F_1)

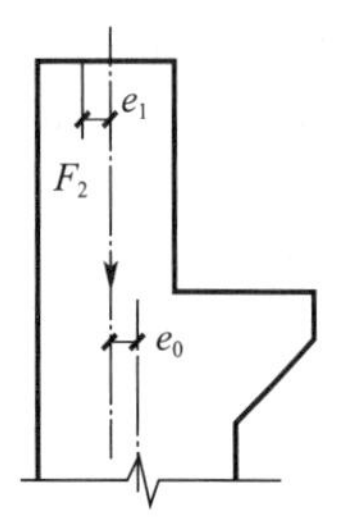

Fig. 2-18 Action point of weight of the upper column (F_2)

2) Weight of upper column (F_2)

The weight of the upper column (F_2) is calculated according to the cross-section size and height of upper column. F_2 acts on the center line of the upper column section on the top surface of the corbel, as shown in Fig. 2-18.

3) Weight of lower column (F_3)

The weight of the lower column (F_3) is calculated according to the cross-section size and height of the lower column. F_3 acts on the top surface of the foundation along the center line of the lower column.

4) Weight of crane beams and track parts (F_4)

Weight of crane beams and track parts (F_4) can be calculated according to standard drawing of crane beam and track connection part. F_4 acts on the top of the corbel along the center line of the crane beam. In general, the distance from the center line of the crane beam to the longitudinal positioning axis of the column is 750 mm, and the eccentricity from the center line of the lower column section is e_4 as shown in Fig. 2-19(a).

5) Weight of maintenance structure (F_5)

When the coupling beams are set to support the wall, the weight of the wall within the range of calculation unit is transmitted to the top surface of the corbel (supporting the coupling beam) in the form of a vertical concentrated force F_5. F_5 acts at center line of the coupling beam or wall section. Its distance from the geometric center line of the lower column section is e_5, as shown in Fig. 2-19(b).

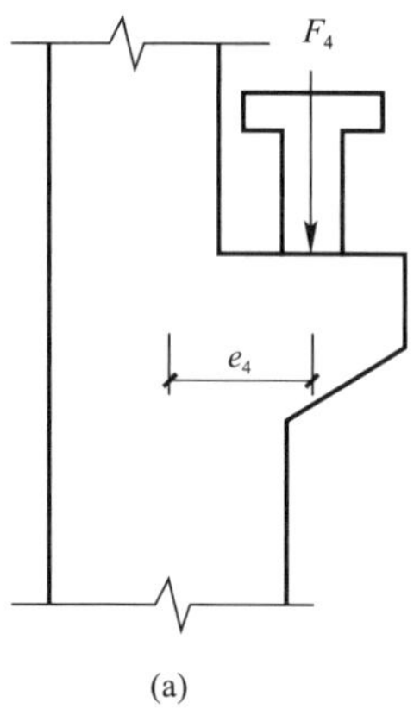

(a)

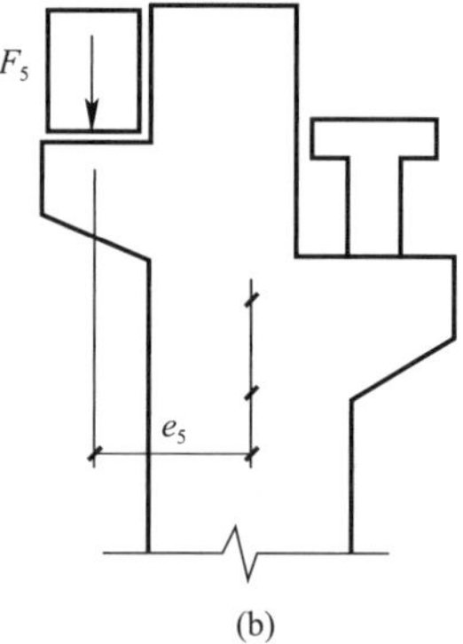

(b)

Fig. 2-19 Action points of F_4 and F_5

2. Live loads on roof

Live loads on roof include roof distributed live load, snow load and ash load. They are all calculated on the basis of the horizontal projection area of the roof.

1) Roof distributed live load

The characteristic value of the roof distributed live load on the horizontal projection of the roof is determined as follows.

For the inaccessible roof, it is 0.5 kN/m^2; for the accessible roof, it is 2.0 kN/m^2.

2) Snow load on roof

The snow pressure on the top surface of the building is referred as snow load on roof. The characteristic value of snow load S_k(kN/m^2) on the horizontal projection of the roof is calculated as follows:

$$S_k = \mu_r S_0 \tag{2-1}$$

Where,

μ_r—— the distribution coefficient of snow load on roof, which is determined according to different roof forms and considered as the uniform distribution for the bent frame;

S_0—— the basic snow pressure (kN/m^2), which is determined by the largest snow pressure on the general local open and flat ground in 50 years.

μ_r and S_0 are both determined according to ***Load Code for the Design of Building Structures***(GB 50009—2012).

3) Ash load on roof

The ash load should be considered in design and production when an industrial building or its neighboring building has a large amount of ash discharge. For machinery, metallurgy, and cement plants which have certain dust removal facilities and ash removal system, the ash load area on the horizontal projection of the industrial building roof is calculated in accordance with the relevant regulations in the ***Load Code for the Design of Building Structures*** (GB 50009—2012). For the places where ash piles are easy to form on the roof, when designing roof slabs and purlins, the characteristic value of ash load is multiplied by amplifying coefficient according to the following.

① When ash load is within the distribution width that is twice the height difference of the roof between high and low spans but not more than 6 m, the amplifying coefficient of ash load is taken as 2.0.

② When ash load is within the distribution width that is not more than 3 m at the gutter, the amplifying coefficient of ash load is taken as 1.4.

Considering the possibility of the simultaneous occurrence of the above loads on roof, the ***Load Code for the Design of Building Structures*** (GB 50009—2012) stipulates that the roof distributed live load shall not be considered with the snow load at the same time and take the larger value of the two values to calculate. When there is ash load, the ash load and the larger value of the snow load and the roof distributed live load should be considered at the same time.

After the characteristic value of live load on roof is determined, the concentrated force on the column top Q_1 can be calculated according to the loading area of the calculation unit. The acting position of Q_1 is the same as F_1. The calculation diagram under the live load on roof is shown in Fig. 2-20.

3. Wind load

1) Characteristic value of wind load on building surface

The pressure or suction generated on the surface of theindustrial building by the wind is called wind load. The characteristic value of wind load on the surface of the building w_k(kN/m^2) is calculated as follows:

$$w_k = \beta_z \mu_s \mu_z w_0 \tag{2-2}$$

Where,

w_0—— the basic wind pressure value (kN/m^2);

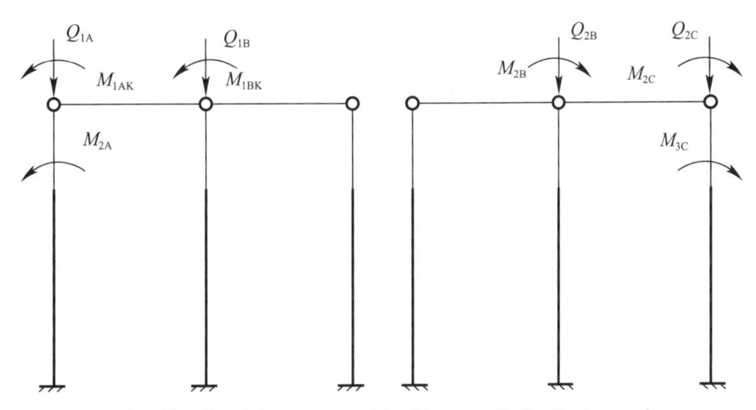

Fig. 2-20 Calculation diagram of bent frame under live load on roof

μ_s—— the shape coefficient of wind load;

μ_z—— the altitude variation coefficient of wind pressure;

β_z—— the wind vibration coefficient at height of z.

2) Calculation diagram

(1) Wind load below the column top

When calculating the bent frame structure, the wind load acting on the wall below the column top is approximately considered as the uniform distributed load as shown in Fig. 2-21(a), and μ_z can be determined according to the distance from the column top to the outdoor ground. q_{1k} is the characteristic value of windward load on the structure windward side, and q_{2k} is the characteristic value of wind load on the structure leeward side.

(2) Wind load above the column top

The wind load from the column top to the roof top is still taken as the uniform distributed load which is perpendicular to the roof (Fig. 2-21b). Its effect on the bent is considered as the horizontal concentrated wind load characteristic value $\overline{W}_k$ acting on the column top:

$$\overline{W}_k = \overline{W}_{1k} + \overline{W}_{2k} \tag{2-3}$$

Where,

$\overline{W}_{1k}$—— the characteristic value of wind load on the vertical surface of roof;

$\overline{W}_{2k}$—— the characteristic value of wind load on the sloping roof.

$$\overline{W}_{1k} = (\mu_{s1} + \mu_{s2})\mu_z \omega_0 h_1 B \tag{2-4}$$

$$\overline{W}_{2k} = F_2 - F_1 = (\mu_{s2} - \mu_{s1})\mu_z \omega_0 h_2 B \tag{2-5}$$

Where,

μ_{s1}, μ_{s2}—— the shape coefficient of wind load, taken as absolute value because the force direction has been considered in formula (2-3);

B —— the width of calculation unit.

μ_z can be determined according to the following conditions. When there is a rectangular skylight, it is determined according to the height from the skylight eave to the outdoor ground; when there is no rectangular skylight, it is determined according to the height from the building's roof ridge to the outdoor ground.

The wind load may change the direction, so the left and right wind situations need tobe considered when calculating the bent frame structure as shown in Fig. 2-22.

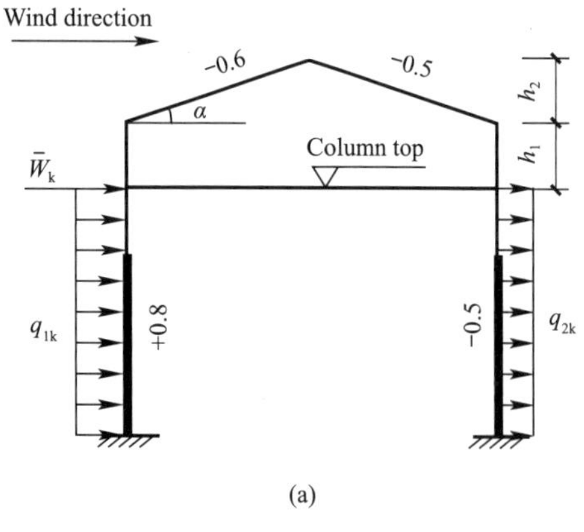

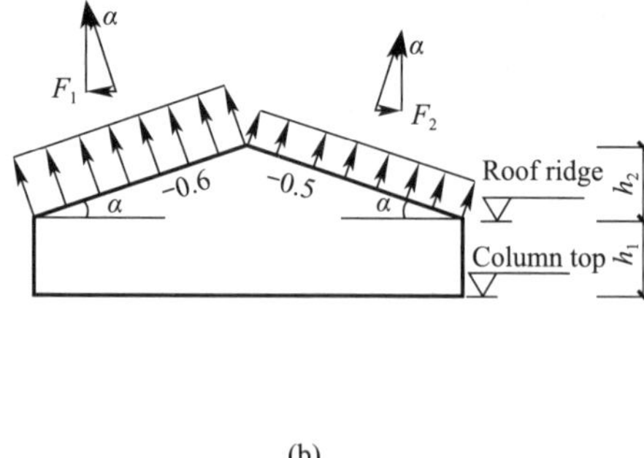

Fig. 2-21 Calculation of wind load on roof

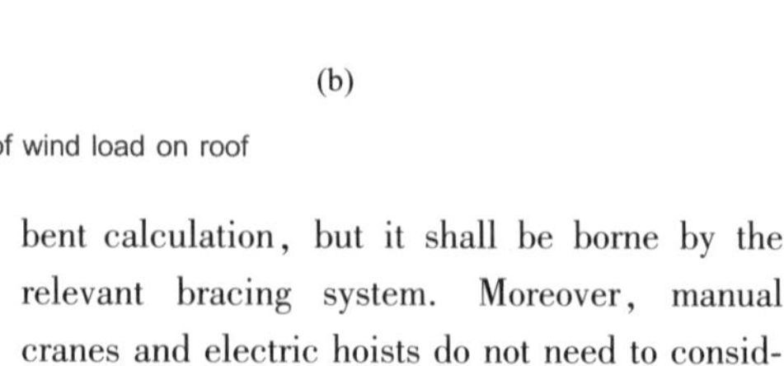

4. Crane load

The cranes commonly used in single-storey industrial building include suspended cranes, manual cranes, electric hoists and bridge cranes. Among them, the horizontal load of the suspended cranes may not be included in the bent calculation, but it shall be borne by the relevant bracing system. Moreover, manual cranes and electric hoists do not need to consider the horizontal load. The calculation of bridge crane load is introduced in this section.

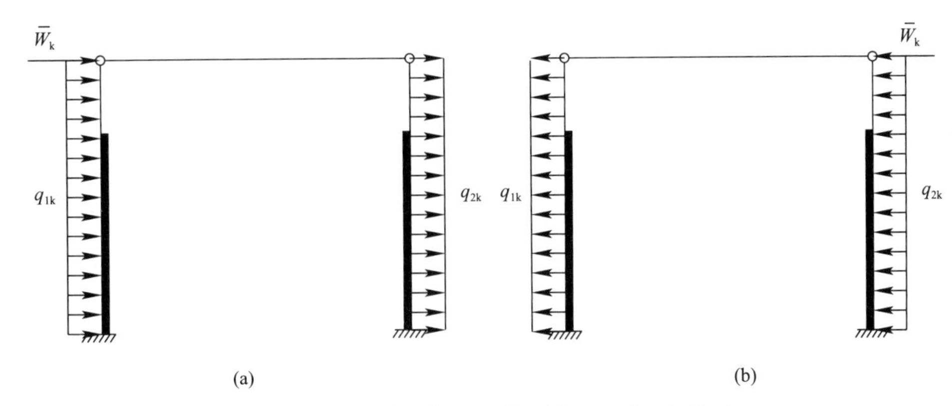

Fig. 2-22 Calculation diagram of bent frame under wind load

In the design of theindustrial building structure, the calculation parameters of the crane load are determined by the working levels of the crane.

The working levels of the crane according to the service levels and load condition grade required by the crane during the usage period are divided into 8 working levels, i.e., A1—A8 as shown in Appendix 1 as the basis for the design of the crane. The service levels are divided into 10 utilization levels according to the total number of working cycles required by the crane during the service period. The load condition grade refers to the ratio of each of crane's representative hoisting load to the crane's rated hoisting load and the corresponding lifting times to the total lifting times in the crane's design working life. The working levels of the crane reflect the frequency that the crane reaches its rated hoisting load. The working levels of the crane include light level (A1—A3), intermediate level (A4, A5) and heavy level (A6—A8).

Generally, for plants with low operating speed and few full-load run opportunities, such as hydropower stations and mechanical equipment maintenance stations, the cranes belong to A1—A3 working levels. The cranes in the mechanical processing industrial building and the assembly industrial building belong to A4 and A5 working levels; the cranes in the smelting industrial building and the cranes directly participating in continuous production belong to A6—A8 working levels.

For most bridge cranes, the crane loads acting on the transversal bent frame include the crane vertical load and the crane transversal horizontal load. The crane load acting on the longitudinal bent frame structure of the industrial building is crane longitudinal horizontal load.

1) Crane vertical load

The bridge crane consists of the bridge operating mechanism (big cart) and the operating part of lifting mechanism (small cart). The bridge operating mechanism runs along the longitudinal direction of the industrial building on the track of the crane beam, and the operating part of lifting mechanism runs transversely on the track of the bridge operating mechanism. The hoisting winch with lifting hook is installed on the operating part of lifting mechanism. As observed in Fig. 2-23, when the operating part of lifting mechanism with the rated hoisting load drives to a certain limit position of the bridge operating mechanism, each large wheel pressure on this side is called the characteristic value of maximum wheel pressure of the crane $P_{max,k}$. The wheel pressure on the other side is called the characteristic value of minimum wheel pressure of crane $P_{min,k}$. $P_{max,k}$ and $P_{min,k}$ occur simultaneously.

$P_{max,k}$ and $P_{min,k}$ are usually obtained according to the product specification of the crane manufacturer. For cranes of common specification, Appendix 1 lists the basic parameters and main dimensions. For the four-wheel bridge crane:

$$P_{\min,k}=\frac{G_{1,k}+G_{2,k}+G_{3,k}}{2}-P_{\max,k} \quad (2\text{-}6)$$

Where,

$G_{1,k}$, $G_{2,k}$—— the characteristic value of weight of the bridge operating mechanism and the operating part of lifting mechanism, respectively (kN);

$G_{3,k}$—— the rated hoisting load of crane(kN).

The vertical load generated by the crane acting on the bent can be calculated by the wheel pressure of each wheel of the crane (maximum or minimum wheel pressure) according to the influence line of support reaction force, as shown in Fig. 2-24. In Fig. 2-24, B is the width of big cart (i.e., bridge width); K is the track width. The characteristic value of maximum reaction force $D_{\max,k}$ is generated by $P_{\max,k}$ which acting on the crane beam. Meanwhile, the characteristic value of minimum reaction force $D_{\min,k}$ is generated by $P_{\min,k}$ which acting on the crane beam of the other side of the bent column. $D_{\max,k}$ and $D_{\min,k}$ are the characteristic value of crane vertical load acting on the bent frame.

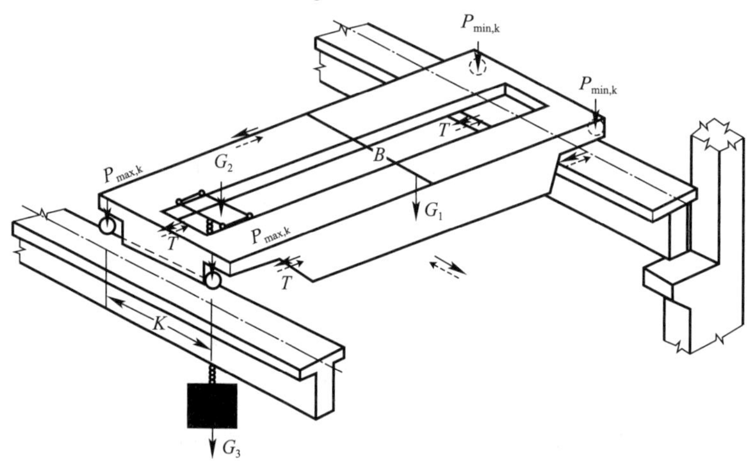

Fig. 2-23 Schematic diagram of crane load

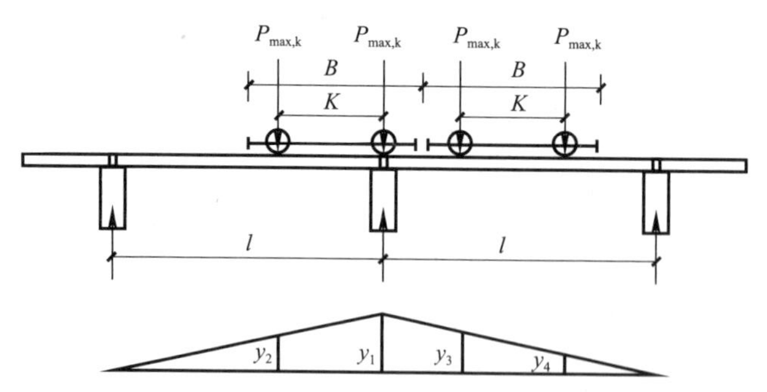

Fig. 2-24 Influence line of crane beam brace reaction

$$\left.\begin{aligned} D_{\max,k}&=\beta[P_{1\max,k}(y_1+y_2)+P_{2\max,k}(y_3+y_4)] \\ D_{\min,k}&=\beta[P_{1\min,k}(y_1+y_2)+P_{2\min,k}(y_3+y_4)] \end{aligned}\right\} \quad (2\text{-}7)$$

Where,

$P_{1\max,k}$, $P_{2\max,k}$—— the characteristic value of maximum wheel pressure of crane 1 and crane 2, $P_{1\max,k}>P_{2\max,k}$;

$P_{1\min,k}$, $P_{2\min,k}$——the characteristic value of minimum wheel pressure of crane 1 and crane 2, additionally $P_{1\min,k}>P_{2\min,k}$;

y_1, y_2, y_3, y_4—— the ordinate values on the influence line of reaction force corresponding to the wheels of crane 1 and crane 2;

β—— the load reduction co-effi-cient of multiple cranes, as shown in Table 2-2.

Because $D_{max,k}$ can occur on both the left column or the right column, two situations in Fig. 2-25 should be considered for a single-span industrial building under the crane vertical load. D_{max} and D_{min} are the design values which are the eccentric pressures acting on the lower column. $M_{max} = D_{max} \cdot e_4$ and $M_{min} = D_{min} \cdot e_4$. The method of internal force analysis is the same as that of the dead load.

2) Crane horizontal load

The horizontal loads generated by the crane on the bentframe are divided into two types, i.e., transversal horizontal load and longitudinal horizontal load.

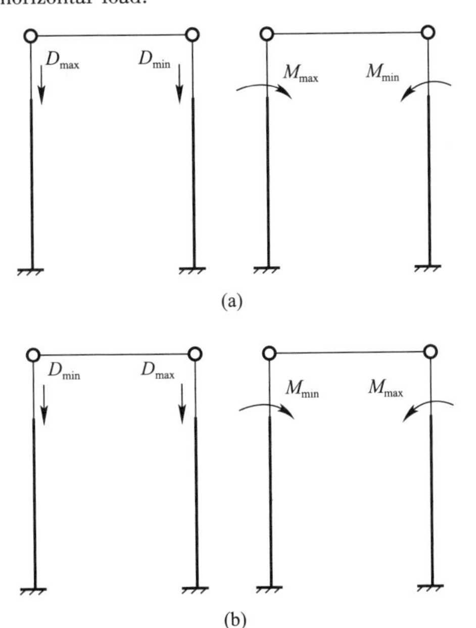

Fig. 2-25 Calculation diagram of single-span bent under D_{max} and D_{min}

(1) Crane transversal horizontal load

The crane transversal horizontal load is the transversal horizontal inertia force caused by the braking system when the operating part of lifting mechanism is hoisting the goods. It is transmitted to the bridge operating mechanism through the friction between the brake wheel of operating part of lifting mechanism and the track. Then, it is transferred to the crane beam through the wheels of small cart. Furthermore, it is transferred to the bent column through the connecting steel slab between the column and the crane beam, as shown in Fig. 2-26. Therefore, the crane transversal horizontal load acts on the horizontal direction of the top surface of the crane beam.

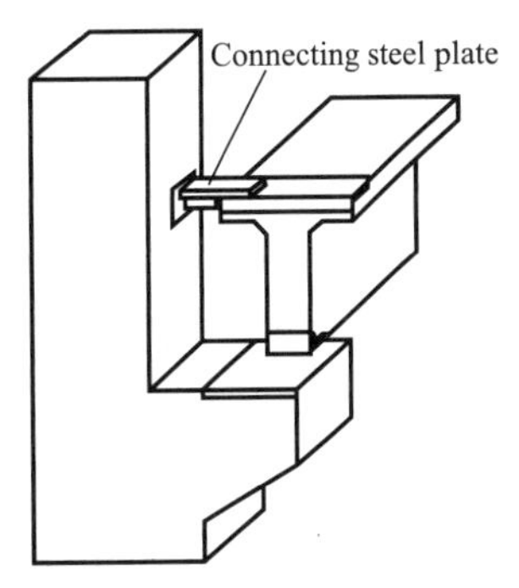

Fig. 2-26 Connecting between the column and the crane beam

The characteristic value of the crane transversal horizontal load on the bent should be the sum of the characteristic value of the gravity of the operating part of lifting mechanism and the characteristic value of the rated hoisting load multiplied by the transversal braking force coefficient α. Therefore, the characteristic value of the total transversal horizontal load of crane on the bent $\sum T_{i,k}$ can be expressed as:

$$\sum T_{i,k} = a(G_{2,k} + G_{3,k}) \tag{2-8}$$

Where,

α—— the crane transversal braking force coefficient is determined by ***Load Code for the Design of Building Structures*** (GB 50009—2012); for hard hook crane, $\alpha = 0.2$; for soft hook crane, it is determined according to Table 2-1.

Transversal braking force coefficient of soft hook crane

Table 2-1

$Q(t)$	≤10	16—50	≥75
α	0.12	0.10	0.08

For the soft hook crane, the hoisting weight is transmitted to the operating part of lifting mechanism through the steel wire rope. While the hard hook crane is a kind of crane that transfers the hoisting weight to the operating part of lifting mechanism through the rigid structure (such as clamps, claws, etc.). The hard hook crane works frequently and runs at a high speed. The rigid cantilever structure attached to the operating part of lifting mechanism prevents the hoisting weight from swinging freely, so that the transversal horizontal inertia force generated by braking is relatively large. In addition, the jamming phenomenon of the hard hook crane is also relatively serious. Therefore, the transversal horizontal load coefficient of the hard hook crane is relatively high.

The crane transversal horizontal load should be equal at the two ends of the cable tray, which is evenly transmitted to the wheels on the track respectively. Generally, the bridge operating mechanism has four wheels. That is, the number of wheels at each side is two. Therefore, the characteristic value of crane transversal horizontal load is transmitted through a bridge operating mechanism wheel, which is calculated as follows:

$$T_{k} = \frac{1}{4}\sum T_{i,k} = \frac{1}{4}\alpha(G_{2,k} + G_{3,k}) \quad (2\text{-}9)$$

The crane position when the characteristic value of maximum crane transversal horizontal load $T_{max,k}$ is generated on the bent is the same as the crane position when the vertical load of crane $D_{max,k}$, $D_{min,k}$ are generated. $T_{max,k}$ can also be calculated by influence line.

Considering load reduction coefficient of multiple cranes β:

$$T_{max,k} = \beta T_{k}\sum y_{i} = \frac{1}{4}\alpha\beta(G_{2,k} + G_{3,k})\sum y_{i} \quad (2\text{-}10)$$

The calculation diagrams of bent frame under T_{max} are shown in Fig. 2-27.

(2) Crane longitudinal horizontal load

The longitudinal horizontal load generated by the crane on the bent frame is the longitudinal horizontal braking force caused by the bridge operating mechanism when braking. The characteristic value of crane longitudinal horizontal load should be the maximum wheel pressure P_{max} of all brake wheels acting on one side of the track multiplied by the sliding friction coefficient α'. When $\alpha'=0.1$, $T_0=0.1\ P_{max}$.

The crane longitudinal horizontal load acts on the contacting point between the brake wheel and the track whose direction is the same as the track direction. When there is column bracing in the industrial building, the longitudinal horizontal load is borne by the column bracing; when there is no column bracing in the industrial building, the longitudinal horizontal load is borne by the longitudinal columns.

3) Load reduction coefficient of multiple cranes

When calculating the vertical loads of multiple cranes, for each bent of a single-span industrial building, cranes participating in the combination should not be more than two cranes; for each bent of multi-span industrial building, cranes participating in the combination should not be more than four cranes.

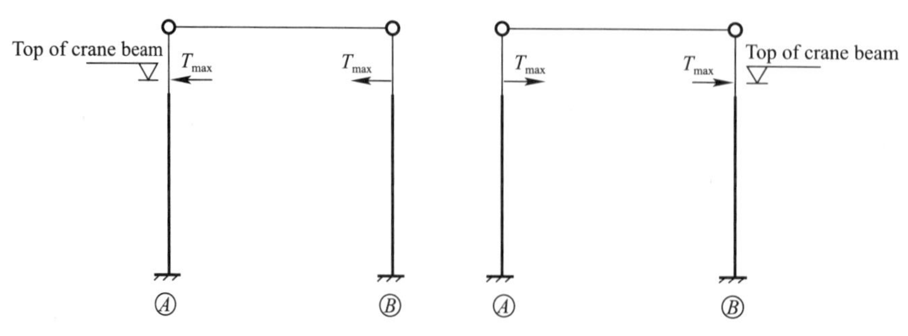

Fig. 2-27 Calculation diagrams of bent frame under T_{max}

When calculating the horizontal load of multiple cranes, the number of cranes participating in the combination should not be more than 2 for each bent of a single-span or multi-span industrial building.

When calculating the bent, for the vertical load and horizontal load of multiple cranes, the-load reduction coefficient is adopted according to Table 2-2.

Load reduction coefficient of multiple cranes β Table 2-2

The number of crane participating in the combination	Crane working level	
	A1—A5	A6—A8
2	0.90	0.95
3	0.85	0.80
4	0.80	0.85

【Example 2-1】

Conditions: There is a single-storey industrial building whose span is 18 m, column spacing is 6 m. Two cranes with intermediate load whose specification is 10 t are considered in the design (bridge width is 5.55 m, track width is 4.40 m, weight of the operating part of lifting mechanism is 3.8 t, the maximum wheel pressure characteristic value of crane $P_{max,k}$ is 115 kN, weight of the bridge operating mechanism is 18 t).

Question: Calculate D_{max}, D_{min}, T_{max}.

【Solution】

① Calculation of D_{max}, D_{min}

The influence line of reaction of crane beam is shown in Fig. 2-28.

$$P_{min,k}=\frac{G_{1,k}+G_{2,k}+G_{3,k}}{2}-P_{max,k}$$

$$=\frac{18+3.8+10}{2}\times 10-115$$

$$=44\ \text{kN}$$

$$D_{max}=\gamma_Q \beta P_{max,k}\sum y_i$$

$$=1.5\times 0.9\times 115\times\left(1+\frac{1.6+4.85+0.45}{6}\right)$$

$$=333.79\ \text{kN}$$

$$D_{min}=D_{max}\frac{P_{min,k}}{P_{max,k}}$$

$$=311.54\times\frac{44}{115}$$

$$=127.71\ \text{kN}$$

② Calculation of T_{max}

$$T_k=\frac{1}{4}\alpha\beta(G_{2,k}+G_{3,k})$$

$$=\frac{1}{4}\times 0.12\times 0.9\times(3.8+10)\times 10$$

$$=3.73\ \text{kN}$$

$$T_{max}=D_{max}\frac{T_k}{P_{max,k}}$$

$$=311.54\times\frac{3.73}{115}$$

$$=10.82\ \text{kN}$$

2.2.3 Internal force calculation of equal-height bent frame by shear force distribution method

The equal-height bent refers to a bent whose the horizontal displacements of the all

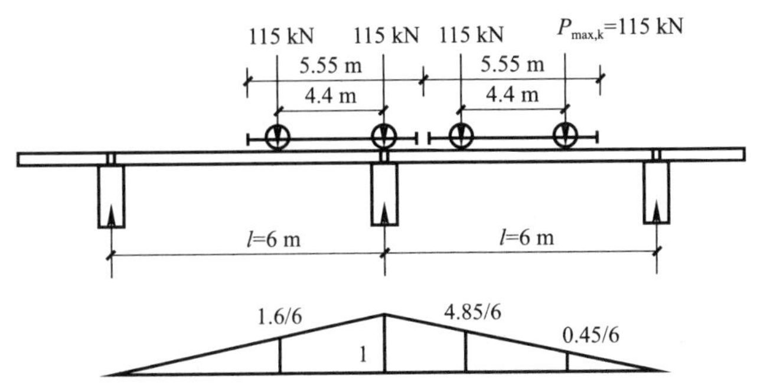

Fig. 2-28 Influence line of reaction of crane beam

column tops of each bent are equal under the loads. The equal-height bent has two situations. One situation is that it has the same column top elevation (Fig. 2-29a). The other situation is that the columns with different elevations are connected by inclined beams (Fig. 2-29b).

The unequal-height bent with different column top displacements can be solved by "force method" in structural mechanicals. Here a simple method is introduced for calculating equal-height bent, that is, the shear force distribution method.

1) Lateral stiffness of single-step cantilever column

According to structural mechanicals, when a unit horizontal force acts on a single-step column top (Fig. 2-30), the horizontal displacement Δu of the column top is:

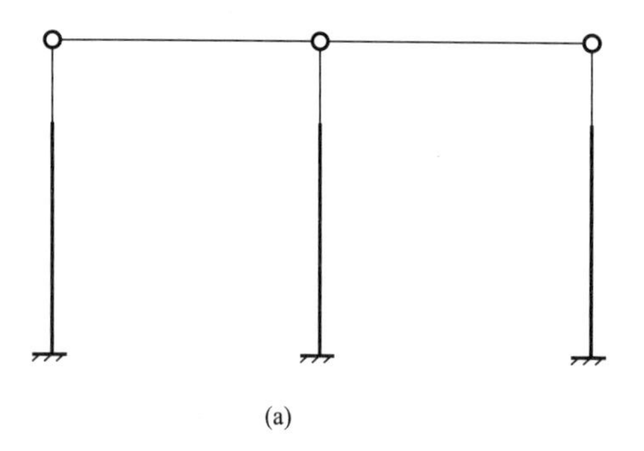
(a)

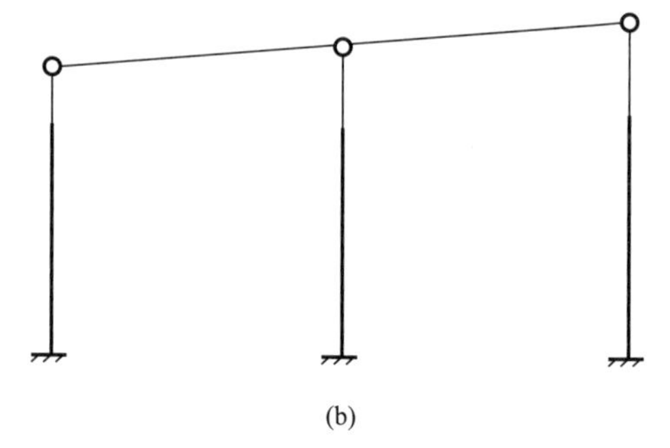
(b)

Fig. 2-29 Two situations of equal-height bent calculation

$$\Delta u=\frac{H^3}{C_0 E_c I_l} \quad (2\text{-}11)$$

$$\left|\begin{array}{l} C_0=\dfrac{3}{1+\lambda^3\left(\dfrac{1}{n}-1\right)} \\ \lambda=\dfrac{H_u}{H};\ n=\dfrac{I_u}{I_l} \end{array}\right|$$

Where,

H_u, H —— the upper column height and total column height, respectively;

I_u, I_l —— the moment of inertia of the upper column and lower column, respectively.

When a unit horizontal displacement occurs at the single-step cantilever column top, the horizontal force applied at the column top is called the lateral stiffness of the single-step cantilever column (Fig. 2-30):

$$D_0=\frac{1}{\Delta u}=\frac{C_0 E_c I_l}{H^3} \quad (2\text{-}12)$$

2) Internal force of equal-height bent under horizontal concentrated force on the column top

When there is a horizontal concentrated force acting on the column top (Fig. 2-31a), supposing that there are n columns, the horizontal displacement of each column top is u, and the lateral stiffness of any column is $D_{0,i}$. Then, the shear force of column top V_i can be obtained according to the force balance condition and the deformation compatibility condition. According to the definition of lateral stiffness, $V_i=D_{0,i}\cdot u$.

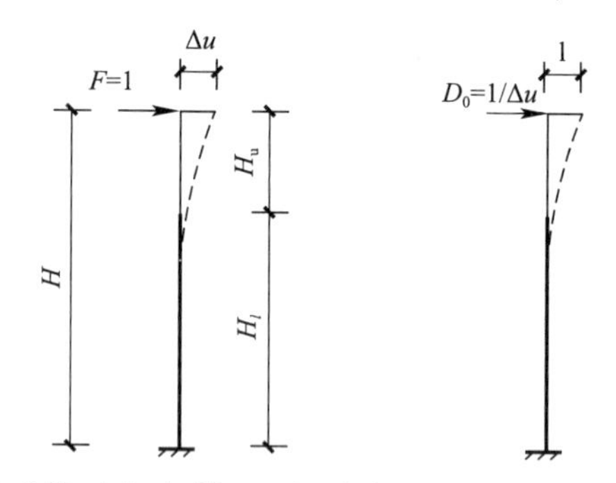

Fig. 2-30 Lateral stiffness of a single-step cantilever column

$$F=\sum V_i=\sum D_{0,i}\cdot u \quad (2\text{-}13)$$

$$u=\frac{F}{\sum D_{0,i}} \quad (2\text{-}14)$$

So,
$$V_i=\frac{D_{0,i}}{\sum D_{0,i}}F=\eta_i\cdot F \quad (2\text{-}15)$$

Where,

η_i —— the distribution coefficient of shear force for the i-th column;

$D_{0,i}$ —— the lateral stiffness of the i-th column is determined by formula (2-12).

After calculating the shear force on the column top, the internal force diagram of the column can be drawn as the cantilever column.

3) Internal force of equal-height bent under the arbitrary load

As shown in Fig. 2-32, the calculation process can be divided into four steps in order to make use of the above shear force distribution coefficient when the equal-height bent is subjected to the arbitrary load (such as crane horizontal load T_{max}).

(1) First, the horizontal hinged support is added to the bent column top to prevent the horizontal displacement of the column top. The horizontal reaction force R of the horizontal hinged support is calculated to obtain V_{A1} and V_{B1}, as shown in Fig. 2-32(b).

(2) Then, remove the additional fixed hinge support, the reverse effect of $R=R_A+R_B$ on the bent column top is added. Then, V_{A2} and V_{B2} are got by shear force distribution method, as shown in Fig. 2-32(c).

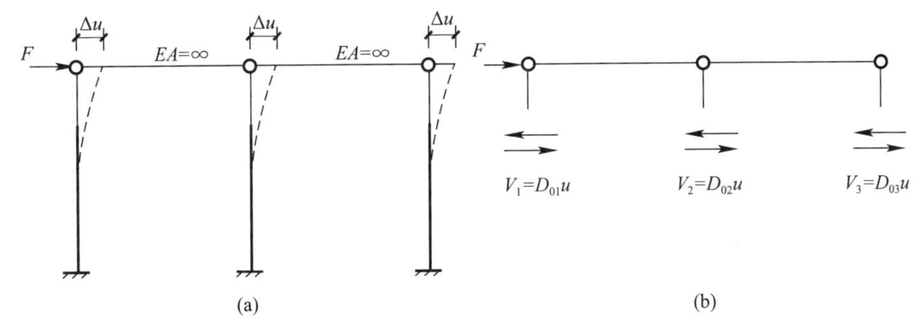

Fig. 2-31 Shear force of column top under horizontal load on the column top

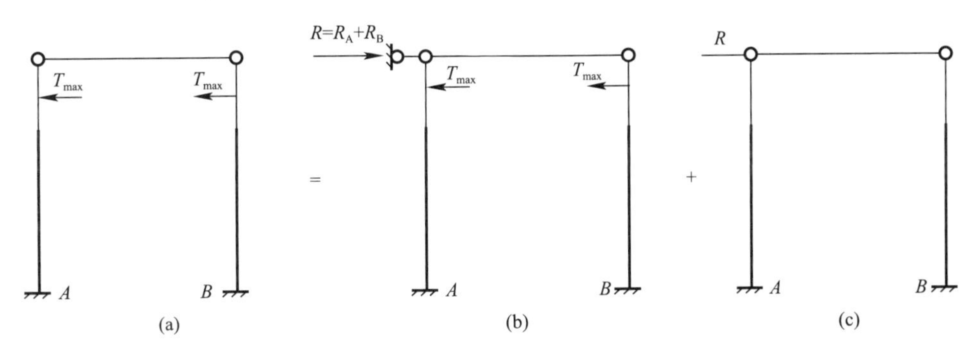

Fig. 2-32 Calculation of equal-height bent under arbitrary load

(3) Superimpose the above two states to restore the original state of the bent frame, that is, superimpose the above two steps to obtain the shear force of each column top, $V_A=V_{A1}+V_{A2}$, $V_B=V_{B1}+V_{B2}$.

(4) After obtaining the shear force at the column top, the internal force of the bent column can be obtained.

The reaction force R of the fixed hinge support on the single-step column top under various loads can be calculated according to the Appendix 2.

【Example 2-2】

Conditions: The calculation diagram of a metalworking shop's bent is shown in Fig. 2-33. The shape and cross-section size of the columns

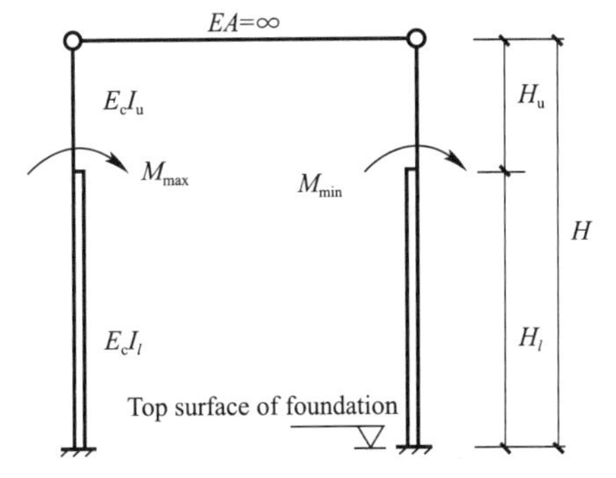

Fig. 2-33 Calculation diagram of the bent frame structure

A and B are the same (Fig. 2-34). $H_u = 3.3$ m and $H = 11$ m. The moments of inertia of the upper and lower columns are 2.13×10^9 mm^4 and 5.85×10^9 mm^4, respectively.

Question: Calculate the shear force at the column top according to the shear distribution method under the action of $M_{max} = 103$ kN · m and $M_{min} = 35.9$ kN · m.

【Solution】

① Calculate parameters n and λ

Moment of inertia of upper column: $I_u = \frac{1}{12} \times 400 \times 400^3 = 2.13 \times 10^9$ mm^4

Moment of inertia of lower column:

$$I_l = \frac{1}{12} \times 400 \times 600^3 - \frac{1}{12} \times 300 \times 350^3 - 2 \times \frac{1}{2} \times 25 \times 300 \times \left(170 + \frac{25}{3}\right)^2 = 5.85 \times 10^9 \text{ mm}^4$$

$$n = \frac{I_u}{I_l} = \frac{2.13 \times 10^9}{5.85 \times 10^9} = 0.36;\ \lambda = \frac{H_u}{H} = \frac{3.3}{11} = 0.3$$

② Add a virtual horizontal hinged support on the column top (Fig. 2-35)

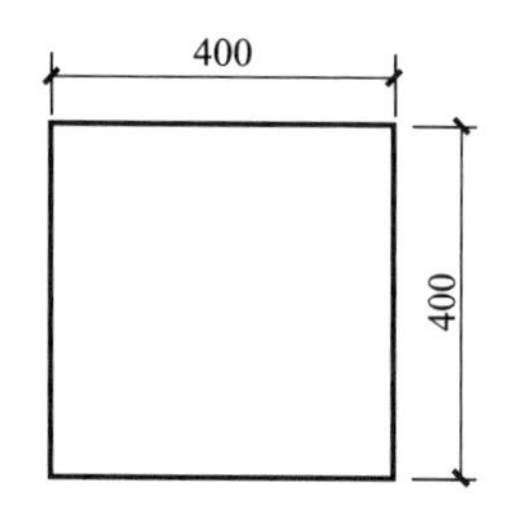

(a) The upper column section

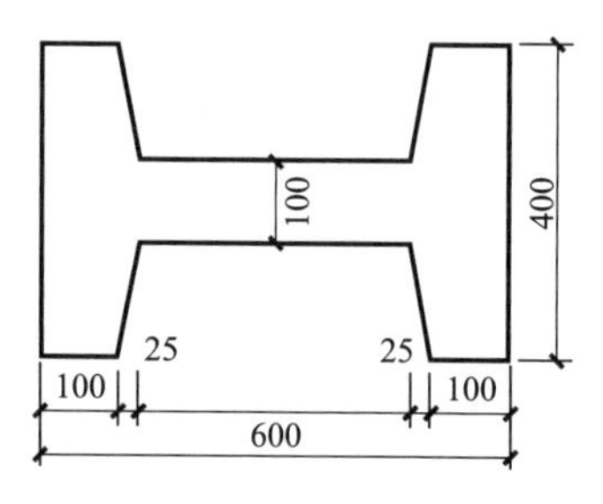

(b) The lower column section

Fig. 2-34 Schematic diagram of upper and lower column cross-sections

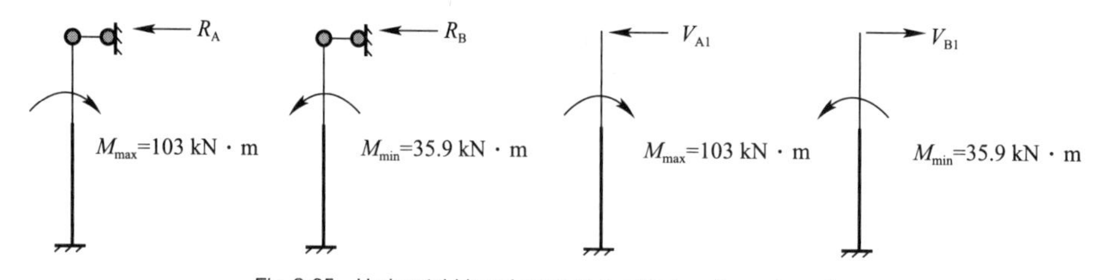

Fig. 2-35 Horizontal hinged support is added on the column top

According to Appendix 2, $C_0 = 1.5 \times \frac{1-\lambda^2}{1+\lambda^3\left(\frac{1}{n}-1\right)} = \frac{1.5 \times (1-0.3^2)}{1+0.3^3\left(\frac{1}{0.36}-1\right)} = 1.3$

$$R_A = -\frac{M_{max}}{H} C_0 = -\frac{103}{11} \times 1.3 = -12.17 \text{ kN};$$

$$R_B = \frac{M_{min}}{H} C_0 = \frac{35.9}{11} \times 1.3 = 4.24 \text{ kN}$$

So, $V_{A1} = R_A = -12.17$ kN; $V_{B1} = R_B = 4.24$ kN

③ Remove the virtual hinged support and reversely add the support reaction (Fig. 2-36)

$$V_{A2} = -\eta_A (R_A + R_B) = 0.5 \times (12.17 - 4.24) = 3.97 \text{ kN}$$

$$V_{B2} = 3.97 \text{ kN}$$

④ Superimpose the above two steps to solve the total shear force of column top

$$V_A = V_{A1} + V_{A2} = -12.17 + 3.97 = -8.20 \text{ kN}$$

$$V_B = V_{B1} + V_{B2} = 4.24 + 3.97 = 8.21 \text{ kN}$$

R_B R_A

Fig. 2-36 Adding support reaction after removing the virtual hinge support

2.2.4 Internal force combination

1. Control section of single-step bent column

In the general single-step bent column shown in Fig. 2-37, the reinforcements of each cross-section of the upper column are usually the same. The internal force of the bottom section of the upper column is the largest, so section Ⅰ-Ⅰ is the control section of the upper column. In the lower column, the reinforcements of each section are also usually the same. The internal forces of section Ⅱ-Ⅱ and section Ⅲ-Ⅲ are relatively larger, so section Ⅱ-Ⅱ and section Ⅲ-Ⅲ are the control sections of the lower column. In addition, the internal force of section Ⅲ-Ⅲ is also used for designing the foundation under column. Although section Ⅰ-Ⅰ and section Ⅱ-Ⅱ are at one place, which represent the upper and lower column section, respectively. In the design of section Ⅱ-Ⅱ, the influence of corbel on the cross-section bearing capacity is ignored.

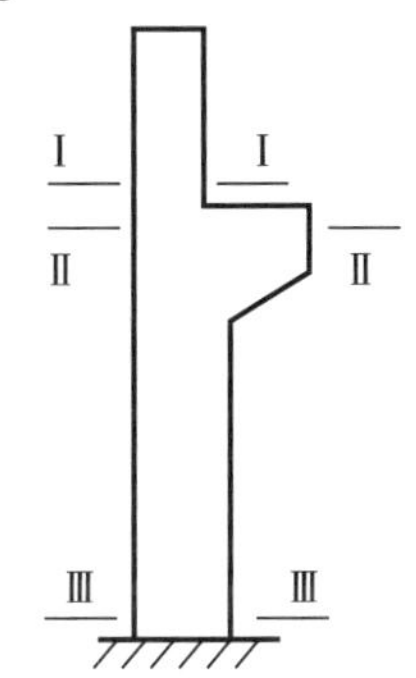

Fig. 2-37 Control sections of single-step bent column

2. Load effect combination

1) Basic combination of load effects

We can take the simplified load combination rule. Unfavorable values are determined according to the following load effect combination:

$$S=\gamma_G S_{Gk}+\gamma_{Q1} S_{Q1k} \tag{2-16}$$

$$S=\gamma_G \sum_{j=1}^{m} S_{Gjk}+0.9\sum_{i=1}^{n}\gamma_{Qi} S_{Qik} \tag{2-17}$$

Where,

γ_G—— the partial factor for the permanent load; when its effect is unfavorable to the structure, 1.3 should be taken for the combination; when the combination effect is beneficial to the structure, γ_{Gj} should be taken as 1.0;

γ_{Qi}—— the partial factor for the i-th live load. In general, 1.5 should be taken;

S_{Gjk}—— the load effect value calculated according to the characteristic value of permanent load G_{jk};

S_{Qik}—— the load effect value calculated according to the characteristic value of live load Q_{ik};

ψ_{ci}—— the combination value coefficient of live load Q_i as shown in Table 2-3;

m—— the number of permanent loads participating in the combination;

n—— the number of live loads participating in the combination.

2) Standard combination of load effects

For the checking calculation of the subgrade bearing capacity of bent frame structure, the standard combination of load effect should be adopted.

$$S_d=\sum_{j=1}^{m} S_{Gjk}+S_{Q1k}+\sum_{i=2}^{n}\psi_{ci} S_{Qik} \tag{2-18}$$

The meaning of other symbols is the same as before.

3) Quasi-permanent combination of load effects

When checking the crack width of bent column and foundation deformation, the quasi-permanent combination of load effect should be taken as:

$$S_d=S_{G1k}+\sum_{i=1}^{n}\psi_{qi} S_{Qik} \tag{2-19}$$

Where,

ψ_{qi}—— the quasi-permanent value coefficient of variable load Q_i as shown in Table 2-3.

The meaning of other symbols is the same as before.

3. Internal forces combinations

The internal forces of the control section of bent column include axial force, bending moment and shear force.

Bent column is an eccentric compression member, and the calculation of its longitudinal load-bearing reinforcement depends on the axial force N and bending moment M, and the influence of N and M on the bearing capacity is related. Generally, the following four most unfavorable internal force combinations can be considered.

Basic combination value coefficient ψ_c、quasi-permanence coefficient ψ_q

Table 2-3

Serial number	Type of live load	Combination value coefficient ψ_c	Quasi-permanent value coefficient ψ_q
1	Live load on roof	$\psi_c=0.7$	The roof without function, take $\psi_q=0$; a functional roof, take $\psi_q=0.4$
2	Snow load on roof	$\psi_c=0.7$	According to the different snow load zones Ⅰ, Ⅱ and Ⅲ, ψ_q is taken as 0.5, 0.2 and 0 respectively
3	Ash load on roof	Generally take $\psi_c=0.9$, roof of a single-storey industrial building near blast furnace, $\psi_c=1.0$	Generally take $\psi_q=0.8$, roof of a single-storey industrial building near blast furnace, $\psi_q=1.0$
4	Wind load on roof	$\psi_c=0.6$	$\psi_q=0$
5	Crane load	Soft hook crane: $\psi_c=0.7$; hard hook crane and working crane of A8 level, $\psi_c=0.95$	When designing bent, take $\psi_q=0$; when designing crane beam, for soft hook crane, working crane of A1—A3 level, $\psi_q=0.5$; working crane of A4 and A5 level, $\psi_q=0.6$; working crane of A6 and A7 level, $\psi_q=0.7$; for hard hook crane and working crane of A8 level, $\psi_q=0.95$

① $+M_{max}$ and corresponding N and V;

② $-M_{max}$ and corresponding N and V;

③ N_{max} and corresponding M and V;

④ N_{min} and corresponding M and V.

When the cross-section of the column adopts symmetrical reinforcement, the internal force combination of ① and ② are combined into one, namely $|M_{max}|$ and corresponding N and V. Generally, the above four kinds of internal force combination can meet the design requirements, but in some cases, they may not be the most unfavorable. For example, when the eccentricity $e_0=M/N$ is larger (i.e., M is larger, N is smaller), the reinforcement is more often. Therefore, M is not the maximum value but slightly smaller than the maximum value, and its corresponding N is much smaller, then the reinforcement required by this group of internal forces will be larger.

Attention should be paid to the internal force combination as follows.

1) Each combination must include the in-

ternal forces generated by the dead load.

2) Each combination takes one kind of internal force as the target to decide the choice of load items. For example, when considering the first internal force combination, we must take $+M_{max}$ as the goal, and then get the corresponding N and V values.

3) When N_{max} or N_{min} is taken as the combined target, the corresponding absolute value of M should be as large as possible. Therefore, the bending moment values in the load items which do not generate axial force (wind load and crane horizontal load) should also be combined.

4) There are two kinds of wind load items, i.e., left wind and right wind. Only one of them can be selected for each combination.

5) The following three points should be paid attention to for crane load items.

(1) Pay attention to the relationship between D_{max} (or D_{min}) and T_{max}. It is impossible to separate the horizontal load of crane from its vertical load. Therefore, when the internal force generated by T_{max} is used, the internal force generated by D_{max} (or D_{min}) in the same span should be combined, that is, "if there is T, there must be D". On the other hand, the vertical load of the crane can be separated from the horizontal load of the crane. Considering that T_{max} can act along both the left or the right direction, the vertical load of the crane can exist independently. If the internal force produced by D_{max} (or D_{min}) is used, T_{max} must be used at the same time to obtain the larger unfavorable internal force of M. Therefore, when combining the internal forces of crane load, the internal force of D and T need to be considered at the same time.

(2) Since the possibility of multiple cranes fully loaded at the same time is very small, the internal force should be multiplied by the corresponding load reduction factor when multiple cranes participate in the combination.

(3) The crane transversal horizontal load T_{max} acts on two columns along the left or the right direction in the same span. Only one situation can be selected. Also, the vertical load of crane, D_{max} (or D_{min}) acts on the left or right columns in the same span. Only one situation can be selected.

6) Since the horizontal shear force at the column bottom will produce the bending moment on the bottom of foundation, the corresponding horizontal shear force value should be calculated when combining the internal force of control section Ⅲ-Ⅲ.

2.2.5 Calculation of bent frame structure considering spatial behavior

1. The concept of spatial effect of single-storey industrial building

The single-storey industrial building is actually a space structure. Fig. 2-38 indicates schematic diagram of four types of horizontal displacements in a single-span industrial building with different structures or load conditions under the horizontal load on the column top. In Fig. 2-38(a), the horizontal displacement of each bent is the same. Therefore, it is actually the same as the bent without longitudinal members, which belongs to the planar bent frame structure. In Fig. 2-38(b), the lateral stiffness of the industrial building with gables is very large, and the horizontal displacement is small, every bent gets different restraints from the gables at both ends. Therefore, the horizontal displacement of the column top is curved, and $u_b < u_a$. In Fig. 2-38(c), the bent that is not directly loaded will also produce horizontal displacement due to the impact of the directly loaded bent. In Fig. 2-38(d), the horizontal displacement of each bent is smaller than that in Fig. 2-38(c) due to the gable, and $u_d < u_c$. Therefore, in the last three cases, each bent or gable wall cannot be deformed individually, but they are restrained each other. The overall effect of the interrelated relationship between the bent and the bent or between the bent and the gable wall is called the overall spatial function of the industrial building.

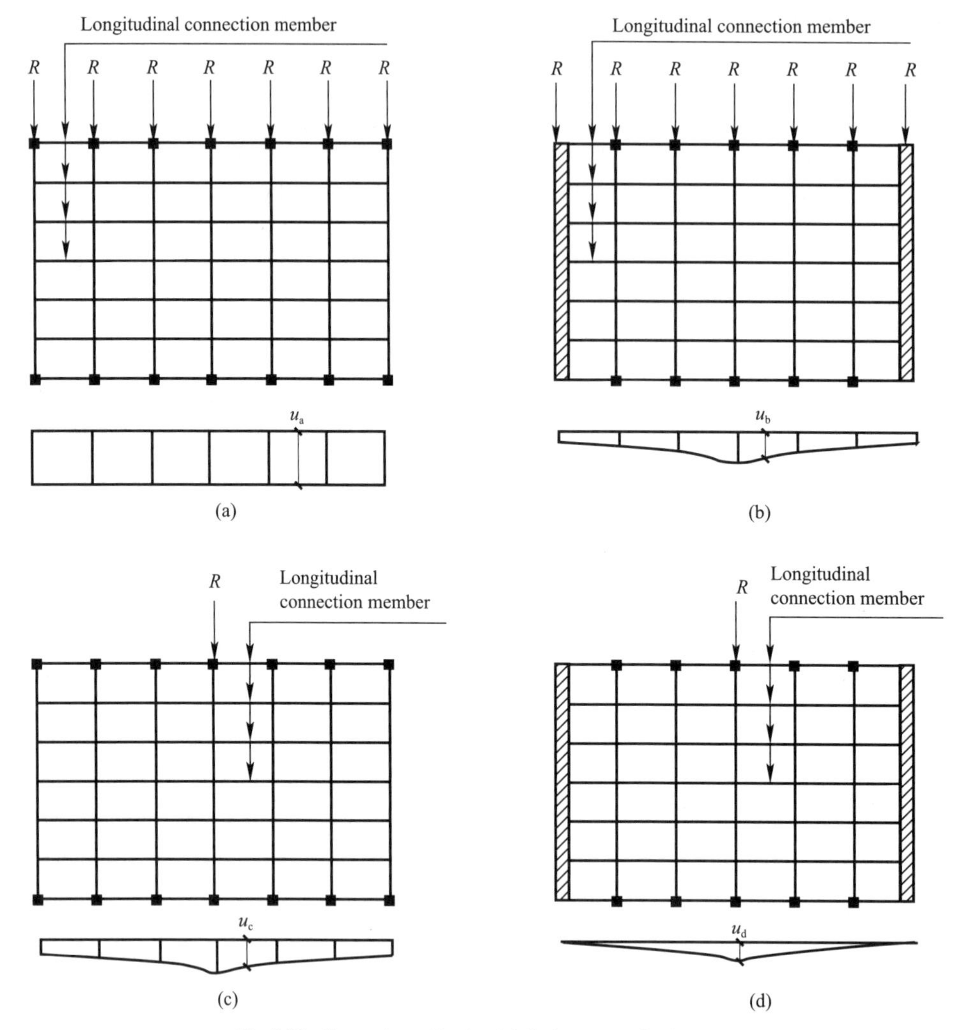

Fig. 2-38 Comparison of horizontal displacement of column top

There are two conditions for the overall spatial function of a single-storey industrial building. One is that there must be longitudinal members between the horizontal bents to connect them. The other is that the horizontal bents are different from each other, that is, the structure is different or the load is different.

2. Calculation method of bent frame under crane load considering spatial behavior

The crane load only acts on a few bents, which are a kind of local load. Therefore, when calculating the internal force of the structure according to the planar bent under the crane load, the overall spatial effect of the bent may be considered.

When the bent is subjected to a single horizontal load, due to the overall spatial function of the single-storey industrial building is shown in Fig. 2-38 (c) or (d). When a horizontal concentrated force R acts on a column top of a bent, due to the spatial function of the industrial building, the horizontal concentrated force R is not only borne by the directly load-bearing bent, but also a part of R will be transmitted to other adjacent bents through longitudinal connection members. The horizontal concentrated force of the bent is reduced, its value is R', and $R' < R$. The ratio of R' to R is called the spatial distribution coefficient, denoted by η.

For an elastic structure, the force is proportional to the horizontal displacement, so the spatial distribution coefficient can also be expressed as:

$$\eta=\frac{R'}{R}=\frac{u'}{u}\leqslant 1 \tag{2-20}$$

Where,

u'—— the displacement of the column top of the directly loaded bent when considering spatial function. In Fig. 2-38 (c), $u'=u$. In Fig. 2-38(d), $u'=u_d$. u is the displacement of column top calculated according to the planar bent.

η reflects the special working effect of industrial building. The smaller the η value, the greater the spatial function of the industrial building, and vice versa.

F_x is the reaction force of the elastic support (Fig. 2-39). According to the proportional relationship between the force and the displacement, u: $(1-\eta)u=R$, so, $F_x=(1-\eta)R$. Therefore, the internal force calculation steps of the bent frame structure when considering the overall spatial function of industrial building under the crane load are as follows according to Fig. 2-39.

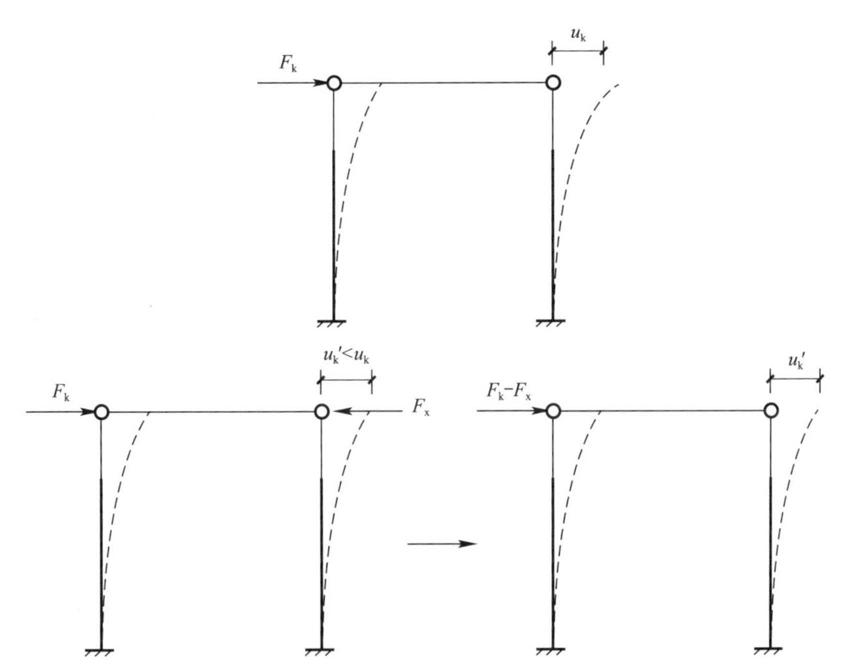

Fig. 2-39 Calculation method of bent in industrial building space

1) Apply a horizontal hinged support on the column top and calculate the reaction force R of the support by Appendix 2.

2) Add adversely the reaction force R on the column top which is superimposed with the reaction force $(1-\eta)R$, that is, $R-(1-\eta)R=\eta R$ on the column top. According to the shear distribution method, the shear force at the top of each column is calculated according to the shear force distribution method.

3) The shear force at the column top calculated by superimposing the above two steps is the shear force at the column top of the bent frame structure considering the whole spatial function. According to the shear force at the column top and the load on the column, the bending moment of each column can be obtained according to the cantilever column.

2.3 Design of column of single-storey industrial building

The design of the column generally includes: ① determining the column shape and cross-section size; ② designing the section bearing capacity according to the most unfavorable internal force of each control section and meeting the structural requirements; ③ checking bearing capacity and crack width during construction hoisting and transportation; ④ designing connecting structure with roof truss, crane beam and other members and drawing construction drawings, etc. Corbel designing is required when there is the crane.

2.3.1 Column forms

The columns in the single-storey industrial building are mainly composed of two types, i. e. , the bent column and the wind-resisting column.

1) Bent column

The bent column of a single-storey industrial building is generally composed of an upper column, a lower column and a corbel. There are many types of bent columns. At present, the commonly used upper columns are mainly rectangular or circular, and the lower columns include rectangular column, I-shaped column, latticed column, pipe column, etc. as shown in Fig. 2-40.

Rectangular column shown in Fig. 2-40(a) is the most commonly used which has the advantages of simple shape, convenient construction and good seismic performance. However, the amount of concrete is large and the economic property is poor. The column with I-shaped cross-section shown in Fig. 2-40(b) is relatively reasonable, which can give full play to the load-bearing effect of concrete and has good integrity and is currently widely used in single-storey industrial building, but its concrete consumption is more than that of double-legged columns.

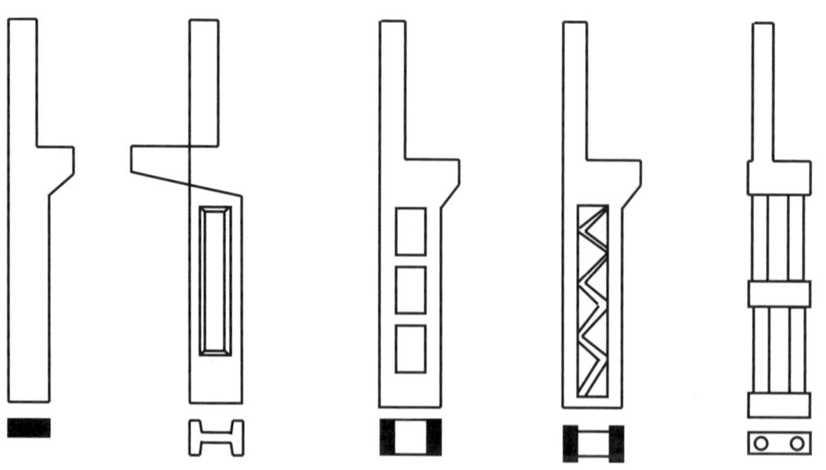

(a) Rectangular column (b) I-shaped column (c) Latticed column with horizontal web (d) Latticed column with diagonal web (e) Double-legged pipe column

Fig. 2-40 Forms of the lower columns of the single-storey industrial building

The latticed column includes horizontal flat web rods shown in Fig. 2-40(c) and diagonal web rods shown in Fig. 2-40(d). The structure of the horizontal web bar is simpler and more convenient for manufacture.

When bearing a large horizontal load, it is advisable to adopt a latticed columnwith diagonal web rods. However, its construction and production are more complicated. Compared with I-shaped column, the latticed column has less concrete consumption and lighter weight, especially when the column is tall, but its integral rigidity is lower.

The pipe column consists of round pipe column and square pipe column, which can be made into single-legged, double-legged or four-legged column. Currently, double-legged column shown in Fig. 2-40(e) is more commonly used.

According to engineering experience, the section type of columns can be determined according to the section height h.

When $h \leqslant 600$ mm, rectangular section column shall be adopted;

When h = 600—800 mm, I-shaped or rectangular section column shall be adopted;

When h = 900—1400 mm, I-shaped section column shall be adopted;

When $h > 1400$ mm, battened column shall be adopted.

2) Wind-resisting column

The wind-resisting column is generally composed of an upper column and a lower column. The upper column is generally rectangular in cross-section, and the lower column is generally I-shaped cross-section.

2.3.2 Section size of column

In addition to ensure that the column has a certain bearing capacity, the column should also be sufficiently rigid, so as not to cause excessive horizontal and vertical deformation of the building, or cause cracks in the wall and roof and the normal use of the plant.

The cross-section form and size of the column depend on the height of the column and the lifting capacity of the crane. According to the rigidity requirements, the minimum cross-section size of the industrial building column and the outdoor trestle column with 6 m column spacing can be determined according to Table 2-4.

2.3.3 Effective length of column

Before calculating the reinforcement of bent column, it is necessary to determine the effective length of the column. The effective length l_0 of the column of a single-storey industrial building with rigid roof can be adopted according to Table 2-5.

Cross-section size of column in a single-storey industrial building with 6 m column spacing **Table 2-4**

Project	Schematic diagram	Situation	Section height h	Section width b
Industrial building without crane		Single span	$\geqslant H/18$	$\geqslant H/30$ and $\geqslant 300$; pipe column $r \geqslant H/105$, $D \geqslant 300$ mm
		Multi-span	$\geqslant H/20$	

Continued

Project	Schematic diagram	Situation		Section height h	Section width b
Industrial building with crane		$Q \leqslant 10$ t		$\geqslant H_K/14$	$\geqslant H_l/20$ and $\geqslant 400$; pipe column $r \geqslant H_l/85$, $D \geqslant 400$ mm
		$Q=$ (15—20) t	$H_K \leqslant 10$ m	$\geqslant H_K/11$	
			10 m $< H_K \leqslant 12$ m	$\geqslant H_K/12$	
		$Q=$ 30 t	$H_K \leqslant 10$ m	$\geqslant H_K/9$	
			$H_K > 12$ m	$\geqslant H_K/10$	
		$Q=$ 50 t	$H_K \leqslant 11$ m	$\geqslant H_K/9$	
			$H_K > 13$ m	$\geqslant H_K/11$	
		$Q=$ (75—100) t	$H_K \leqslant 12$ m	$\geqslant H_K/9$	
			$H_K > 14$ m	$\geqslant H_K/8$	
Outdoor trestle		$Q \leqslant 10$ t		$\geqslant H_K/10$	$\geqslant H_l/20$ and $\geqslant 400$; pipe column $r \geqslant H_l/85$, $D \geqslant 400$ mm
		$Q=$ (15—30) t	$H_K \leqslant 12$ m	$\geqslant H_K/9$	
		$Q=$ 50 t	$H_K \leqslant 12$ m	$\geqslant H_K/8$	

Notes:

1. In the table, Q is the lifting capacity of the crane; H is the total height of the column from the top of the foundation to the top of the column; H_l is the height from the top of the foundation to the bottom of the crane beam; H_K is the height from the top of the foundation to the top of the crane beam; r is the single pipe rotation of the pipe column; D is the outer diameter of the single pipe of the pipe column;

2. When adopting latticed column with horizontal web, h should be multiplied by 1.1; when adopting latticed column with oblique web, h should be multiplied by 1.05;

3. The height of the cross-section of the industrial building with crane is considered according to the heavy and extra heavy load conditions, such as medium and light load condition, should be multiplied by a factor of 0.95;

4. When the column spacing of the plant is 12 m, the cross-section size of the column should be multiplied by a factor of 1.1.

2.3.4 Reinforcement calculation of bent column

Calculation of reinforcement of each control section of the bent column is based on unfavorable internal force combination values (M, N, V). In general, the bent direction of the column is calculated according to the eccentric compression member. The calculated longitudinal steel bars are symmetrically arranged on both sides of the bending moment direction of the column. The calculation method of eccentric compression members is detailed in the textbook ***Principles of Concrete Structure Design***.

Effective length l_0 of column **Table 2-5**

Type of column		Bent transversal direction	Bent longitudinal direction	
			With inter-column support	No inter-column support
Column of no crane industrial building	Single span	$1.5H$	$1.0H$	$1.2H$
	Two-span and multi-span	$1.25H$	$1.0H$	$1.2H$
Column of industrial building with crane	Upper column	$2.0H_u$	$1.25H_u$	$1.5H_u$
	Lower column	$1.0H_l$	$0.8H_l$	$1.0H_l$
Outdoor crane column and trestle column		$2.0H_l$	$1.0H_l$	—

Notes:

1. In the table, H is the height of the column from the top of the foundation; H_l is the height of the lower column from the top of the foundation to the bottom of the assembled crane beam or the top surface of the cast-in-place crane beam; H_u is the height of the upper column.

2. In the table, the effective length of the upper column in the bent direction is only applicable when $H_u/H_l \geqslant 0.3$; when $H_u/H_l < 0.3$, $2.5H_u$ should be adopted.

2.3.5 Construction detailing of column

1. Material

The concrete strength grade of column is generally C20—C40. When adopting steel bars with strength of 400 MPa and above, concrete strength grade should not be lower than C25. The concrete strength grade has a great influence on the bearing capacity of the column under compression, so higher strength grade concrete should be adopted.

Longitudinal steel bars of columns generally adopt HRB400, HRB335. Constructional reinforcement usually adopts HPB300 or HRB335. Column stirrups usually adopt HRB335, HRB400 or HPB300.

2. Longitudinal reinforcement

1) The diameter of longitudinal steel bars for column should not be less than 12 mm. The reinforcement ratio of all longitudinal steel bars should not be greater than 5%. When the concrete grade is less than or equal to C50, the reinforcement ratio of all longitudinal steel bars should not be less than 0.5%; when the concrete grade is greater than C50, it should not be less than 0.6%.

2) When the section height of the column is large than 600 mm, longitudinal construction steel bars with a diameter of 10—16 mm should be installed on the sides, and set compound stirrups or tie bars accordingly.

3) The clear distance of longitudinal steel bars of the column should not be less than 50 mm, and should not be greater than 300 mm. For horizontally precast columns, the minimum clear distance should not be less than 25 mm and the diameter of the longitudinal steel bars. Middle distance of longitudinal bars perpendicular to the bending moment plane should not be larger than 350 mm.

4) In eccentric compression column, for longitudinal steel bars on the sides perpendicular to the action plane of the bending moment, and the longitudinal steel bars on each side of the axial compression column, the middle distance should not be greater than 300 mm.

3. Stirrup

1) The stirrup diameter is $d_{sv} \geqslant d/4$, and greater than or equal to 6 mm. Here, d is the maximum diameter of the longitudinal reinforcement.

2) The stirrup spacing should not be grea-

ter than 400 mm, larger than the short side dimension of the member section and greater than $15d$ (d is the minimum diameter of the longitudinal steel bars).

3) When the reinforcement ratio of all longitudinal steel bars in the column is greater than 3%, diameter of the stirrup $d_{sv} \geq 8$ mm, the spacing should not be greater than $10d$ (d is the minimum diameter of longitudinal tensile bar) and shouldn't be greater than 200 mm. The end of the stirrup is made into a 135° hook, and the length of the straight section at the end of the hook should not be less than 10 times of the diameter of the stirrup.

4) When the size of the short side of the column section is greater than 400 mm and the number of longitudinal bars on each side is more than 3, or when the short side of the column is not more than 400 mm but the number of longitudinal bars is more than 4, compound stirrups should be set.

5) The stirrups should be made into a closed form, and no inner corners are allowed. The structure of the stirrups of the I-shaped and L-shaped column sections is shown in Fig. 2-41.

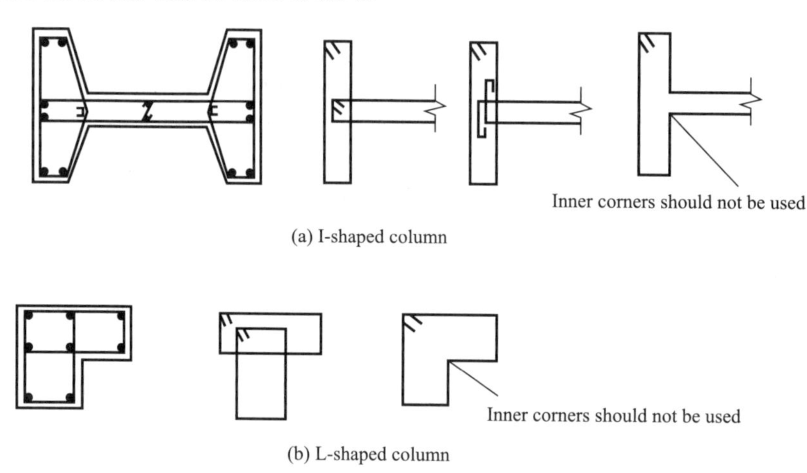

Fig. 2-41 Stirrup of I-shaped or L-shaped sections

2.3.6 Hoisting checking of column

The precast column should be checked according to the actual strength of the concrete during transportation and hoisting. Considering turning over lifting (Fig. 2-42a) or flat lifting (Fig. 2-42b), the most unfavorable position and the corresponding calculation diagram are shown in Fig. 2-42(c). In the Fig. 2-42(c), g_1 is the weight of the upper column; g_2 is the weight of the corbel part; g_3 is the weight of the lower column. The bearing capacity and crack width are checked respectively according to the control sections 1-1, 2-2 and 3-3 in Fig. 2-42 (c). Pay attention to the following issues when checking calculations.

1. Considering construction vibration, the weight of the column should be multiplied by the dynamic coefficient 1.5. (it can be increased or decreased according to the force during hoisting)

2. The hoisting checking calculation is temporary. Therefore, the safety level of the member can be reduced by one level compared with the safety level of the using stage.

3. The concrete strength of the column is generally considered at 70% of the design strength. When hoisting, the concrete strength shall not be less than 70% of the design strength.

4. Single point binding is generally adopted for hoisting, and the hoisting point should be set at the bottom of the corbel. When multiple lifting points are required, the lifting method shall be negotiated with the construction company and

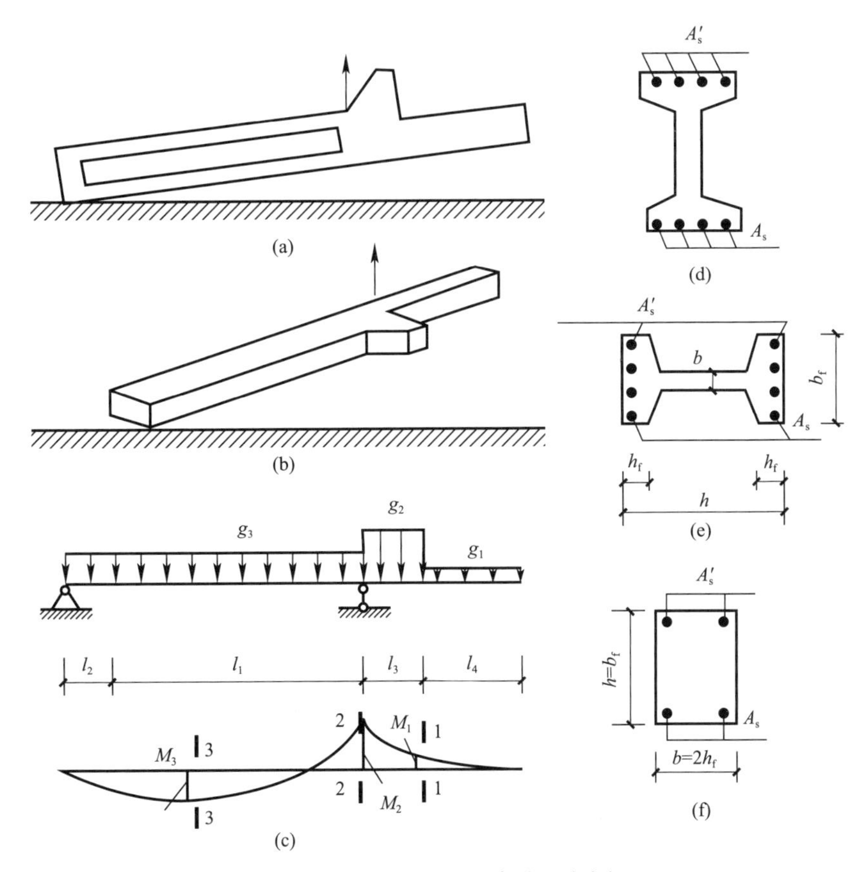

Fig. 2-42 Checking diagram of column hoisting

the corresponding checking calculation shall be carried out.

5. When the reinforcement of the cross-section at the variable step of the column is insufficient, short steel bars can be added in the local section.

6. When lifting by turning over method, the cross-section of the lower column is checked according to the I-shaped cross-section (Fig. 2-42d). When using flat lifting, the cross-section of the lower column is checked according to the rectangular cross-section (Fig. 2-42e). At this time, the width of the rectangular section is $2h_f$, the stressed steel bars only consider the steel bars at the upper and lower edges (Figs. 2-42e and f).

2.3.7 Design of corbel

In a single-storey industrial building, the corbel protruding from the column side is often used to support crane beam, roof truss (roof beam) and bracket. Although the corbel is relatively small, the load is large or has a dynamic effect, it is a relatively important structural member. The main contents of the corbel design are to determine the cross-section size, bearing capacity calculation and reinforcement of the corbel.

According to the horizontal distance a from the point of action of the vertical force F_v of the corbel to the edge of the lower column, the corbel is generally divided into two categories.

When $a \leqslant h_0$, it is a short corbel (Fig. 2-43), and when $a > h_0$, it is a long corbel (Fig. 2-44). Here h_0 is the effective height of the vertical section of the corbel at the junction of the corbel and the lower column. The mechanical properties of the long corbel are similar to the cantilever beam and the long corbel can be designed according to the cantilever beam. The corbel supporting crane beams or other members is a short corbel and it is a variable cross-section deep beam. Its mechanical property is different from that of ordinary cantilever beam.

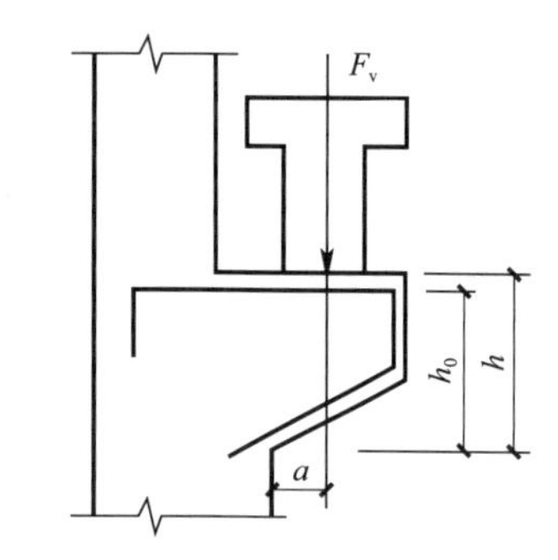

Fig. 2-43 Short corbel($a \leq h_0$)

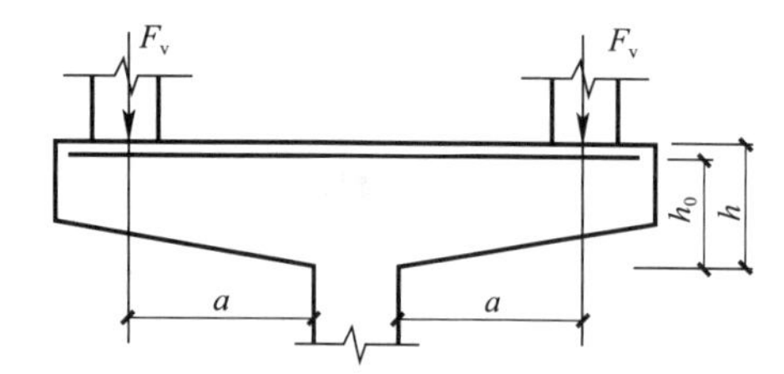

Fig. 2-44 Long corbel ($a>h_0$)

This section will clarify the design method of the corbel on the basis of the results of the corbel's experimental investigation.

1. Experimental investigation

1) Stress distribution in elastic stage

Fig. 2-45 shows the principal stress traces obtained from the photo elastic experiment on the model of epoxy corbel with $a/h_0 = 0.5$. As illustrated, on top of the corbel, the main tensile stress trace is basically parallel to the upper edge of the corbel and relatively uniform along the length; the main compressive stress traces of the corbel are roughly parallel to the line ab (from the loading point b to the lower corner of the corbel); there is stress concentration at the junction of the corbel and the upper column; the main tensile stress trace in the middle and lower part of the corbel is inclined, which explains why the cracks starting from the loading slab have the trend downward sloping.

2) Appearance and spread of cracks

The experiment of reinforced concrete corbel under the action of vertical force shows that the initial vertical cracks generally appear when the load reaches 20%—40% of the ultimate load, but at this time the cracks are small and develop slowly, which has little effect on the performance of the corbel. As the load continues to increase to reach 40% to 60% of the ultimate load, the first diagonal crack ① appears near the inside of the loading slab (Fig. 2-46). After that, as the load increases, apart from the continuous development of this diagonal crack, the second diagonal crack almost no longer appears. Until approach to the destruction (about 80% of the ultimate load), a second crack ② suddenly appears, which indicates that the corbel is about to break. During use stage of corbel, the so-called allowable diagonal cracks refer to crack ①, which is the main factor to control the size of the corbel section.

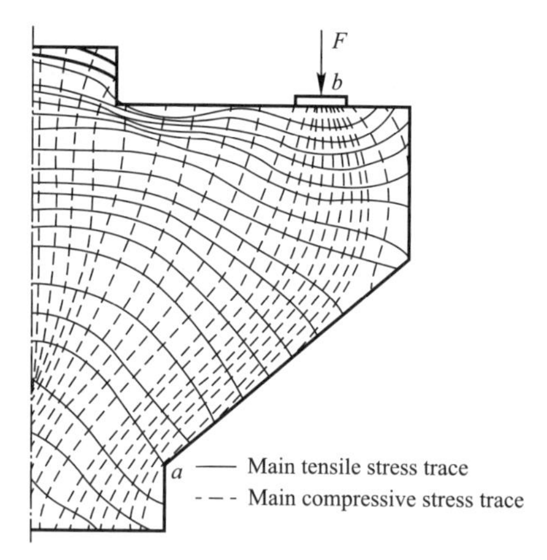

Fig. 2-45 Photo elastic experiment

The experiment shows that a/h_0 is the main parameter which affects the appearance of diagonal cracks. As a/h_0 increases, the load of diagonal cracks continues to decrease. This is because as the a/h_0 increases, the horizontal stress also increases, and the vertical stress decreases. Then the main tensile stress increases and diagonal cracks appear earlier.

3) Failure modes of corbel

The failure pattern of the corbel mainly depends on the a/h_0 value, there are three main failure modes (Fig. 2-47).

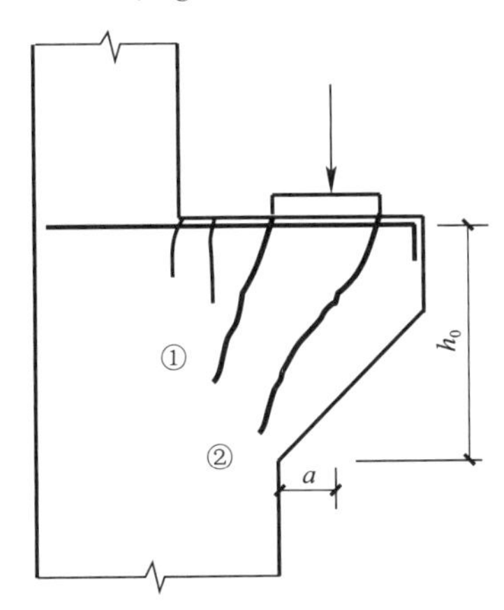

Fig. 2-46 Schematic diagram of corbel crack

(1) Bending failure

When $a/h_0 > 0.75$ and the reinforcement ratio of longitudinal tensile bar is low, bending failure generally occurs. Its characteristic is that when diagonal crack ① appears, the crack continues to extend to the compressive region as the load increases; the tensile stress of the horizontal longitudinal tensile bars gradually increases and reaches the yield strength. At this time, the outer part of the diagonal crack ① rotates around the junction point between the lower part of the corbel and the column, causing the concrete in the compression zone to crush damage as shown in Fig. 2-47(a).

(2) Shear failure

Shear failure is divided into three types, that is, pure shear failure, diagonal compression failure and diagonal tension failure.

Pure shear failure close to the vertical section along the inner side of the loadingslab may occur (Fig. 2-47b) when the value of a/h_0 is smaller than 0.1 or the value of a/h_0 is large but the edge height h_1 is small. It is characterized by a series of short diagonal cracks on the interface between the corbel and the lower column. Finally the corbel is cut from the column along the cracks and is destroyed. At this time, the stress of longitudinal steel bar in the corbel is lower.

Diagonal compression failure (Fig. 2-47c) occurs when $0.1 \leqslant a/h_0 \leqslant 0.75$ and there are many horizontal stirrups. As the load increases, a batch of short and fine diagonal cracks ② appear on the outside of the diagonal crack ①. These diagonal cracks gradually develop, when the main compressive stress of concrete between diagonal cracks exceeds its compressive strength, the concrete peels off and the corbel is damaged.

Diagonal tension failure (Fig. 2-47d) occurs when $0.1 \leqslant a/h_0 \leqslant 0.75$ and there are few horizontal stirrups. Due to the large main tensile stress, a full-length diagonal crack ③ suddenly appears under the loading slab, and the corbel is damaged along the crack.

(3) Local compression failure

When the loading slab is too small or the concrete strength is too low, the concrete under the loading slab will be partially crushed and damaged due to large local compressive stresses as shown in Fig. 2-47(e).

In addition, when the corbel's longitudinal tensile bar is insufficiently anchored, the steel bar will be pulled out and other damages will occur.

4) The corbel under the action of vertical force and horizontal tension

The experiment results of corbels with vertical force F_v and horizontal tension F_h at the same time show that the load when the diagonal

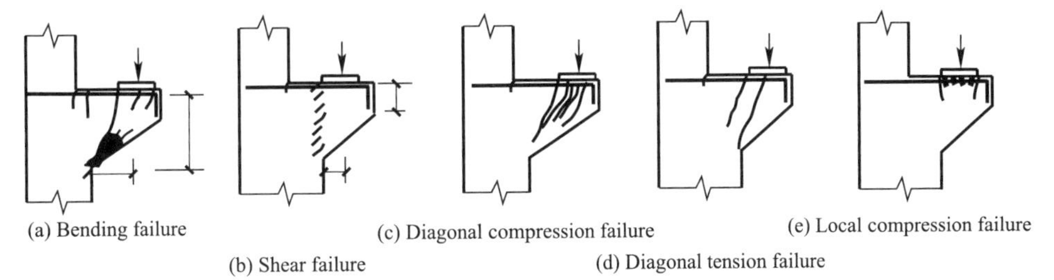

(a) Bending failure (b) Shear failure (c) Diagonal compression failure (d) Diagonal tension failure (e) Local compression failure

Fig. 2-47 Destruction form of corbel

cracks occur in the corbel with vertical and horizontal loads is lower than that of the corbel with only vertical force. When $F_v/F_h = 0.2$—0.5, the cracking load is reduced by 36%—47%. At the same time, the bearing capacity of the corbel is also reduced. The experiment results also show that the damage laws of the corbel with horizontal tension and the corbel without horizontal tension are similar.

2. Corbel design

1) Cross-section size of corbel

The cross-section dimensions of the column corbel (including the width of the corbel, the length of the top surface, the height of the outer edge and the inclination angle of the bottom surface, etc.) can be determined with reference to the structural requirements of Fig. 2-48.

Determining the length of the top surface of the corbel according to the width of the crane beam and the distance from the outer edge of the crane beam to the outer edge of the corbel (greater than or equal to 100 mm).

According to the structural requirements of corbel outer edge height $h_1 \geqslant h/3$ and $h_1 \geqslant$ 200 mm (h_1 shall not be too small, otherwise shear failure will occur), and the inclination angle α should not be greater than 45°, generally 45° (in order to prevent the occurrence of diagonal cracks that may cause serious stress concentration at the intersection of the bottom surface and the lower column).

The corbel failure occurs after the diagonal cracks appear and propagate according to the experiment results. Therefore, the height of corbel section is determined generally by diagonal crack controlling condition and detailing requirement as formula (2-21) and Fig. 2-48:

$$F_{vk} \leqslant \beta\left(1-0.5\frac{F_{hk}}{F_{vk}}\right)\frac{f_{tk}bh_0}{0.5+\dfrac{a}{h_0}} \tag{2-21}$$

Where,

F_{vk} —— the vertical force acting on the top of the corbel, calculated according to the standard combination of load effect;

F_{hk} —— the horizontal tension acting on the top of the corbel, calculated according to the standard combination of load effect;

β —— the crack control coefficient: 0.65 for corbels supporting crane beams; 0.8 for other corbels;

b —— the width of corbel;

h_0 —— the effective height of the vertical section at the junction of the corbel and the lower column; $h_0 = h_1 - a_s + c \cdot \tan\alpha$. When $\alpha > 45°$, taking $\alpha = 45°$. c is the horizontal length from the edge of the lower column to the outer edge of the corbel.

In formula (2-21), $(1-0.5F_{hk}/F_{vk})$ is to

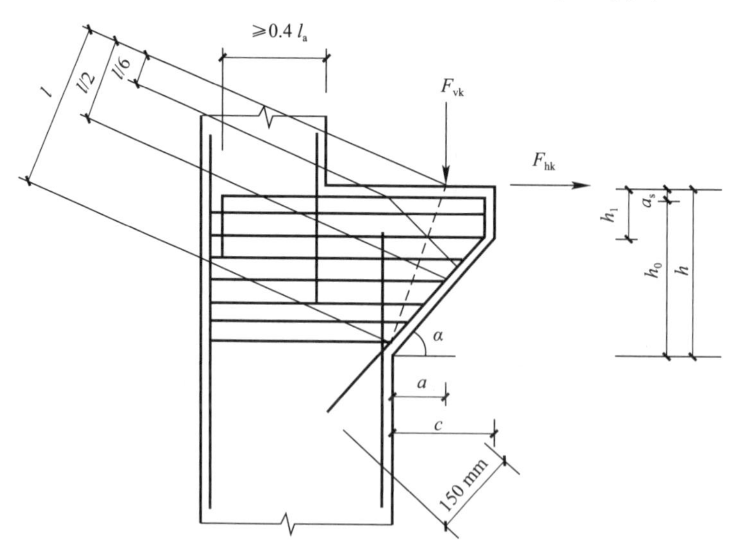

Fig. 2-48 Geometric size and reinforcement of corbel

consider the adverse effect on the crack resistance of the corbel under the simultaneous action of the horizontal tension F_{hk}. The coefficient β takes into account the requirements of different using conditions for the crack resistance of the corbel. When $\beta = 0.80$, most corbels will not have cracks under normal conditions, and some only have slight cracks.

According to the experiment results, the longitudinal reinforcement of the corbel has basically no effect on the appearance of diagonal cracks. Bending bars play an important role in the propagation of diagonal cracks, but it has no obvious effect on the appearance of diagonal cracks. Therefore, in formula (2-21), the parameters related to longitudinal reinforcement and bending reinforcement are not introduced.

The size of the loading slab has a certain influence on the bearing capacity of the corbel. The larger the size of the loading slab (sufficient rigidity) is, the higher the bearing capacity of the corbel will be. If the size is too small, it will reduce the bearing capacity. Therefore, the ***Code for Design of Concrete Structures*** (GB 50010—2010) stipulates that under the action of the vertical force F_{vk}, the local compressive stress on the support surface of the corbel should not exceed $0.75f_c$.

2) Bearing capacity calculation and reinforcement

(1) Calculation diagram

The experiment results showed that the longitudinal steel bars on the top surface of the corbel are under tension. After the diagonal crack appears, the stress of the steel bar increases sharply. The stress of the steel bar is evenly distributed along the full length of the top surface of the corbel during failure.

Steel bars are like horizontal tension rods in a truss. The concrete is compressed inthe range outside the diagonal crack, and the diagonal compressive stress is relatively uniform, just like the compression rod in a truss (Fig. 2-49a). The compressive stress of concrete can reach its compressive strength when it is broken.

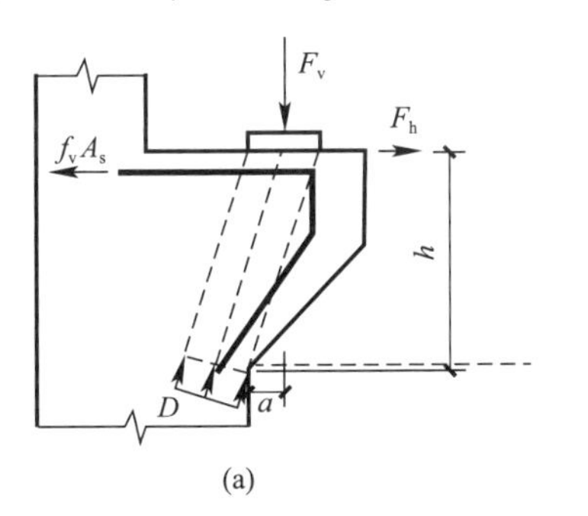

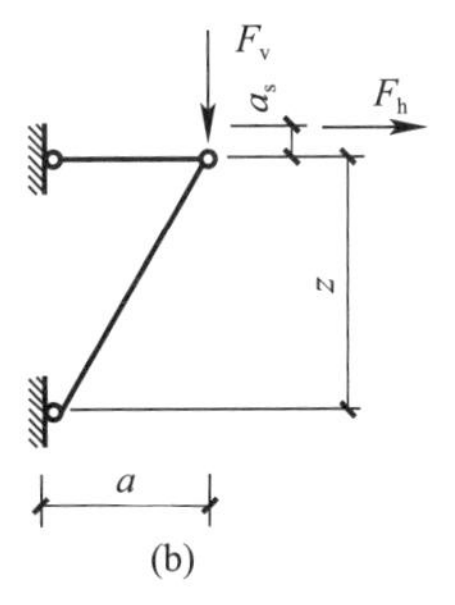

Fig. 2-49 The calculation diagram of corbel

According to the above analysis of the mechanical properties of the corbel, the corbel can be simplified into a triangular truss during the calculation. Steel bars are horizontal tie rod, concrete is compression strut. When the vertical force and horizontal force work together, the calculation diagram is shown in Fig. 2-49(b).

(2) Calculation of longitudinal tensile reinforcement

According to Fig. 2-49(b), taking the balance condition, we can get:

$$f_yA_sZ = F_va + F_h(Z + a_s) \qquad (2\text{-}22)$$

If approximately $z = 0.85h_0$, we can get:

$$A_s = \frac{F_va}{0.85f_yh_0} + \left(1 + \frac{a_s}{0.85h_0}\right)\frac{F_h}{f_y} \qquad (2\text{-}23)$$

In formula (2-23), if approximately taking $a_s/0.85h_0 = 0.2$, then the total area of the longitudinal tensile steel bars bearing the vertical force and the horizontal tensile force is calculated as follows:

$$A_s = \frac{F_va}{0.85f_yh_0} + 1.2\frac{F_h}{f_y} \qquad (2\text{-}24)$$

Where,

F_v—— the design value of the vertical

force acting on the top of the corbel;

F_h—— the design value of the horizontal tension acting on the top of the corbel;

a —— the horizontal distance from the acting point of the vertical force to the edge of the lower column. When $a<0.3h_0$, $a=0.3h_0$.

(3) Detailing of longitudinal tensile reinforcement

The longitudinal tensile bars arranged along the top of the corbel should be HRB335, HRB400 or RRB400 hot-rolled steel bars. All longitudinal reinforcement and bent-up bars should extend 150 mm into the lower column along the outer edge of the corbel and then be cut off (Fig. 2-48). The horizontal section including the curved arc section is not less than $0.4l_a$, while the vertical section is not less than $15d$, and the total length is not less than l_a.

According to formula (2-23), the reinforcement ratio of longitudinal tensile bar bearing vertical force should not be less than 0.2% and $0.45f_t/f_y$, and not be greater than 0.6%. It should not be less than 4, and the diameter should not be less than 12 mm. The horizontal anchor bars that can withstand horizontal tension shall be welded to the embedded parts, and the number shall be no less than 2.

When the corbel is set on the top of the upper column, it is advisable to bend the longitudinally stressed steel bars on the outside of the column opposite the corbel into the corbel horizontally along the top of the column, used as corbel longitudinal tensile bars. When the longitudinal tensile bars on the top surface of the corbel and the longitudinal reinforcement on the outside of the column opposite the corbel are arranged separately, the longitudinal tensile bars on the top surface of the corbel should be bent into the outside of the column and comply with the requirements for steel lap (Fig. 2-50).

3) Detailing of horizontal stirrups and bent-up bars

Because the diagonal crack (control) condition of formula (2-21) is stricter than the shear capacity condition of diagonal section. It is no longer required to calculate the shear capacity of the diagonal section of the corbel when satisfying formula (2-21). However, horizontal stirrups and bent-up bars should be set according to structural requirements.

The diameter of the horizontal stirrups of the corbel is 6—12 mm, and the spacing is 100—150 mm. The total cross-section area of the horizontal stirrups in the upper $2h_0/3$ range should not be less than 1/2 of the cross-section area of the longitudinal tensile bars bearing the vertical force.

Experiment results show that although the bent-up bars have little effect on the anti-cracking of the corbel, it has a significant effect on restricting the propagation of cracks. When the shear span ratio $a/h_0 \geqslant 0.3$, the bent-up bars should be set, which can increase the bearing capacity of the corbel by 10% to 30%. When the shear span is relatively small, and the bent-up bars in the corbels cannot fully work. Bent-up bars should be HRB335 or HRB400 hot-rolled ribbed steel bars, and should be placed in the range between l/6 and l/2 on the upper part of the corbel (Fig. 2-48). The cross-section area should not be less than 1/2 of the cross-section area of the longitudinal tensile bars bearing the vertical force. The number of bent-up bars should not be less than 2, and the diameter should not be less than 12 mm. Longitudinal tensile bars shall not be used as the bent-up bars.

4) Checking calculation of local compressive bearing capacity under the loading slab

The local pressure bearing capacity under the loading slab shall meet the requirements of formula (2-25):

$$\frac{F_{vk}}{A} \leqslant 0.75f_c \tag{2-25}$$

Where,

A —— the local bearing area, $A=ab$, where a and b are the length and width of the loading slab, respectively;

f_c—— the design value of concrete compressive strength.

When the local pressure does not meet the requirements, some measures should be taken. For example, we can increase the local pressure-bearing area or increase the concrete strength grade.

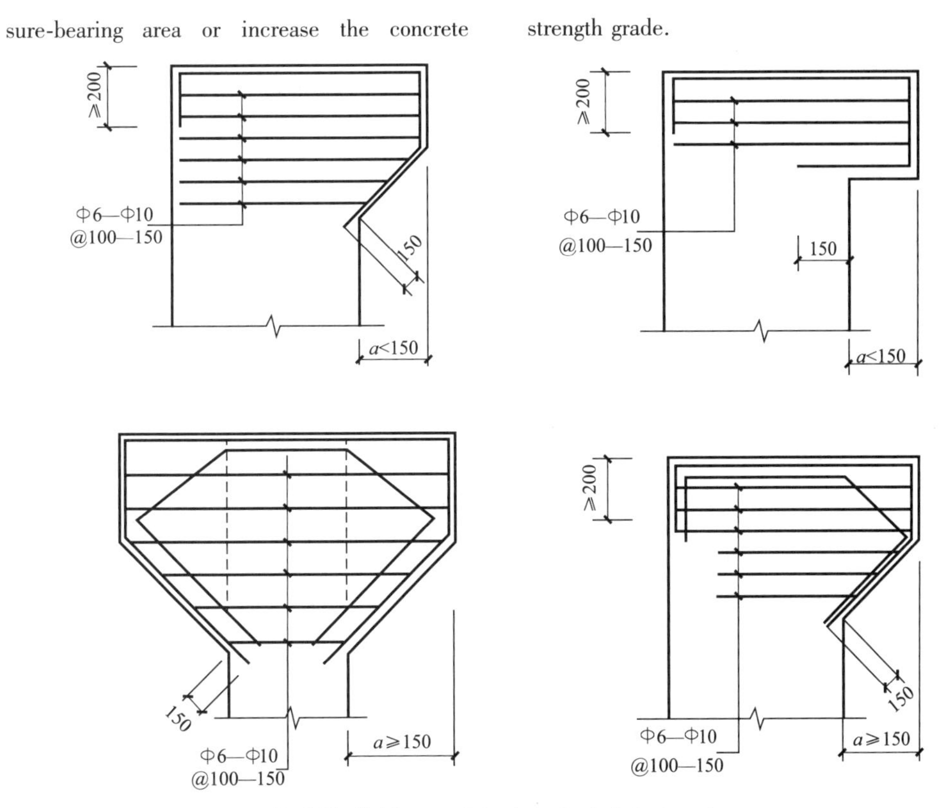

Fig. 2-50 Reinforcement structure of corbel at column top

2.4 Design of independent foundation under column

2.4.1 Design of independent foundation under column

The main design contents of the independent foundation under the column include: ① determination of foundation form; ② determination of embedded depth of foundation; ③ determination of the size of the bottom surface of the foundation; ④ determination of foundation height; ⑤ reinforcement of foundation bottom slab; ⑥ drawing construction drawings of foundation.

1. Forms of independent foundation under column

The independent foundation under column is an important load-bearing member in the single-storey industrial building and the loads from the superstructure are transmitted to the subgrade through the foundation. According to the construction methods, the independent foundation under column can be divided into prefabricated foundation under column and cast-in-situ foundation under column. The common form of the independent foundation under column of single-storey industrial building belongs to the spreading foundation, which is usually designed to stepped or tapered shape Fig. 2-51(a). The prefabricated independent foundation under column is also called cup foundation because the part connected with precast column is made into a cup mouth.

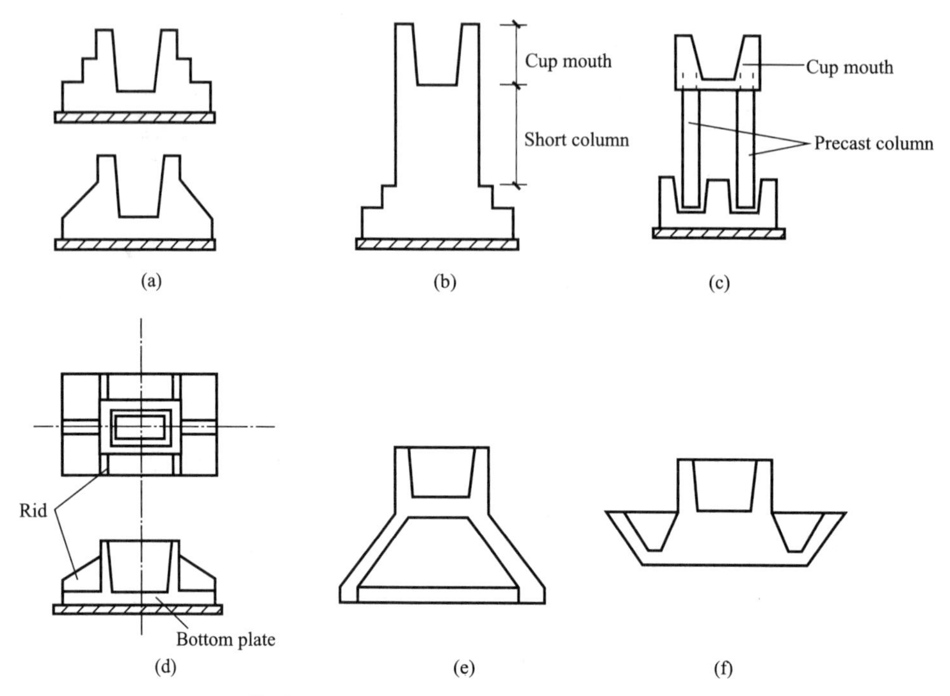

Fig. 2-51 Forms of independent foundation under column

When the foundation must be buried deeper due to geological conditions or there is a deep equipment foundation or a pit nearby, in order not to make the precast column too long and to maintain the uniformity of the length of the precast column, the extended foundation with short column can be made. It is composed of the cup mouth, short column and bottom slab, and is also called the high cup mouth foundation because of the higher position of the cup mouth (Fig. 2-51b). When the short column is very high, it can also be made into a latticed shape to save materials, that is, four prefabricated columns are used instead, and the cup bottom and cup mouth are poured as shown in Fig. 2-51(c).

In order to reduce the amount of concrete pouring on site, save the formwork and speed up the construction progress, semi-assembled ribbed slab foundation can also be used. That is, the cup mouth and ribbed slab are prefabricated, and poured into an integral part with the bottom slab on site, as shown in Fig. 2-51(d).

In practical engineering, shell foundations as shown in Figs. 2-51(e) and (f) are also adopted. They are used in the foundation with little eccentric load, and are often used in the foundation of structures such as chimneys, water towers and television towers. The common forms of shell foundation are conical shell, M-shaped combined shell and combined inner ball and outer cone shell.

When the superstructure load is large, the foundation condition is poor, and industrial building has strict requirements for uneven settlement, the pile foundation is generally adopted.

According to the force form, the independent foundation under column can be divided into two types, that is, axial compression foundation and eccentric compression foundation. In a single-storey industrial building, the independent foundation under column is usually the eccentric compression foundation.

2. Determination of embedded depth of foundation

The embedded depth of foundation refers to the distance from the bottom of the foundation to the designed ground. The purpose of choosing the embedded depth of the foundation is to select the appropriate load-bearing layer of the foundation. The embedded depth of foundation has a great influence on the safety and normal use of buildings, technical measures of foundation construction, construction period and project cost. Therefore, it is important to determine the embedded depth of foundation reasonably. The design must comprehensively consider the building's own conditions (such as construction applications, structural type, size and characteristic of the load, etc.), as well as the geological condition, climate condition, the embedded depth of the foundation of adjacent buildings, frost heaving and thawing of the foundation soil, etc.

3. Determination of size of foundation bottom slab

The dimensions of the foundation bottom slab should meet the requirements of the subgrade bearing capacity and subgrade deformation. According to the embedded depth of foundation and the bearing capacity of foundation, the design is carried out according to the standard combination value of load effect.

1) Size of foundation bottom slab of axial compression foundation

Under axial force, assuming that the pressure on the foundation bottom is uniformly distributed (Fig. 2-52), the design should meet the following requirements:

$$P_k = \frac{N_k + G_k}{A} \leqslant f_a \tag{2-26}$$

Where,

N_k—— the axial force in the standard combination of loads on the top surface of foundation;

f_a—— the characteristic value of the bearing capacity of subgrade modified by correction factor of the width and depth of the foundation;

G_k—— the characteristic value of self-weight of foundation and backfill soil on foundation;

$$G_k \approx \gamma_m dA \tag{2-27}$$

γ_m—— the average weight of foundation and backfill, generally taken as 20 kN/m^2 or 10 kN/m^2 below water level;

d —— embedded depth of foundation.

Substituting formula (2-27) into formula (2-26), we can get:

$$A \geqslant \frac{N_k}{f_a - \gamma_m d} \tag{2-28}$$

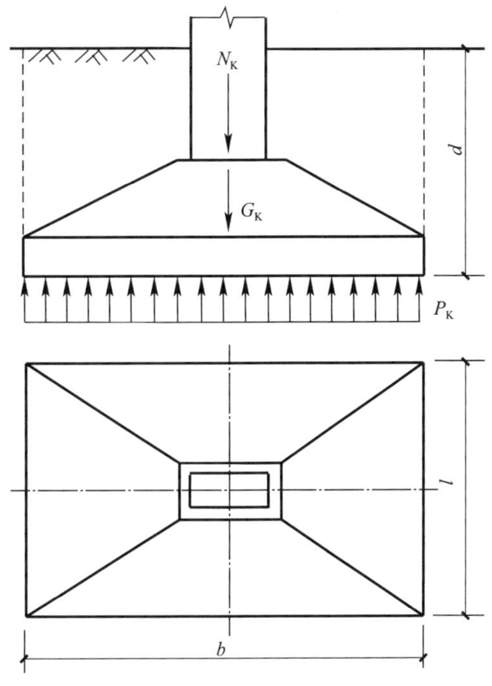

Fig. 2-52 Calculation diagram of axial compression foundation

During the design, A is obtained by the above formula. In general, it is more reasonable to use square bottom slab for axial compression foundation, that is, $b=l=\sqrt{A}$. Alternatively, a rectangle can be used by selecting the ratio of the long side and the short side is generally not greater than 2.

For buildings with foundation design grades of Grade A, Grade B and Grade C in special circumstances, in addition to determine the size of the foundation bottom according to the bearing capacity of the subgrade, the deformation checking calculation of the foundation shall also be carried out. When the foundation subgrade has weak underlying layer, the bearing capacity of the weak underlying layer should also be checked.

2) Size of foundation bottom slab of eccentric compression foundation

When the column is under eccentric compression, it is assumed that the pressure on the foundation bottom is linearly and non-uniformly distributed. The internal force of the top surface of the foundation is first converted to the foundation bottom, and then the maximum and minimum compressive stresses of the foundation bottom are determined (Fig. 2-53).

$$\frac{P_{k,max}}{P_{k,min}}=\frac{N_{bk}+G_k}{A}\pm\frac{M_{bk}}{W} \tag{2-29}$$

Where,

$$W=\frac{lb^2}{6},\ M_{bk}=M_k+N_{wk}e_w,\ e=\frac{M_{bk}}{N_{bk}+G_k}$$

So,

$$\frac{P_{k,max}}{P_{k,min}}=\frac{N_{bk}+G_k}{bl}\left(1\pm\frac{6e}{b}\right) \tag{2-30}$$

Where,

$P_{k,max}$ —— the maximum compressive stress at the edge of the foundation bottom corresponding to the standard combination of load effect;

$P_{k,min}$ —— the minimum compressive stress at the edge of the foundation bottom corresponding to the standard combination of load effect;

M_{bk} —— the value of moment in the standard combination of moments on the foundation bottom, $M_{bk}=M_kN_{wk}e_w+V_{kh}$;

N_{bk} —— the value of axial force in the standard combination of moments on the foundation bottom, $N_{bk}=N_k+N_{wk}$;

N_{wk} —— the characteristic value of vertical force from the foundation beam;

W —— the resistance moment of the foundation bottom, $W=\frac{1}{6}lb^2$.

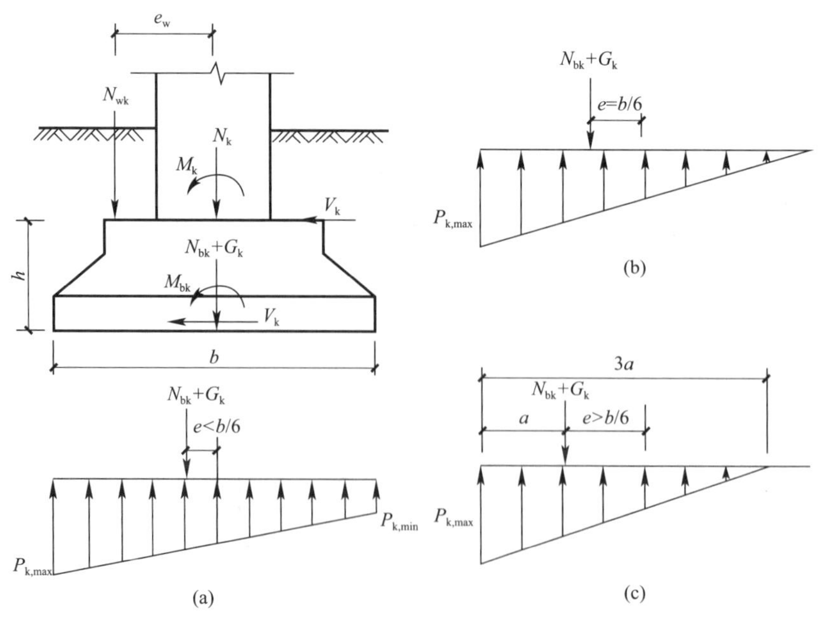

Fig. 2-53 Calculation diagram of eccentric compression foundation

According to formula (2-30), when $e<b/6$, $P_{k,min}>0$, the foundation reaction graph is trapezoidal as shown in Fig. 2-53(a); when $e_k=b/6$, $P_{k,min}=0$, the foundation reaction graph is triangular as shown in Fig. 2-53(b); when $e_k>b/6$, $P_{k,min}<0$, the foundation pressure graph is shown in Fig. 2-53(c), which indicates that a part of the area of the foundation will generate tensile stress. However, since the contacting stress between the foundation and the foundation is impossible to be tensile stress, this part of the foundation bottom is actually separated from the foundation. In other words, at this time, the foundation area bearing the foundation reaction is not bl but $3al$, where a is the distance between the eccentric load acting point and the edge with the maximum compressive stress $P_{k,max}$. Therefore, $P_{k,max}$ cannot be calculated according to formula (2-30). $P_{k,max}$ can be calculated according to the equilibrium condition of the force as shown in Fig. 2-53(c).

By $N_{bk}+G_k=\frac{1}{2}P_{k,max}\cdot 3a\cdot l$,

$$P_{k,max}=\frac{2(N_{ck}+G_k)}{3al} \tag{2-31}$$

Where,

a —— the distance from the point of eccentric load to the edge with $P_{k,max}$, $a=\frac{b}{2}-e$;

Under the action of eccentric load, the bearing capacity of foundation should meet the following requirements:

$$P_k=\frac{P_{k,max}+P_{k,min}}{2}\leqslant f_a \tag{2-32}$$

$$P_{k,max}\leqslant 1.2f_a \tag{2-33}$$

The reason why the characteristic value of subgrade bearing capacity is increased by 20% in formula (2-33) is that $P_{k,max}$ only appears in the local scope of the foundation edge, and most of $P_{k,max}$ is generated by the live loads.

The trial algorithm is generally used to determine the foundation bottom slab size of eccentrically compressed foundation. Firstly, the bottom area required by axial compression foundation is increased by 20%—40%. The bottom surface of eccentric compression foundation is generally rectangular (the ratio of the long side to the short side should not exceed 1.5—2.0). Secondly, the dimensions of long and short sides are selected, and then the values under eccentric load should meet the requirements of formulas (2-32) and (2-33). For foundations with low compressibility subgrade, the requirement for eccentricity e may be relaxed when considering the standard combination of load effects. But e should also be within $b/4$.

4. Determination of foundation height

After the size of foundation bottom slab is determined, the foundation height is estimated according to the structural requirements, and then the foundation height size is checked according to the requirement of anti-punching bearing capacity of concrete at the junction of column and foundation. For the stepped foundation, the height at the variable step should also be checked according to the same principle.

The test result shows that when the foundation height (or the height at the variable step) is not large enough, the load transferred from the column to the foundation will cause the foundation to occur the shear failure as shown in Fig. 2-54(a), that is, the section along the column side in the direction of 45° will be pulled apart to generate the pyramid failure as shown in shaded part of Figs. 2-54 (b) and (c). The same damage can also occur at the variable step. In order to prevent the punching damage, the punching force F_l produced by the foundation reaction outside the anti-punching surface must be less than or equal to the anti-punching bearing capacity of concrete at the punching surface.

The checking calculation of the foundation height includes two control sections: ① the junction of the column and the foundation; ② the variable step of the foundation. For rectangular foundation, the failure of short side of column is more dangerous than that of long side of column. Therefore, the height of foundation is generally determined according to the punching failure conditions of short side of column (Fig. 2-55). The anti-punching bearing capacity

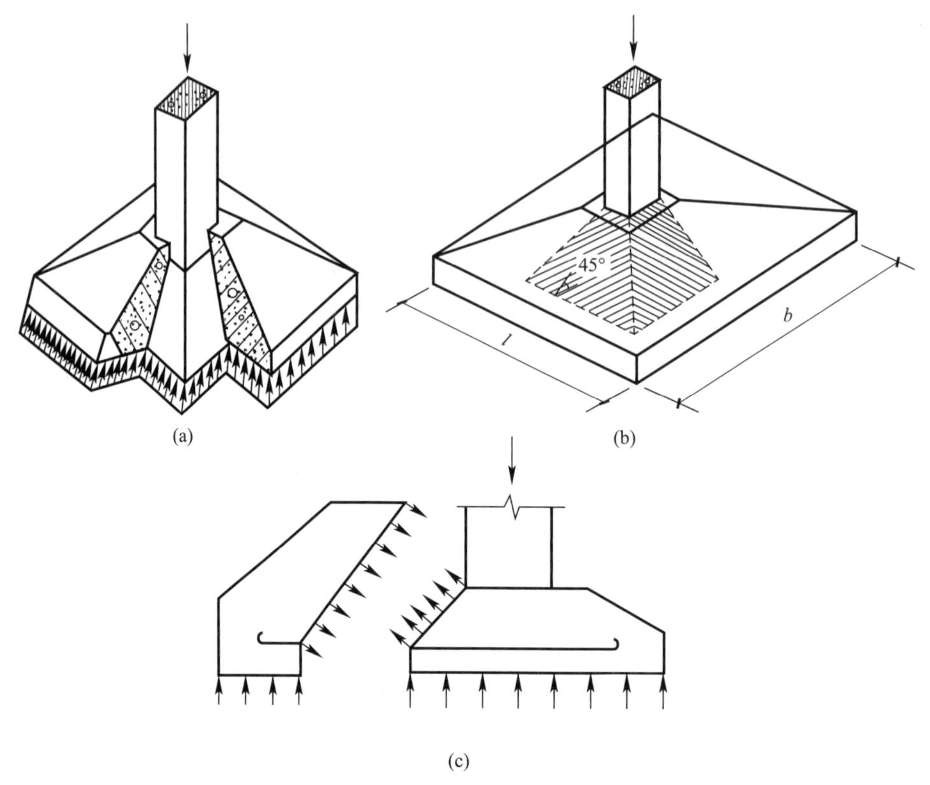

Fig. 2-54 Schematic diagram of foundation punching failure

of the foundation shall meet the requirements of formula (2-33).

$$F_l \leqslant 0.7\beta_{hp} f_t a_m h_0 \quad (2\text{-}34)$$

$$F_l = P_s A_l \quad (2\text{-}35)$$

$$a_m = \frac{a_t + a_b}{2} \quad (2\text{-}36)$$

Where,

β_{hp}—— the coefficient of cross-section height; when $h \leqslant 800$ mm, $\beta_{hp} = 1.0$; when $h \geqslant 2000$ mm, $\beta_{hp} = 0.9$; when h is in the middle value between 800 mm and 2000 mm, β_{hp} is taken by linear interpolation;

f_t—— the design value of concrete tensile strength;

a_m—— the average value of the upper side length b_t and the lower side length b_b of the cross-section of the cone that is destroyed by punching;

a_t—— the length of the upper side of the oblique section of a cone that is destroyed by punching; when calculating the punching bearing capacity at the junction between the column and the foundation, a_t is the column width; when calculating the punching bearing capacity at the variable step of the foundation, a_t is the width of upper step;

a_b—— the length of the lower side of the oblique section of a cone that is destroyed by punching; when the bottom side of the punching failure cone is within the foundation, as shown in Figs. 2-55 (a) and (b), that is, when $l \geqslant a_t + 2h_0$, $a_b = a_t + 2h_0$; when the bottom side of the punching failure cone is outside the foundation in the direction of l, as shown in Fig. 2-56, that is, when $l < a_t + 2h_0$, $a_b = l$;

h_0—— the effective height of the punching failure cone of the foundation;

A_l—— the part of the foundation area used for punching checking is the polygon shadow area $ABCDEF$ in Figs. 2-55 (a) and (b) or the shadow area $ABCD$ in Fig. 2-56;

P_s—— the net reaction force of the foundation bottom.

When the bottom side of the punching failure cone is outside the foundation as shown in Fig. 2-56, the punching bearing capacity do not need to be calculated. When the punching bearing capacity calculation does not meet the requirements, the height of the foundation should be adjusted until it meets the requirements.

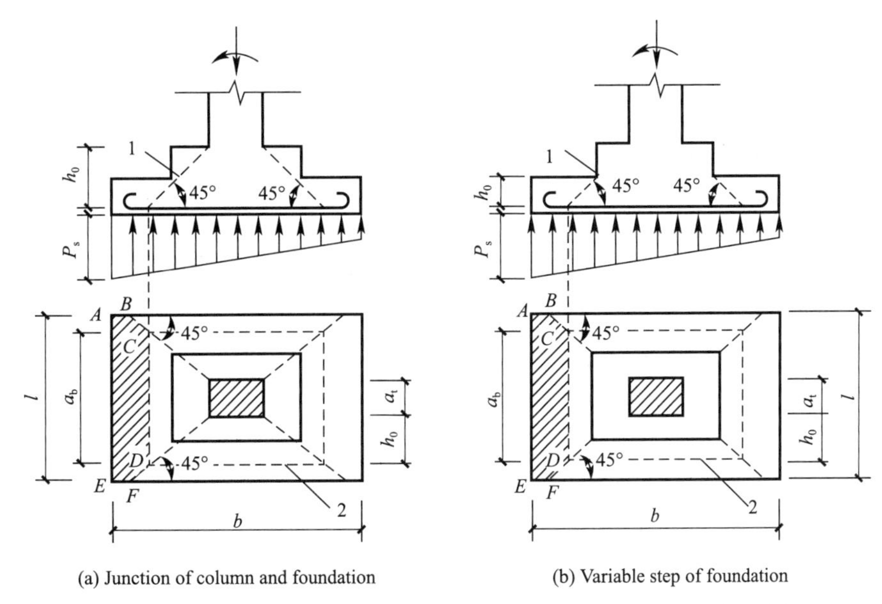

(a) Junction of column and foundation (b) Variable step of foundation

Fig. 2-55 Calculation of section position of punching bearing capacity of stepped foundation

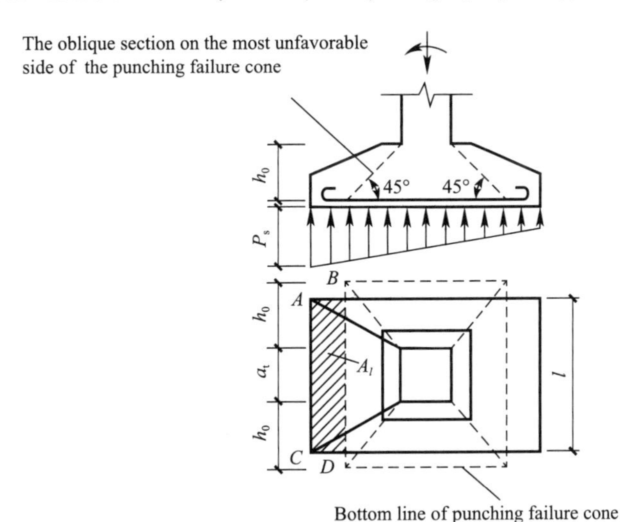

Fig. 2-56 Bottom of the punching failure cone falls outside the foundation in the direction l

5. Reinforcement of foundation bottom slab

The purpose of reinforcement calculation of foundation slab is to make the foundation meet the requirements of flexural bearing capacity. Under the action of net reaction force of foundation, the foundation bottom slab will bend upward along two directions. Therefore, the foundation slab needs to be equipped with reinforcing bars in both directions. The bottom slab can be regarded as a cantilever slab fixed on the side of the column or at the variable step. And the calculated section is taken at the side of the column or the variable step. The control section for reinforcement calculation of foundation slab is taken at the junction of column and foundation or at the variable step (for the stepped foundation) as shown in Fig. 2-57. The reinforcement of the foundation slab is calculated by the internal force design value.

1) Axial compression foundation

For the axial compression foundation, the bending moment M_{I} at the section Ⅰ-Ⅰ is equal to the net reaction force P_s of the foundation acting on the area of the trapezoid *ABCD* multiplied by the distance from the centroid of the area to the cross-section Ⅰ-Ⅰ.

$$M_{\mathrm{I}}=\frac{1}{24}P_s(b-b_t)^2(2l+a_t) \quad (2\text{-}37)$$

The tension steel bars $A_{s\mathrm{I}}$ along the long side can be calculated approximately by the following formula:

$$A_{s\mathrm{I}}=\frac{M_{\mathrm{I}}}{0.9f_y h_{0\mathrm{I}}} \quad (2\text{-}38)$$

Where,

$h_{0\mathrm{I}}$ —— the effective height of the cross-section Ⅰ-Ⅰ. When there is cushion layer, $h_{0\mathrm{I}}=h-45$; when there is no cushion layer, $h_{0\mathrm{I}}=h-75$.

For the axial compression foundation, the bending moment M_{II} at the section Ⅱ-Ⅱ along the short side direction is equal to the net reaction force P_s of the foundation acting on the area of the trapezoid *BCFE* multiplied by the distance from the centroid of the area to the cross-section Ⅱ-Ⅱ.

$$M_{\mathrm{II}}=\frac{1}{24}P_s(l-a_t)^2(2b+b_t) \quad (2\text{-}39)$$

The reinforcement along the short side is generally placed on the top of the reinforcement along the long side, and if the diameter of reinforcement in both directions is d, then the effective height of section Ⅱ-Ⅱ is $h_{0\mathrm{II}}=h_{0\mathrm{I}}-d$, therefore, the sectional area of reinforcement along the short side direction is $A_{s\mathrm{II}}$ is as follows:

$$A_{s\mathrm{II}}=\frac{M_{\mathrm{II}}}{0.9f_y h_{0\mathrm{II}}} \quad (2\text{-}40)$$

After the reinforcement of the foundation slab is determined, the appropriate diameter and spacing of the reinforcement can be selected.

2) Eccentric compression foundation

When eccentricity is less than or equal to 1/6 of foundation width b, the design values of the bending moment M_{I} and M_{II} corresponding to the basic combination of load effect at Ⅰ-Ⅰ of any section along the direction of bending moment and at Ⅱ-Ⅱ of any section perpendicular to the direction of bending moment can be calculated according to the following formulas (2-41) and (2-42), respectively:

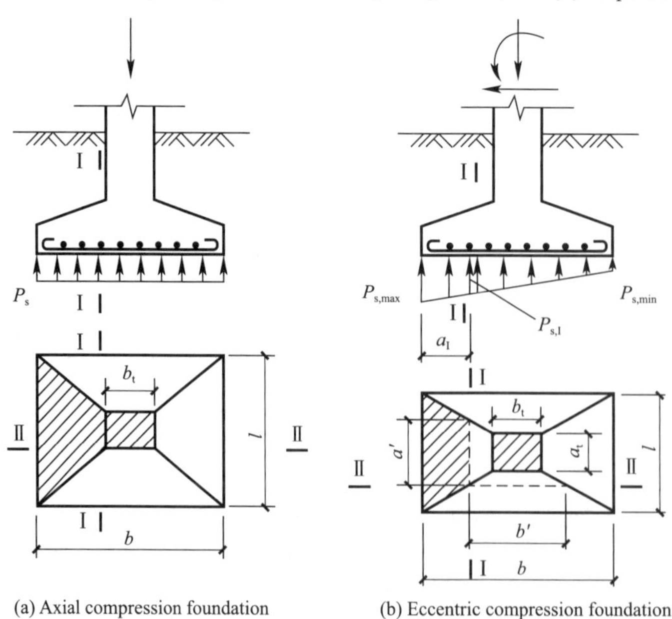

(a) Axial compression foundation (b) Eccentric compression foundation

Fig. 2-57 Calculation diagram of rectangular foundation

$$M_{\text{I}}=\frac{1}{12}a_1^2[(2l+a')(P_{s,max}+P_{s,\text{I}})+(P_{s,max}-P_{s,\text{I}})l] \quad (2\text{-}41)$$

$$M_{\text{II}}=\frac{1}{48}(l-a')^2(2b+b')(P_{s,max}+P_{s,min}) \quad (2\text{-}42)$$

Where,

a_1 —— the distance from any cross-section I - I to the maximum reaction at the foundation edge;

$P_{s,max}$, $P_{s,min}$ —— the design values of maximum and minimum net reaction forces on the bottom edge of foundation corresponding to the basic combination of load effects, respectively;

$P_{s,\text{I}}$ —— the net reaction design value of foundation bottom at any section I - I corresponding to the basic combination of load effect.

When the eccentricity is greater than 1/6 of foundation width b, stress will not appear in some part of the foundation bottom along the bending moment direction because the foundation soil does not bear tensile force, and the reaction force distribution takes on triangular (Fig. 2-53c). At this time, along the direction of bending moment, the design value of bending moment M_{I} corresponding to the basic combination of load effect at any section I - I can still be calculated according to formula (2-37). In the direction perpendicular to the action of bending moment, the design value of bending moment corresponding to the basic combination of load effect at any section should be calculated according to the actual stress distribution. In order to simplify the calculation, it can also be taken safely as $P_{s,min}=0$ and be calculated according to formula (2-42).

2.4.2 Structural requirements for independent foundation under column

1. General structural requirements

The edge height of the tapered foundation shall not be less than 300 mm; the slope in both directions shall not be greater than 1:3, and the height of each step is generally 300—500 mm.

The concrete strength grade of foundation shall not be lower than C20. The plain concrete cushion is usually made under the foundation, the strength grade generally is C10. The thickness of the cushion is generally 100 mm. The area of the cushion is larger than that of the foundation, and it usually extends 100 mm from the foundation.

The steel bars of foundation bottom slab generally adopt HRB400, HRB335 or HPB300, and the bearing reinforcement of foundation shall meet the requirement of minimum reinforcement ratio of 0.15%. The minimum diameter of the reinforced bar shall not be less than 10 mm, and the spacing shall not be greater than 200 mm. When there is a cushion, the thickness of the concrete protective layer shall not be less than 35 mm, and when there is no cushion, it shall not be less than 70 mm.

When the side length of foundation bottom slab is greater than 2.5 m, the reinforcing bars can be shortened by 10% and should be staggered (Fig. 2-58).

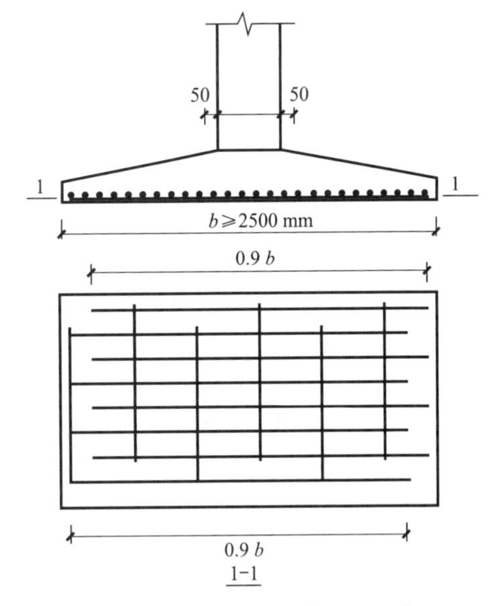

Fig. 2-58 Staggered reinforcement of foundation bottom slab

For the cast-in-situ foundation under column, if it is not poured at the same time with the column, the number, diameter and type of reinforcement shall be the same as the longitudinal stressed reinforcement inside the column (Fig. 2-59). The anchorage bars and the length of lapping with longitudinal stressed bars of columns shall comply with the provisions of ***Code for Design of Concrete Structure*** (GB 50010—2010).

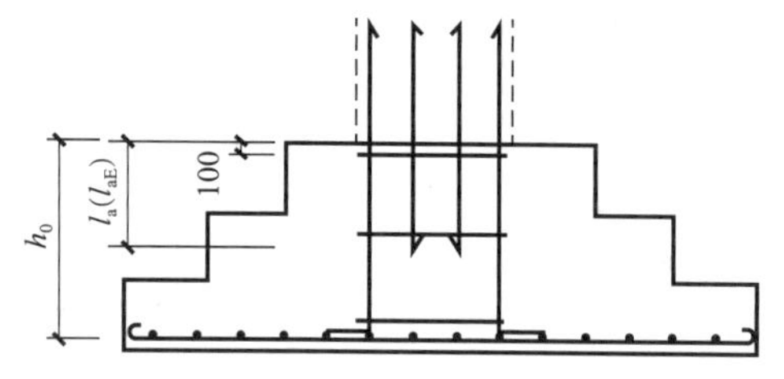

Fig. 2-59 Schematic diagram of reinforcement in cast-in-situ foundation

2. Cup mouth form of prefabricated foundation and insertion depth of column

When the section of the prefabricated column is rectangular or I-shaped, the foundation under column adopts the type of single cup mouth. When it is a two-limb column, a double cup mouth or a single cup mouth can be used. The structure of the cup mouth is shown in Fig. 2-60.

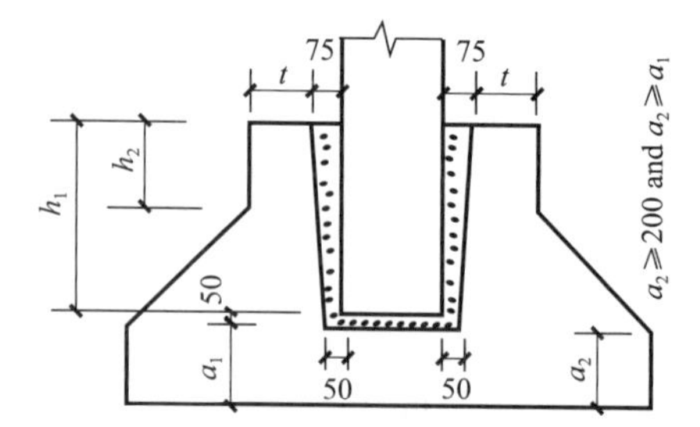

Fig. 2-60 Construction of the cup mouth of the prefabricated column

The prefabricated column shall be inserted into the cup mouth of the foundation with sufficient depth to make sure that the column is securely embedded in the foundation. The insertion depth h_1 shall meet the requirements of Table 2-6. At the same time, it shall also meet the requirements of the anchorage length of longitudinal stressed reinforcement of the column and the stability of the column during hoisting, that is, $h_1 \geqslant 0.05$ times of the column length (the column length during hoisting).

Cup bottom thickness a_1 and cup wall thickness t of the foundation can be determined according to Table 2-7.

Insert depth h_1 of the column (mm) **Table 2-6**

Rectangular or I-shaped column				Double-limb column
$h<500$	$500 \leqslant h<800$	$800 \leqslant h \leqslant 1000$	$h>1000$	
h—$1.2h$	h	$0.9h$ and $\geqslant 800$	$0.8h$ and $\geqslant 1000$	$(1/3$—$1/2)h_a$ $(1.5$—$1.8)h_b$

Notes:

1. h is the long side size of the column section; h_a is the long side size of the whole section of the double-limb column; h_b is the short side size of the whole section of the double-limb column.

2. When the column is under axial compression or small eccentric compression, h_1 can be reduced; when the eccentricity is greater than $2h$, h_1 can be appropriately increased.

Cup bottom thickness and cup wall thickness of the foundation **Table 2-7**

Long side size of column section h (mm)	Thickness of the cup bottom a_1 (mm)	Thickness of the cup wall t (mm)
$h<500$	$\geqslant 150$	150—200
$500 \leqslant h<800$	$\geqslant 200$	$\geqslant 200$
$800 \leqslant h<1000$	$\geqslant 200$	$\geqslant 300$
$1000 \leqslant h<1500$	$\geqslant 250$	$\geqslant 350$

Continued

Long side size of column section h (mm)	Thickness of the cup bottom a_1 (mm)	Thickness of the cup wall t (mm)
$1500 \leqslant h < 2000$	$\geqslant 300$	$\geqslant 400$

Notes:

1. The cup bottom thickness of the double-legged column can be increased appropriately.

2. When there is a foundation beam, the thickness of the cup wall under the foundation beam shall meet the requirements of its supporting width.

3. The surface where the column is inserted into the cup should be gouged to be rough, and the gap between the column and the cup should be filled with fine stone concrete which has higher grade than the strength grade of the foundation concrete. When the strength of the concrete reaches more than 70% of the designed value, the superstructure can be lifted.

3. Reinforcement of cup mouth of the foundation without short column

When the column is under axial or small eccentric compression and $t/h_2 \geqslant 0.65$, or when the column is under large eccentric compression and $t/h_2 \geqslant 0.75$, the cup wall may be unreinforced. When the column is under axial or small eccentric compression and $0.5 \leqslant t/h_2 < 0.65$, the cup wall can be reinforced according to the requirements of Table 2-8, and the reinforcement is placed at the top of the cup mouth (Fig. 2-61a). In other cases, the reinforcement shall be calculated.

When the width of the middle separator on the foundation with the double cups is less than 400 mm, Φ12@200 longitudinal reinforcement and Φ8@300 transversal reinforcement should be arranged for the separator (Fig. 2-61b).

Constructional reinforcement of cup wall **Table 2-8**

Long side size of column section (mm)	$h < 1000$	$1000 \leqslant h < 1500$	$1500 \leqslant h < 2000$
Diameter of steel bar (mm)	8—10	10—12	12—16

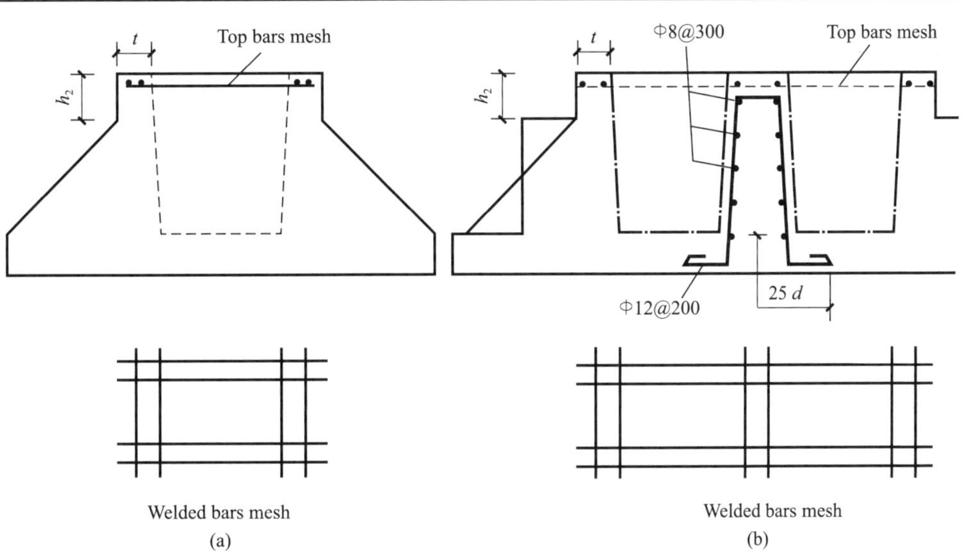

Fig. 2-61 Cup mouth constructional reinforcement of foundation without short column

2.5 Design points of reinforced concrete crane beam and roof truss

2.5.1 Design points of reinforced concrete crane beam

Crane beam is one of the main load-bearing members in a single-storey industrial building which directly bears the crane load. The functions of the crane beam are as follows: ① bearing the various loads generated when the crane is lifting, running and braking; ② transmitting longitudinal loads of industrial building to the longitudinal columns; ③ strengthening the longitudinal rigidity of the single-storey industrial building.

1. The mechanical characteristics of the crane beam

Assembled crane beam is a simply-supported beam supported on the columns, and its mechanical characteristics depend on the characteristics of the crane load including the following four points.

1) Crane load is the moving load

The crane load is two groups of moving concentrated loads. One group is the moving vertical load F, the other is the moving horizontal load T (for the crane beam in open air, the horizontal wind load should also be considered simultaneously with the crane transversal horizontal load). Here "a group" refers to the crane wheels that may act on the crane beam. Therefore, both the self-weight and the vertical bending under the action of F must be considered, as well as the two-way bending under the combined action of the self-weight, F and T. By the knowledge of structural mechanics, the envelopes of bending moment and shear force of the crane beam are obtained as shown in Fig. 2-62.

2) Dynamic characteristics of crane load

The crane load has impacting and vibration effects. When calculating the strength of the crane beam and its connection, the dynamic characteristics of the crane load should be considered, and the vertical load of the crane should be multiplied by the dynamic factor μ. For suspended cranes and soft hook cranes of working levels A1—A5, the dynamic factor μ can be taken as 1.05; for soft hook cranes, hard hook cranes and other special cranes with working levels A6—A8, the dynamic factor μ can be taken as 1.1.

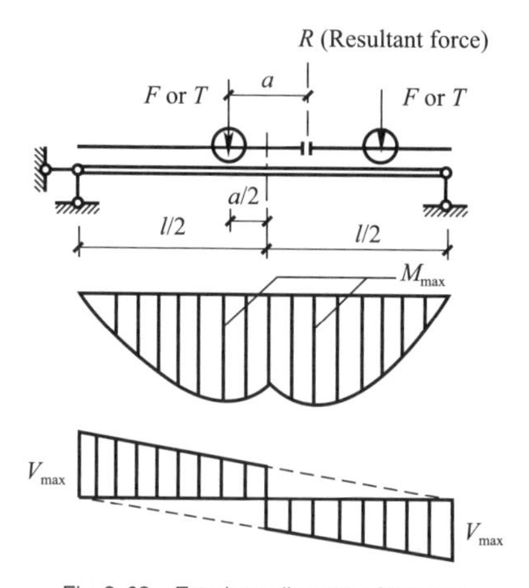

Fig. 2-62 Envelope diagram of bending moment and shear force of crane beam

3) Crane load is the repeated load

Actual investigations show that if the industrial building is used for 50 years, the number of repetition of the crane load with A4 and A5 working levels can generally reach 2×10^6, and the repetition number of the crane load with A6 and A7 working levels can reach 4×10^6 to 6×10^6. For structures or members that directly bear such repeated loads, the material strength will decrease due to fatigue. Therefore, in addition to static calculations, fatigue strength checking calculations are also required for members that directly bear the crane load.

4) Crane load is the eccentric load

The crane vertical load μF_{max} and the lateral horizontal load T are eccentric to the bending center of the cross-section of the crane beam (Fig. 2-63). The torque generated by each crane wheel is calculated according to two conditions: ① when calculating the static force, two cranes are considered; ② when checking the fatigue strength, only one crane is considered, and the influence of the transversal horizontal load of the crane is not considered.

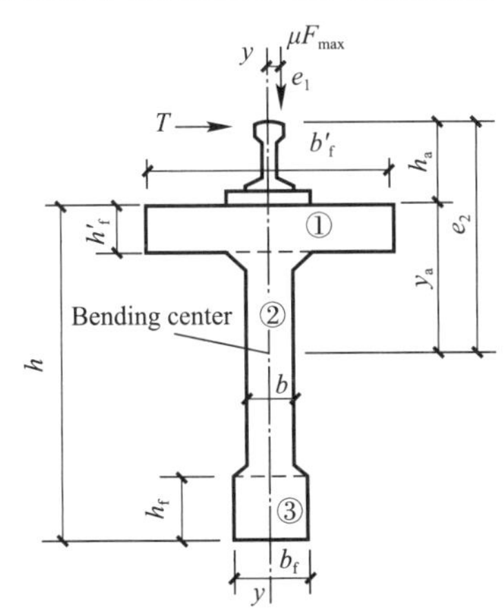

Fig. 2-63 Eccentric load of crane

2. Construction points of crane beam

1) Material

Prestressed steel wire, steel strand or prestressed ribbed bar can be used for prestressed reinforcement in crane beam; HRB400 or HRB500 reinforcement shall be used for non-prestressed reinforcement. The strength grade of concrete can be C30—C50. Generally, C40 shall be used for prestressed concrete crane beam, and C50 shall be used if necessary.

2) Section form and size

The cross-section of crane beam is generally designed as an I-shape or T-shape to reduce its self-weight and also facilitate the arrangement of the steel bars.

The height of crane beam section is related to the lifting capacity of the crane. The beam height can be taken as 1/12—1/5 of the span, generally including 600 mm, 900 mm, 1200 mm and 1500 mm; the thickness of the web, determined by the shear resistance and reinforcement construction requirements, can be taken as 1/7—1/4 of the web height, generally including 140 mm, 160 mm and 180 mm, and gradually thicker to 200 mm, 250 mm and 300 mm at the beam end. The width of the upper flange can be 1/15—1/10 of the span, not less than 400 mm, generally including 400 mm, 500 mm and 600 mm. The lower flange of the I-shaped section should be smaller than the upper flange, which is determined by the construction of the prestressed tendons.

3) Constructional reinforcement

As shown in Fig. 2-64 for reinforcement diagram of crane beam, because longitudinal steel bars are directly subjected to repeated loads, lashing joints or welded joints are not allowed for longitudinally stressed steel bars, and no attachments (except for end anchor) shall be welded. For the prestressed concrete crane beam, except for special anchoring measures, smooth carbon steel wire should not be used. In the prestressed crane beam, the upper and lower prestressed steel bars should be placed symmetrically. In order to prevent cracks in the pre-tensioned area due to the application of prestress and reduce the principal tensile stress of the section near the support, it is advisable to bend a part of the prestressed steel bars near the support.

Open stirrup is not allowed for the beam with section height greater than 800 mm. The stirrup diameter should not be less than 8 mm; for the beam with section height not greater than 800 mm, it should not be less than 6 mm. When the beams are equipped with longitudinal compression bars required for calculation, the diameter of the stirrups should not be less than $d/4$ (d is the maximum diameter of the bearing steel bars). The spacing of stirrups shall meet the requirements of shear-torsion bearing capacity and conform to the construction regulations.

4) Connection

The connection between the track and the crane beam and the connection between the crane beam and the column can be checked in the relevant standard. Fig. 2-65 is a general construction drawing. The connecting angle steel or steel slab connected with the column on the upper wing edge bears the crane transversal horizontal load. The height of all connecting weld seam shall also be determined by calculating and it shall not be less than 8 mm.

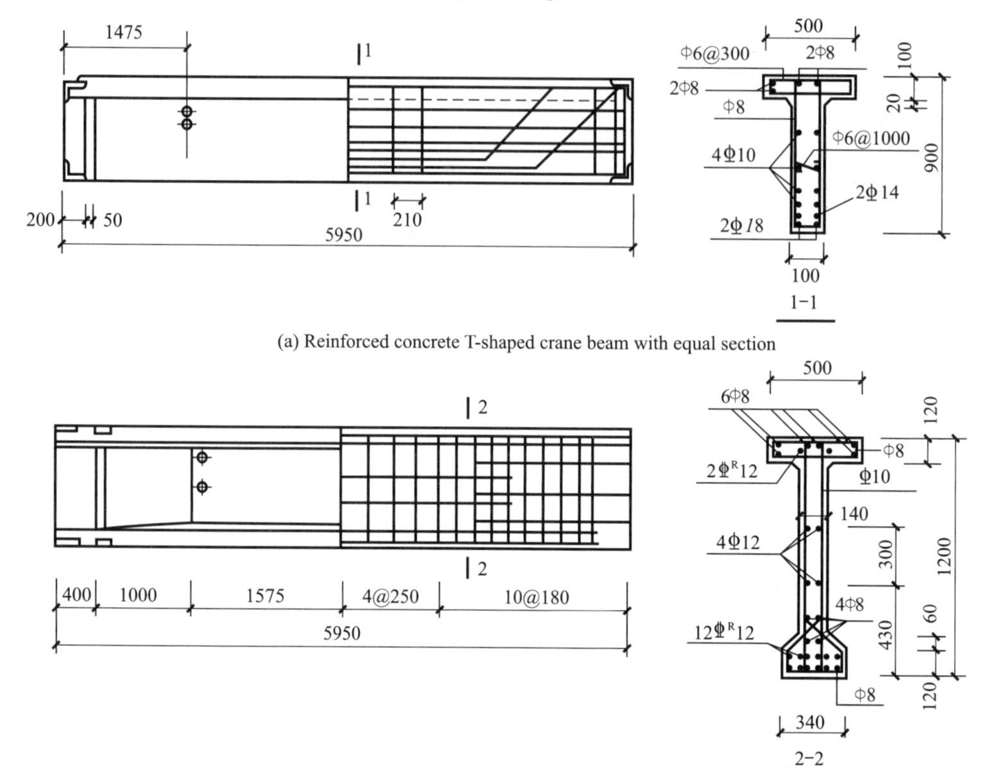

(a) Reinforced concrete T-shaped crane beam with equal section

(b) Pre-tensioned concrete I-shaped crane beam with equal section

Fig. 2-64 Cross-section of crane beam

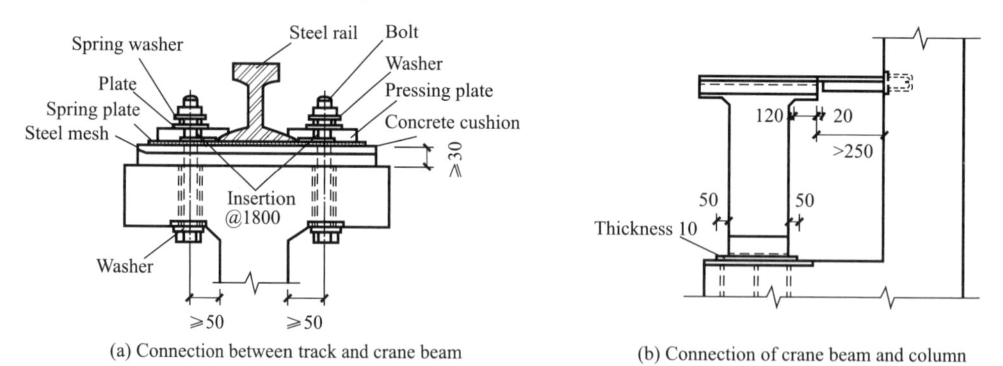

(a) Connection between track and crane beam

(b) Connection of crane beam and column

Fig. 2-65 Connection of crane beam

2.5.2 Design points of reinforced concrete roof truss

The roof truss is an important member of a single-storey industrial building. The roof truss (or girder) bears the roof loads and transmits them to the bent column. It also acts as a beam in the bent frame structure to connect the columns so that they can work together under various loads.

1. General requirements

The type of roof truss (or girder) should be selected according to the factors of technology, building, materials, construction, etc. When the column distance is 6 m and the span is 15—30 m, the prestressed concrete roof truss should generally be preferred; when the span is 9—15 m, the reinforced concrete roof girder can be used.

The shape of the roof truss should be compatible with the use requirements of the industrial building, the span and the roof structure, and the shape should be as close to the bending moment diagram of the simply supported beam as possible to make the internal forces of the members more uniform. The height-span ratio of the roof truss is generally 1/10—1/6. The top chord slope of the trapezoidal roof truss can be 1/7.5 for non-coiled material waterproof roof, 1/10 for coiled waterproof roof. The top chord slope of the single-slope polyline roof truss can be 1/7.5, which is suitable for both coiled waterproof roof and non-coiled waterproof roof. The top chord slope of the double-slope polyline roof truss can be 1/5 (end) and 1/15 (middle).

The top and bottom chords and end oblique rod of the roof truss shall have the same section width to facilitate construction and production. The cross-section width of the top chord should meet the construction requirements of supporting roof slabs and skylight frames. Generally, it should not be less than 200 mm, and the height should not be less than 160 mm (9 m roof truss) or 180 mm (12—30 m roof truss). The cross-sectional width of the bottom chord is generally not less than 200 mm and the height is not less than 140 mm; the cross-section size of the bottom chord of the prestressed roof truss should meet the construction requirements of the prestressed tendon channel. The section width of web members (referring to the out-of-plane section size of the roof truss) should generally be smaller than the chord, and the section height (referring to the section size in the in-plane direction of the roof truss) should be less than or equal to the section width.

The concrete strength grade of the roof truss is generally C30—C50; the prestressed steel bar adopts medium-strength prestressed steel wire, steel stranded wire, stress-relieved wire, etc. The non-prestressed steel bar adopts HRB400 or HRB500 hot-rolled steel bar.

The longitudinal reinforcement of the top chord and the non-prestressed reinforcement of the prestressed bottom chord are generally not less than 4Φ12; the longitudinal reinforcement of the web rods is not less than 4Φ10; the stirrups are all closed, and the diameter is not less than 4 mm. The spacing is no more than 200 mm in the upper and bottom chords, and no more than 250 mm in the web.

2. Loads and load combination

The loads acting on the roof truss include the loads from the roof slab, the self-weight of the roof truss, the concentrated load from the skylight frame, sometimes the load of hanging crane or other suspension equipment, and suction or pressure on roof truss caused by the wind load.

Roof live loads (including construction loads) act on the full span or half-span; and the half-span load may maximize the internal force of web members of the roof truss and even change the sign of the internal force. Therefore, the following three load combinations should be considered (Fig. 2-66).

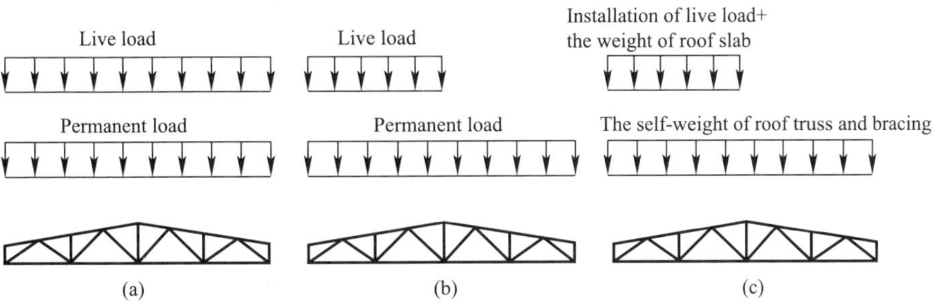

Fig. 2-66 Load combination diagram of roof truss

① Full span permanent load + full span live load;

② Full span permanent load + half span live load;

③ Roof truss (including roof support) weight + half-span roof slab weight + half-span roof installation live load (0.5 kN/m^2).

3. Internal force analysis

Reinforced concrete roof trussis actually statically indeterminate rigidly connected truss due to the joints being poured into a whole. The polyline roof truss is shown in Fig. 2-67(a), and the calculation diagram is shown in Fig. 2-67(b). There are both nodal loads and inter-node loads acting on the top chord, so the top chord of the roof truss will produce bending moment, which is generally in an eccentric compression state. The web rods and bottom chords of the roof truss (ignoring the influence of its self-weight) are the axially stressed members. The internal force can be calculated according to the following methods.

1) The top chord can be assumed to be a polyline continuous beam with a fixed hinge support (Fig. 2-67c), which is calculated by the bending moment distribution method. When the length difference between each section is less than or equal to 10%, it can be approximated as a continuous beam of equal span and calculated by means of the internal force coefficient table.

2) According to the calculation of the axial force of each member according to the hinged truss (Fig. 2-67d), it can be calculated by graphical method, numerical method or the known coefficient table. The node loads R_1, R_2, ⋯, R_5 of the truss are the corresponding support reaction forces of the top chord continuous beam. For the bottom chord, the bending moment caused by its self-weight can generally be ignored.

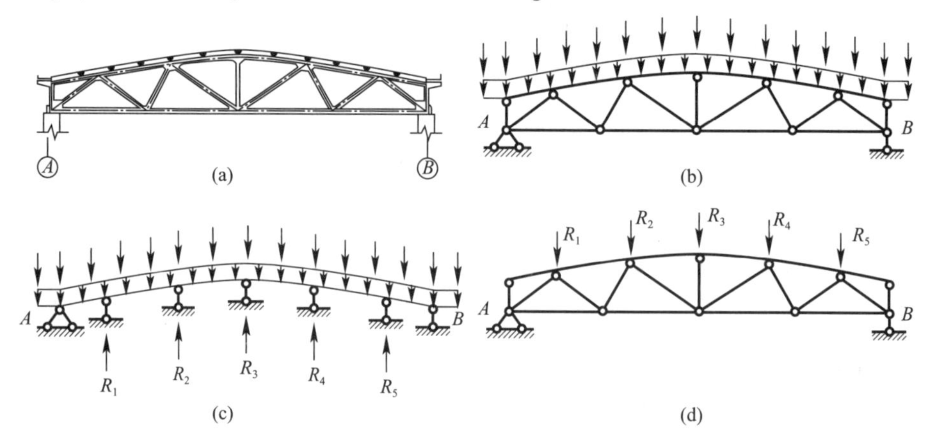

Fig. 2-67 Calculation diagram of roof truss

2.6 Design example of single-storey industrial workshop

2.6.1 Design topic

A metal assembly workshop is reinforced concrete bent frame structure with double-span as shown in Fig. 2-68.

2.6.2 Design contents

1. To calculate the loads on the bent frame.

2. To calculate the internal force of the

bent frame under various loads (the space effect of the factory building is not considered).

3. To design the column, the corbel and the independent foundation under column.

2.6.3 Design resources

1. The workshop is a double-span workshop with the span of 18 m. The total length of the workshop is 60 m, the column spacing is 6 m, and the track top elevation is 8 m. The section of the plant is shown in Fig. 2-68.

2. There are two cranes in each span of the workshop, with A4 working system. The relevant parameters of the cranes are shown in Table 2-9.

3. The basic snow pressure is 0.35 kN/m^2. The basic wind pressure is 0.40 kN/m^2. 0.8 m below the natural floor of the workshop is backfilled soil, 8 m below backfilled soil is uniform cohesive soil, the characteristic value of subgrade bearing capacity is f_a = 240 kPa, and the natural weight of subgrade is 17.5 kN/m^3. The lower layer is coarse sand, the characteristic value of the bearing capacity is f_a = 350 kPa, and the water level is −5.5 m.

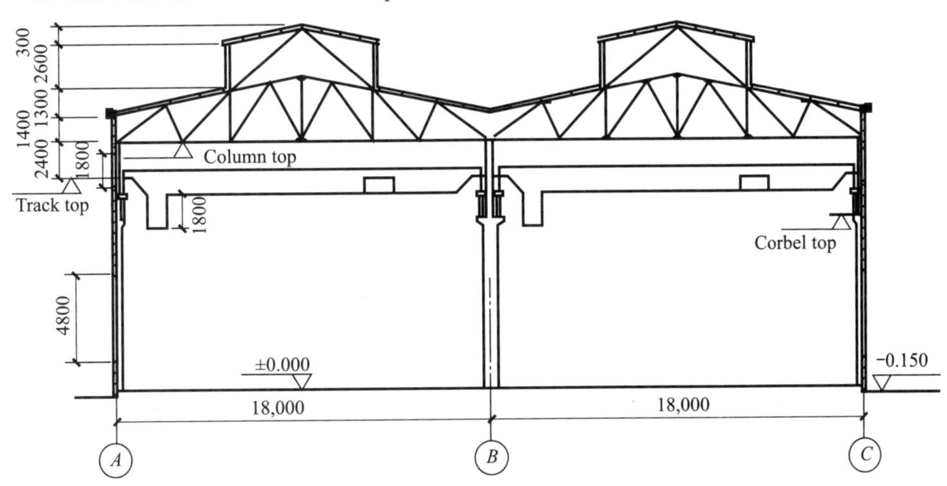

Fig. 2-68 Sectional view of a single-storey workshop (Unit: mm)

Related parameters of the crane **Table 2-9**

Crane position	Lifting capacity (kN)	Bridge span L_K(m)	Trolley weight g(kN)	Maximum wheel pressure $P_{max,k}$(kN)	Wheel K(m)	Cart distance width B(m)	Car height (m)	Total crane weight (kN)
Left span (*AB* span)	300/50	16.5	118	270	4.6	6.05	2.6	340
Right cross (*BC* span)	200/50	16.5	78	195	4.4	5.65	2.2	250

4. Selection of standard components of the workshop and load values

1) A trapezoidal steel roof truss with a span of 18 m is adopted. According to the ***Load Code for the Design of Building Structures*** (GB 50009—2012), the characteristic value of the roof truss self-weight (including support) is $0.12+0.011L$ (L is the span, unit: m), unit is kN/m^2.

2) The crane beam is reinforced concrete bean with equal cross-section. The beam height is 900 mm, the beam width is 300 mm, the characteristic value of dead weight is 39 kN/piece. The dead weight of track and parts is 0.8 kN/m,

and the height of track and cushion structure is 200 mm.

3) A rectangular longitudinal skylight is adopted. The vertical load transmitted to the roof truss on each side of each skylight frame is 34 kN (including self-weight, side panels, sash support, etc.).

4) The characteristic value of gutter board dead load is 2.02 kN/m.

5) Enclosure wall adopts 240 mm thick whitewashed wall, its self-weight is 5.24 kN/m^2. Steel window: self-weight 0.45 kN/m^2, the height of the window is 1800 mm as shown in the section diagram (Fig. 2-68). The cross-section of the foundation beam is 240 mm×450 mm, and the self-weight of the foundation beam is 2.7 kN/m.

5. Material: the concrete strength grade is C25. The longitudinal steel bars of the column are HRB335 grade, and the rest steel bars are HPB300 grade.

6. Roof structure: the 1.5 m×6 m prestressed roof slab is used.

2.6.4 Structural calculation diagram

The load distribution in this assembly shop is even. A representative calculation unit is selected for structural design, and the load range of the bent is shown in Fig. 2-69 (a). The structure calculation diagram is shown in Fig. 2-69(b). Next, the geometric dimensions are determined in the structural calculation diagram.

1. The height of the bent column

1) Determination of the depth of the foundation and the height of the foundation

Considering the freezing depth and backfill soil layer, 1.8 m is selected from the bottom of the foundation to the outdoor ground. The height of the foundation is initially estimated to be 1.1 m. The elevation of the top surface of the foundation is −0.85 m.

2) Determination of the elevation of the top of the upper column

As shown in Fig. 2-68, the top of the upper column is 2.4 m away from the top of the rail, and the elevation of the top of the upper column is 10.4 m.

3) Determination of the elevation of the top surface of the corbel

The elevation of the top of the rail is +8.0 m, the height of the crane beam is 0.9 m, and the height of the track and cushion is 0.2 m. Therefore, the top elevation of the corbel top is 8.0−0.9−0.2 = 6.9 m, which meets the module of 300 mm (allowable deviation of ±200 mm).

4) Determination of the height of the upper and lower columns in the calculation diagram

Upper column height: $H_u = 10.4 - 6.9 = 3.5$ m

Lower column height: $H_l = 6.9 + 0.85 = 7.75$ m

Total height of column: $H = 3.5 + 7.75 = 11.25$ m or $H = 10.4 + 0.85 = 11.25$ m

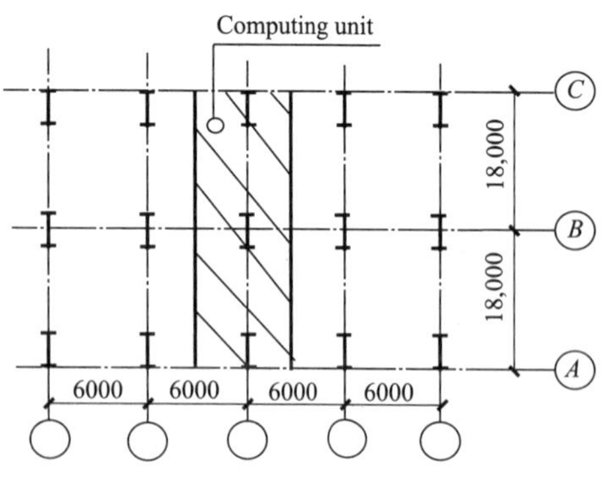

(a)

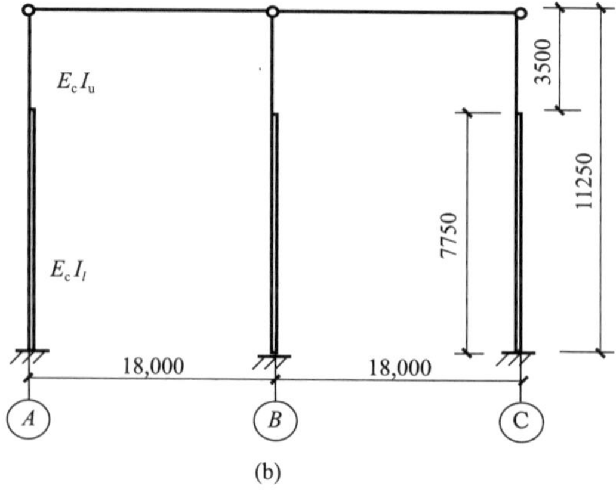

(b)

Fig. 2-69 Determine the structure calculation diagram(mm)

2. Cross-section size of column

Rectangular section for upper column and I-shaped section for lower column are chosen, respectively. The cross-section dimensions of the column are determined from Table 2-10 considering the structural requirements.

For side columns, namely axis A and axis C columns (Fig. 2-70a): cross-section size of the upper column is 400 mm×400 mm. Cross-section size of the lower column is 400 mm×1000 mm×120 mm (Fig. 2-70b). For the center column, that is, axis B column (Fig. 2-70c): cross-section size of the upper column is 400 mm×600 mm. Cross-section size of the lower column is 400 mm×1200 mm×120 mm (Fig. 2-70d).

3. Sectional geometric features and calculation of column weight

1) Geometric features of column sections at axis A and C

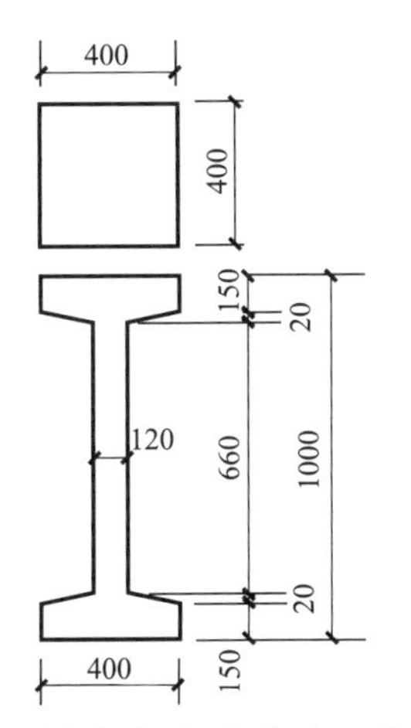

(a) Axis A and axis C column details

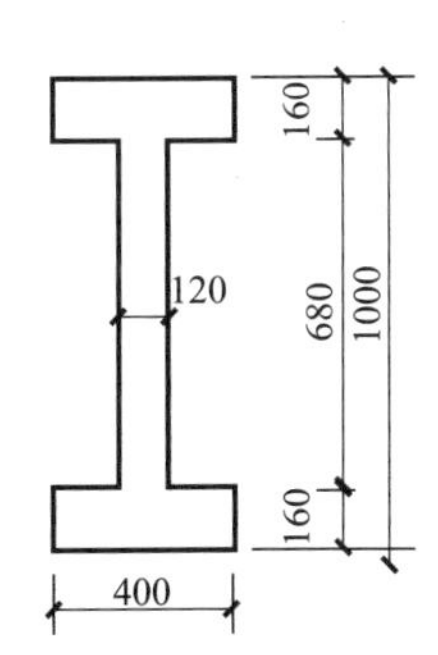

(b) Calculation diagram of lower column of axis A, C

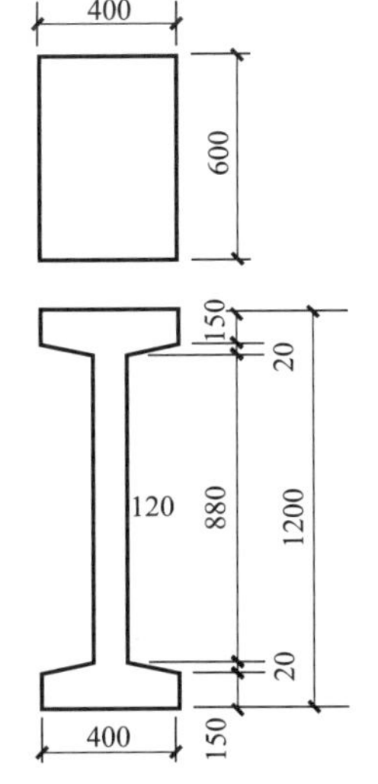

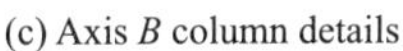
(c) Axis B column details

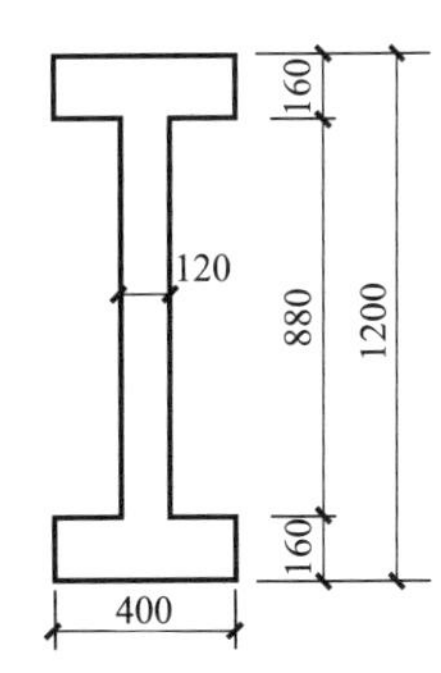

(d) Calculation diagram of lower column of axis B

Fig. 2-70 Detailed cross-sectional view of the column

Upper column (Fig. 2-70a):

$A = 400 \times 400 = 160 \times 10^3$ mm^2; $G = 25 \times 0.16 = 4.0$ kN/m

$I_x = I_y = (1/12) \times 400 \times 400^3 = 21.33 \times 10^8$ mm^4; $i_x = i_y = (I_x/A)^{1/2} = 115.5$ mm

Lower column (Fig. 2-70b):

$A=400\times160\times2+120\times680=209.6\times10^3\ mm^2$; $G=25\times0.2096=5.24\ kN/m$

$I_x=(1/12)\times400\times1000^3-(1/12)\times(400-120)\times680^3=259.97\times10^8\ mm^4$

$I_y=2\times(1/12)\times160\times400^3+(1/12)\times680\times120^3=18.05\times10^8\ mm^4$

$i_x=(I_x/A)^{1/2}=352.18\ mm$; $i_y=(I_y/A)^{1/2}=92.80\ mm$

2) Calculation of the geometric characteristics of the axis B column section of the center column is omitted.

The cross-sectional geometric characteristics of each column are listed in Table 2-10.

Sectional geometric characteristics of each column **Table 2-10**

Column	$A(mm^2)$	$I_x(mm^4)$	$I_y(mm^4)$	$i_x(mm)$	$i_y(mm)$	$G(kN/m)$
Axis A, C upper column	160.0×10^3	21.33×10^8	21.33×10^8	115.50	115.50	4.00
Axis A, C lower column	209.6×10^3	259.97×10^8	18.05×10^8	352.18	92.80	5.24
Axis B upper column	240.0×10^3	72.00×10^8	32.00×10^8	173.20	115.50	6.00
Axis B lower column	233.6×10^3	416.99×10^8	18.33×10^8	422.50	88.58	5.83

Note:

A is the cross-section area; I_x is the moment of inertia of section along the bent direction; I_y is the moment of inertia of section along the direction perpendicular to the bent frame; i_x is the radius of gyration along the bent direction; i_y is the radius of gyration of the section along the direction perpendicular to the bent frame; G is the column weight per unit length.

2.6.5 Load calculation

1. Determination of dead load and calculation diagram

1) Self-weight of roof structure G_{1K}

Roof uniform load:

Polymer modified bitumen membrane waterproof layer: $0.45\ kN/m^2$

25mm thick cement mortar leveling layer: $0.5\ kN/m^2$

100mm thick perlite product insulation layer: $0.4\ kN/m^2$

Cold bottom oil gas barrier (neglected)

25 mm thick cement mortar leveling layer: $0.5\ kN/m^2$

Board seam treatment: paste 300 mm polymer modified asphalt coil
Fine stone concrete pouring
1.5 m×6 m prestressed roof panel } $1.4\ kN/m^2$

Characteristic value of self-weight of roof truss (including supporting):

$0.12+0.011L=0.12+0.011\times18=0.318\ kN/m^2$

Total: $3.568\ kN/m^2$

The self-weight of the roof structure is transmitted from the roof truss, then to the column top of the bent column and it can be calculated according to the loading range in Fig. 2-69(a).

Self-weight of roof structure:

$3.568\times6\times18\times0.5=192.67\ kN$

Skylight frame: 34 kN

Gutter board: $2.02\times6=12.12\ kN$

Total: $G_{1K}=238.79\ kN$

For axis A and C columns: $G_{1A,K}=G_{1C,K}=238.79\ kN$. The eccentricity to the axis of the upper column $e_{1A}=e_{1C}=200-150=50\ mm$ as shown in Fig. 2-71(a).

For axis B column: $G_{1B,K}=2\times238.79\ kN=47758\ kN$. The eccentricity to the axis of the upper column $e_{1B}=0$ as shown in Fig. 2-71 (b).

2) Weight of upper column G_{2K}

For side columns (axis A, C columns):

$G_{2A,K}=G_{2C,K}=4\times3.5=14$ kN; $e_{2A}=e_{2C}=500-200=300$ mm

For the center column (axis B column): $G_{2B,K}=6\times3.5=21$ kN; $e_{2B}=0$

3) Dead weight of the crane beam, track and parts G_{3K} (Fig. 2-72)

Corbel at the side column:

$G_{3A,K}=G_{3C,K}=39+0.8\times6=43.8$ kN; $e_{3A}=e_{3C}=750-500=250$ mm

Corbel at the middle column: $G_{3B,K,left}=G_{3B,K,right}=43.8$ kN, $e_{3B,left}=e_{3B,right}=750$ mm

4) Weight of the lower column G_{4K} (Fig. 2-72)

For side columns: $G_{4A,K}=G_{4C,K}=7.75\times5.24\times1.1=44.67$ kN; $e_{4A}=e_{4C}=0$

For themiddle column: $G_{4B,K}=7.75\times5.83\times1.1=49.70$ kN; $e_{4B}=0$

5) Weight of connecting beam, foundation beam and upper wall G_{5K}, G_{6K} (Fig. 2-72)

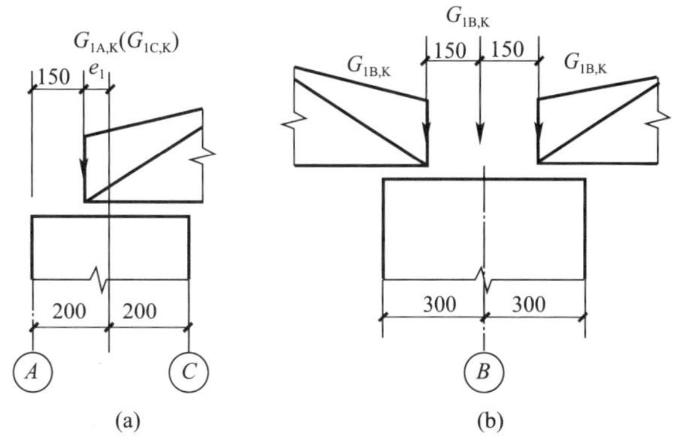

Fig. 2-71 Position of G_{1K}

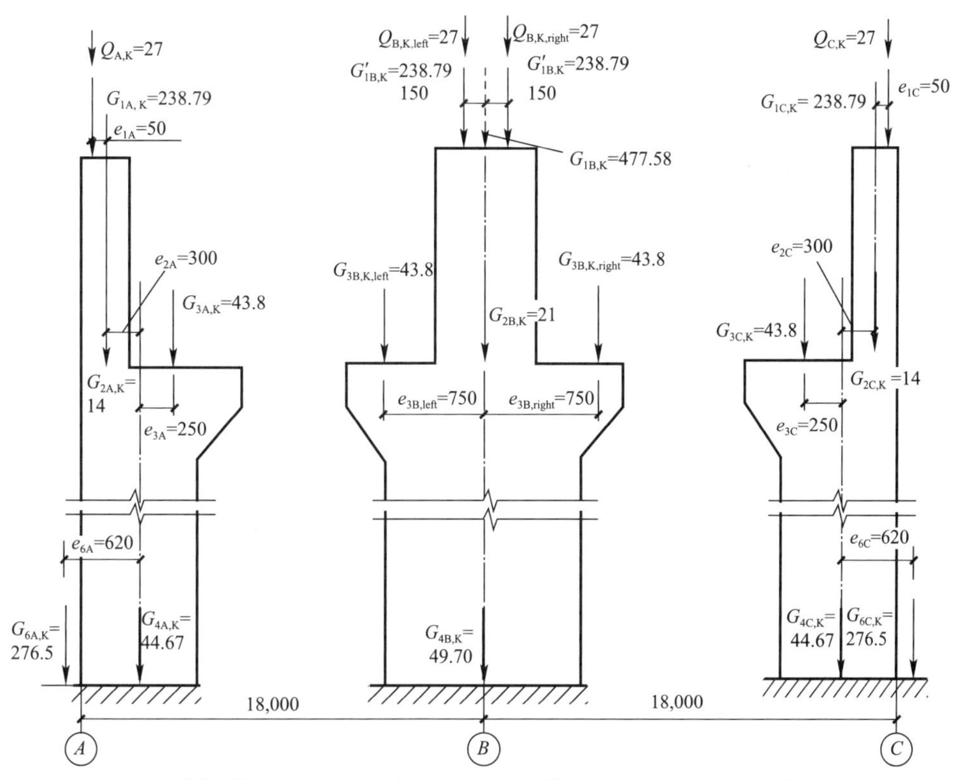

Fig. 2-72 Characteristic value (kN) and acting position (mm) of each permanent load

The net height of the wall is $11.8+0.85-0.45=12.2$ m. The window width is 4 m, and the window height is $4.8+1.8=6.6$ m.

Self-weight of wall on foundation beam:

$5.24\times[12.2\times6-4\times(4.8+1.8)]+0.45\times4\times(4.8+1.8)=260.3$ kN

Self-weight of foundation beam: $2.7\times6=16.2$ kN

Total weight of foundation beam and upper wall: $G_{6A,K}=G_{6C,K}=260.3+16.2=276.5$ kN

G_{6K} acts directly on the top surface of the foundation, $e_{6A}=e_{6C}=120+500=620$ mm

There is no maintenance wall in the center column, $G_{6B,K}=0$

The values and acting positions of each permanent load are shown in Fig. 2-72.

6) Determination of the calculation diagram of dead load (permanent load)

The eccentric pressure on the top of the column not only has an eccentric moment on the top of the column, but also has an eccentric moment on the lower column top because the centroids of the upper and lower column sections of the side columns do not coincide.

$G'_{1A,K}=G_{1A,K}=238.79\ \text{kN}=G'_{1C,K}$

$G'_{2A,K}=G_{2A,K}+G_{3A,K}=14+43.8=57.8\ \text{kN}=G'_{2C,K}$

$G'_{1B,K}=G_{1B,K}=477.58$ kN

$G'_{2B,K}=G_{2B,K}+G_{3B,K,\text{left}}+G_{3B,K,\text{right}}=21.0+43.8\times2=108.6$ kN

$G'_{4A,K}=G_{4A,K}=44.67\ \text{kN}=G'_{4C,K}$

$G'_{4B,K}=G_{4B,K}=49.70$ kN

Calculation of bending moments of each section of axis A and C columns (side columns):

Column top: $M_{1A,K}=M_{1C,K}=238.79\times0.05=11.94\ \text{kN}\cdot\text{m}$

Corbel top: $M_{2A,K}=M_{2C,K}=(238.79+14)\times0.3-43.8\times0.25=64.89\ \text{kN}\cdot\text{m}$

The calculation diagram of bent frame structure under dead load is shown in Fig. 2-73.

2. Determination of the calculation diagram of bent under roof live load

According to ***Load Code for the Design of Building Structures*** (GB 50009—2012), the

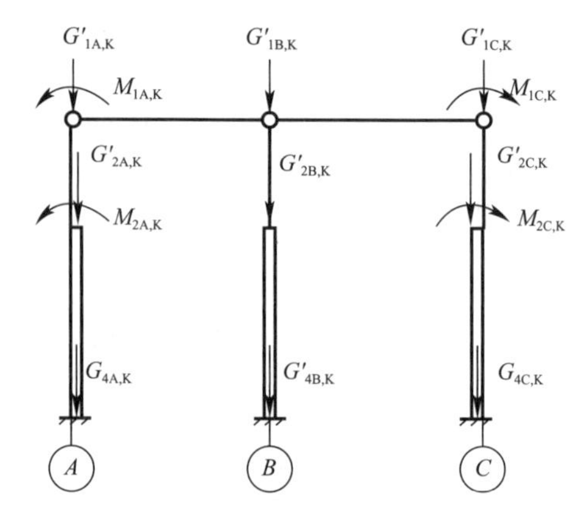

Fig. 2-73 Calculation diagram of double-span bent frame under dead load

characteristic value of roof live load is 0.5 kN/m², and roof snow load is 0.35 kN/m².

The characteristic value of roof live load transmitted from the roof truss to the bent column is calculated according to the load range shown in Fig. 2-69 (a).

$Q_{A,K}=Q_{B,K,\text{left}}=Q_{B,K,\text{right}}=Q_{C,K}=0.5\times6\times18\times0.5=27$ kN

The determination of the calculation diagram of the bent frame structure under live load is shown in Fig. 2-74.

For a double-span single-storey workshop, the impact of live load on the roof of each span should be considered.

There is live load at span AB:

Column top: $Q'_{A,K}=Q_{A,K}=27$ kN; $Q'_{B,K}=Q_{B,K,\text{left}}=27$ kN

$M_{1A,K}=27\times0.05=1.35\ \text{kN}\cdot\text{m}$; $M_{1B,K}=27\times0.15=4.05\ \text{kN}\cdot\text{m}$

Corbel top: $M_{2A,K}=27\times0.3=8.1\ \text{kN}\cdot\text{m}$; $M_{2B,K}=0$

There is live load at span BC:

Column top: $Q'_{C,K}=Q_{C,K}=27$ kN; $Q'_{B,K}=Q_{B,K,\text{right}}=27$ kN

$M_{1C,K}=27\times0.05=1.35\ \text{kN}\cdot\text{m}$; $M_{1B,K}=27\times0.15=4.05\ \text{kN}\cdot\text{m}$

Corbel top: $M_{2C,K}=27\times0.3=8.1\ \text{kN}\cdot\text{m}$; $M_{2B,K}=0$

3. Determination of crane load and calculation diagram

The crane calculation parameters of span AB and span BC are shown in Table 2-9.

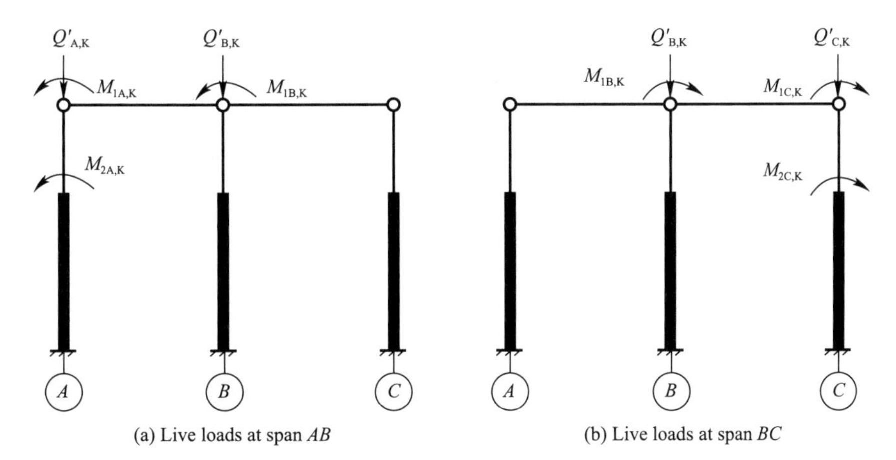

(a) Live loads at span AB (b) Live loads at span BC

Fig. 2-74 Calculation diagram of double-span bent frame under the live load

Span AB: $P_{min,k}=\dfrac{G_{1,k}+G_{2,k}+Q}{2}-P_{max,k}=\dfrac{340+300}{2}-270=50$ kN

Span BC: $P_{min,k}=\dfrac{G_{1k}+G_{2k}+Q}{2}-P_{max,k}=\dfrac{250+200}{2}-195=30$ kN

1) Crane vertical load $D_{max,k}$, $D_{min,k}$

For span AB (Fig. 2-75a):

$D_{max,k}=\beta P_{max,k}\sum y_i=0.9\times270\times\left(1+\dfrac{1.4}{6}+\dfrac{4.55}{6}\right)=483.98$ kN

$D_{min,k}=\dfrac{P_{min,k}}{P_{max,k}}D_{max,k}=\dfrac{50}{270}\times483.92=89.63$ kN

For span BC (Fig. 2-75b):

$D_{max,k}=\beta P_{max,k}\sum y_i=0.9\times195\times\left(1+\dfrac{1.6}{6}+\dfrac{4.75}{6}+\dfrac{0.35}{6}\right)=371.48$ kN

$D_{min,k}=\dfrac{P_{min,k}}{P_{max,k}}D_{max,k}=\dfrac{30}{195}\times371.48=57.15$ kN

2) Crane transversal horizontal load $T_{max,k}$

For span AB:

$T_k=\dfrac{1}{4}\alpha(G_{2k}+Q)=\dfrac{1}{4}\times0.1\times(118+300)$

$=10.45$ kN

$T_{max,k}=\dfrac{T_k}{P_{max,k}}D_{max,k}=\dfrac{10.45}{270}\times483.98$

$=18.73$ kN

For span BC:

$T_k=\dfrac{1}{4}\alpha(G_{2k}+Q)=\dfrac{1}{4}\times0.1\times(78+200)$

$=6.95$ kN

$T_{max,k}=\dfrac{T_k}{P_{max,k}}D_{max,k}=\dfrac{6.95}{195}\times371.48$

$=13.24$ kN

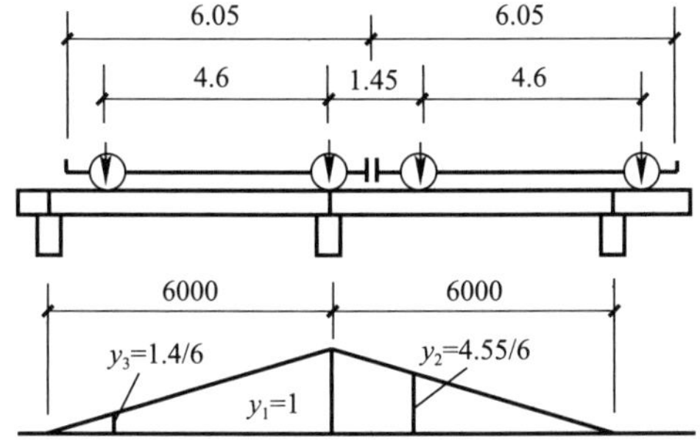

(a) Calculation of crane load at span AB (b) Calculation of crane load at span BC

Fig. 2-75 Calculation of crane load

3) Determination of calculation diagram of bent under crane load

The eccentricity of the vertical crane loads $D_{\max,k}$, $D_{\min,k}$ on the center of the lower column: for the side column: $e_3=0.25$ m; for middle column: $e_3=0.75$ m. When there are cranes on both spans, there are the following four situations (Fig. 2-76).

For span AB (Figs. 2-76a and b):

Right of column A:

$$\begin{cases}M_{\max,k}=D_{\max,k}e_3=483.98\times0.25=121.00\ \text{kN}\cdot\text{m}\\M_{\min,k}=D_{\min,k}e_3=89.63\times0.25=22.41\ \text{kN}\cdot\text{m}\end{cases}$$

Left of column B:

$$\begin{cases}M_{\max,k}=D_{\max,k}e_3=483.98\times0.75=362.99\ \text{kN}\cdot\text{m}\\M_{\min,k}=D_{\min,k}e_3=89.63\times0.75=67.22\ \text{kN}\cdot\text{m}\end{cases}$$

For span BC (Figs. 2-76c and d):

Left of column C:

$$\begin{cases}M_{\max,k}=D_{\max,k}e_3=371.48\times0.25=92.87\ \text{kN}\cdot\text{m}\\M_{\min,k}=D_{\min,k}e_3=57.15\times0.25=15.29\ \text{kN}\cdot\text{m}\end{cases}$$

Right of column B:

$$\begin{cases}M_{\max,k}=D_{\max,k}e_3=371.48\times0.75=278.61\ \text{kN}\cdot\text{m}\\M_{\min,k}=D_{\min,k}e_3=57.15\times0.75=42.86\ \text{kN}\cdot\text{m}\end{cases}$$

Considering $T_{\max,k}$ can either be left or right, the calculation diagrams of a double-span bent frame under the horizontal load of the crane should consider four cases (Fig. 2-77).

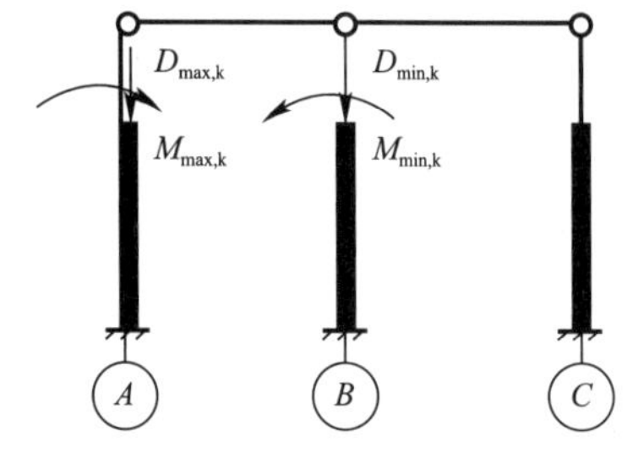

(a) Crane at span AB
($D_{\max,k}$ is on column A)

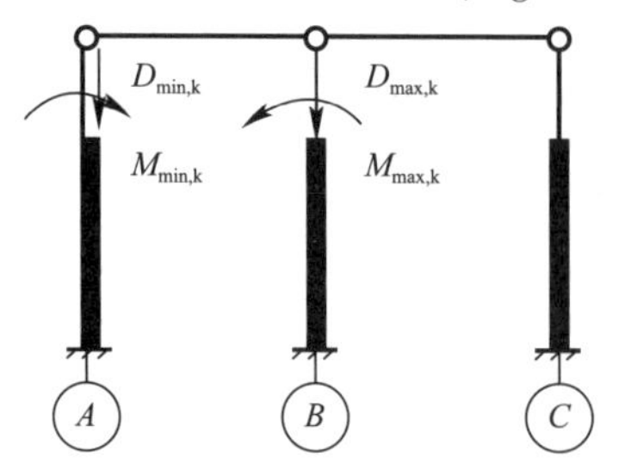

(b) Crane at span AB
($D_{\max,k}$ is on the left of column B)

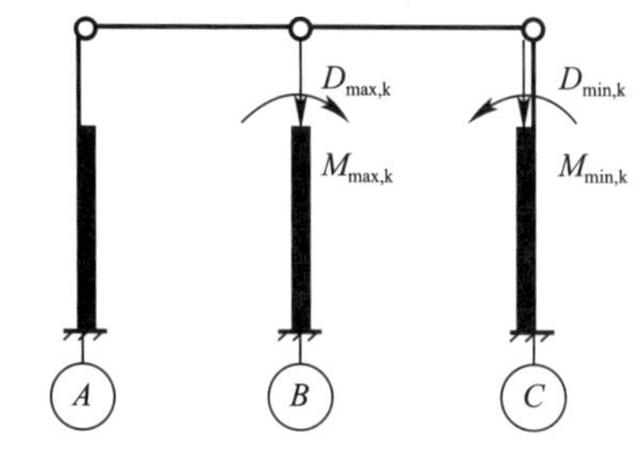

(c) Crane at span BC
($D_{\max,k}$ is on the right of column B)

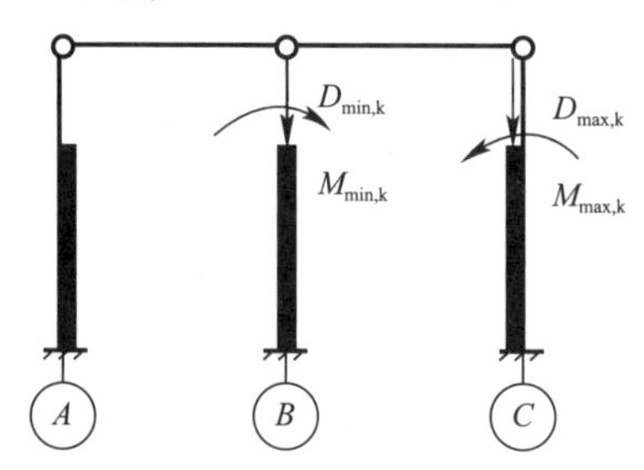

(d) Crane at span BC
($D_{\max,k}$ is on column C)

Fig. 2-76 Calculation diagram of double-span bent frame under $D_{\max,k}$, $D_{\min,k}$

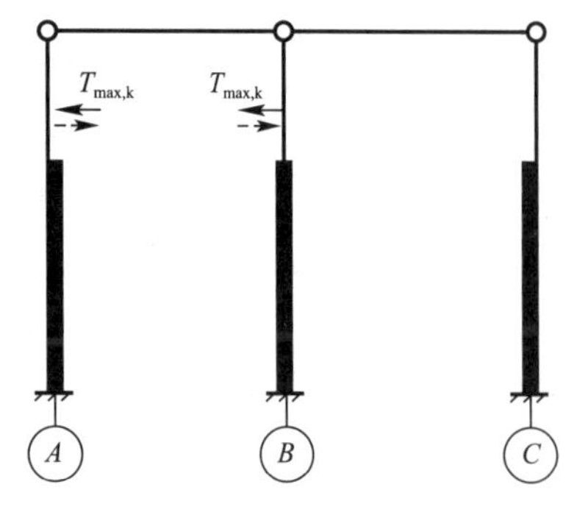

(a) Crane at span AB

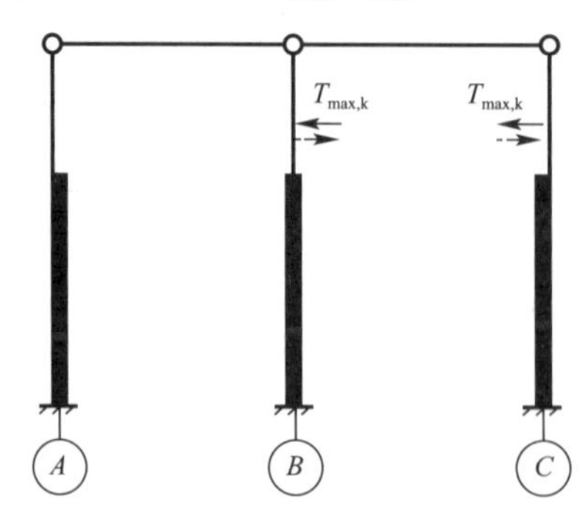

(b) Crane at span BC

Fig. 2-77 Calculation diagram of double-span bent frame under $T_{\max,k}$

4. Determination of wind load and calculation diagram

Shape coefficient of wind load is shown in Fig. 2-78. The basic wind pressure w_0 = 0.40 kN/m^2.

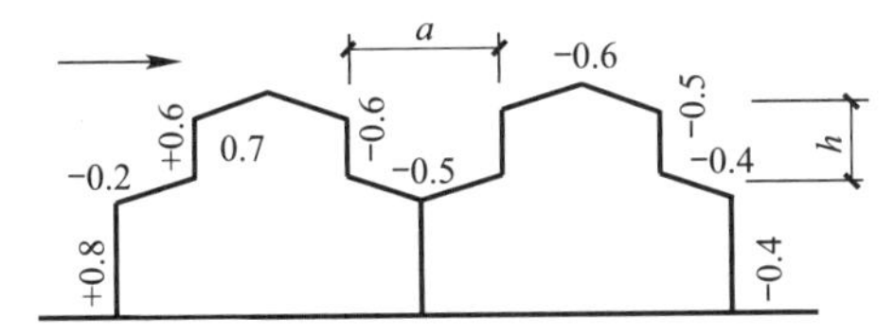

Fig. 2-78 Determination of wind load shape coefficient

In the case of left wind, μ_s at the skylight is determined according to Fig. 2-78, $\mu_s = +0.6$.

The height from the column top to the outdoor ground is: $Z = 10.4 + 0.15 = 10.55$ m

The height from the skylight eave to the outdoor ground is: $Z = 10.55 + 1.4 + 1.3 + 2.6 = 15.85$ m

At column top:

$$\mu_z = 1.0 + \frac{(10.55-10)\times(1.13-1.0)}{15-10} = 1.01$$

At skylight eave:

$$\mu_z = 1.13 + \frac{(15.85-15)\times(1.23-1.13)}{20-15} = 1.15$$

The characteristic value of wind load in the case of left wind (Fig. 2-79a):

$$q_{1k} = \mu_{s1}\mu_z\omega_0 B = 0.8\times1.01\times0.40\times6 = 1.94 \text{ kN/m}$$

$$q_{2k} = \mu_{s2}\mu_z\omega_0 B = 0.4\times1.01\times0.40\times6 = 0.97 \text{ kN/m}$$

$$\begin{aligned} F_{wk} &= \sum\mu_s\mu_z\omega_0 Bh_i = \mu_z\omega_0 B\sum\mu_s h_i \\ &= 1.15\times0.40\times6\times[(0.8+0.4)\times1.4 \\ &\quad +(0.4-0.2+0.5-0.5)\times1.3+ \\ &\quad (0.6+0.6+0.6+0.5)\times2.6+ \\ &\quad (0.7-0.7+0.6-0.6)\times0.3] \\ &= 21.86 \text{ kN} \end{aligned}$$

In the case of right wind, the magnitude of the load remains the same, but the direction is opposite (Fig. 2-79b).

2.6.6 Internal force analysis and internal force combination

1. Shear force distribution coefficient η_i of each column

The geometric characteristics of each column are shown in Table 2-10.

The calculation method of η_i is shown in section 2.2.3.

Shear force distribution coefficient of each column is: $\eta_A = 0.26$; $\eta_B = 0.48$; $\eta_C = 0.26$.

F_{wk}=21.86 kN

q_{1k}=1.94 kN/m

q_{2k}=0.97 kN/m

A B C

(a) Left wind

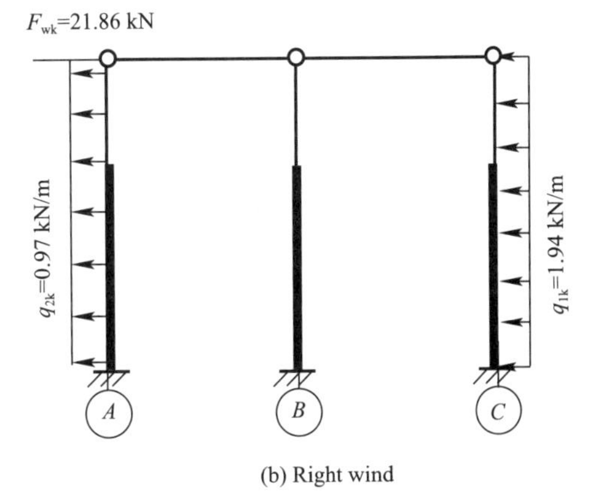

(b) Right wind

Fig. 2-79 Calculation diagram of double-span bent frame under wind load

2. Internal force of bent frame under dead load

The calculation diagram is shown in Fig. 2-73.

The bearing reaction force of the fixed hinge bearing at the top of each column:

Column A: $\lambda = H_u/H = 3.5/11.25 = 0.311$; $n = I_u/I_l = 21.33\times10^8/259.97\times10^8 = 0.082$;

According to Appendix 2:

$$C_1 = 1.5\frac{1-\lambda^2\left(1-\frac{1}{n}\right)}{1+\lambda^3\left(\frac{1}{n}-1\right)} = 1.5\times\frac{1-0.311^2\left(1-\frac{1}{0.082}\right)}{1+0.311^3\left(\frac{1}{0.082}-1\right)} = 2.337$$

$$C_3 = 1.5\frac{1-\lambda^2}{1+\lambda^3\left(\frac{1}{n}-1\right)} = 1.5\times\frac{1-0.311^2}{1+0.311^3\left(\frac{1}{0.082}-1\right)} = 1.01$$

$$R_A = \frac{M_{1A,K}}{H}C_1 + \frac{M_{2A,K}}{H}C_3 = \frac{11.94}{11.25}\times 2.337 + \frac{64.89}{11.25}\times 1.01 = 8.31\ \text{kN}(\rightarrow)$$

Column B: $R_B = 0$

Column C: $R_C = -R_A = -8.31\ \text{kN}(\leftarrow)$

The sum of the reaction forces of the fixed hinge support at the top of the bent frame column is $R = R_A + R_B + R_C = 0$. The actual shear force at the top of each column is:

$V_A = R_A = 8.31\ \text{kN}(\rightarrow)$; $V_B = 0$; $V_C = -8.31\ \text{kN}(\leftarrow)$。

The bending moment, column bottom shear force and axial force of bent frame are shown in Fig. 2-80.

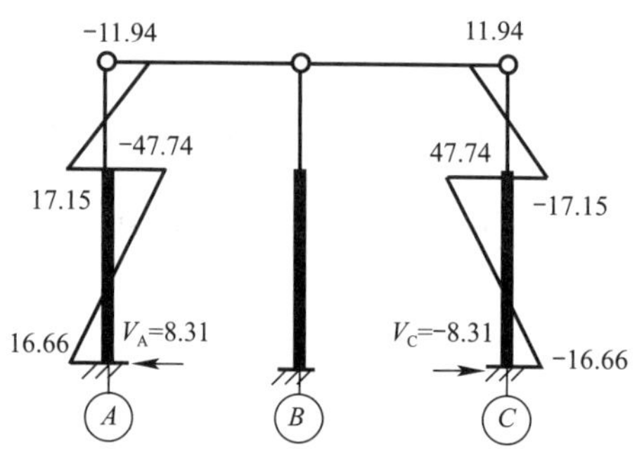

(a) Bending moment (kN · m) and column bottom shear force (kN)

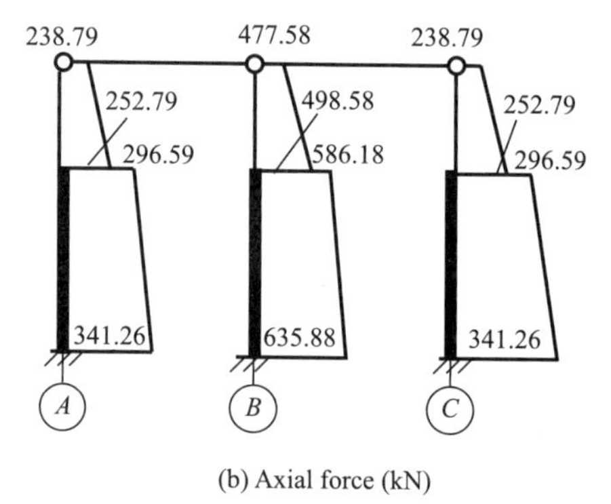

(b) Axial force (kN)

Fig. 2-80 Internal force diagram of the bent frame under dead load

3. Internal force of bent frame under live load

1) There is live load at span AB (Fig. 2-74a)

The bearing reaction force of the fixed hinge bearing at the top of each column:

Column A: $C_1 = 2.331$; $C_3 = 1.02$

$$R_A = \frac{M_{1A,K}}{H}C_1 + \frac{M_{2A,K}}{H}C_3 = \frac{1.8}{11.25}\times 2.337 + \frac{10.8}{11.25}\times 1.01 = 1.344\ \text{kN}(\rightarrow)$$

Column B: $\lambda = H_u/H = 0.311$; $n = I_u/I_l = 0.173$

$$C_1 = 1.5\frac{1-\lambda^2\left(1-\frac{1}{n}\right)}{1+\lambda^3\left(\frac{1}{n}-1\right)} = 1.5\times\frac{1-0.311^2\left(1-\frac{1}{0.173}\right)}{1+0.311^3\left(\frac{1}{0.173}-1\right)} = 1.918$$

$$R_B = \frac{M_{1B,K}}{H}C_1 = \frac{5.4}{11.25}\times 1.918 = 0.921\ \text{kN}(\rightarrow)$$

Column C: $R_C = 0$

The sum of the reaction forces of the fixed hinge support at the top of the bent frame column: $R = R_A + R_B + R_C = 2.265\ \text{kN}(\rightarrow)$.

The shear force at the top of each column:

$$V_A = R_A - \eta_A R = 1.344 - 0.26\times 2.265 = 0.755\ \text{kN}(\rightarrow)$$

$$V_B = R_B - \eta_B R = 0.921 - 0.48\times 2.265 = -0.166\ \text{kN}(\leftarrow)$$

$$V_C = R_C - \eta_C R = 0 - 0.26\times 2.265 = -0.589\ \text{kN}(\leftarrow)$$

The bending moment diagram, column bottom shear force and axial force diagram of bent frame are shown in Fig. 2-81.

2) There is live load at span BC (Fig. 2-74b)

Because the structure is symmetrical, the loads on the span BC and span AB are the same, so you only need to adjust the internal force diagram of Fig. 2-81. The internal force diagram is shown in Fig. 2-82.

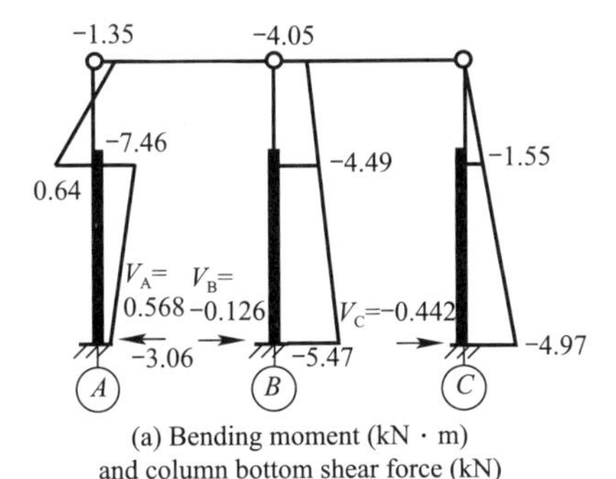

(a) Bending moment (kN · m) and column bottom shear force (kN)

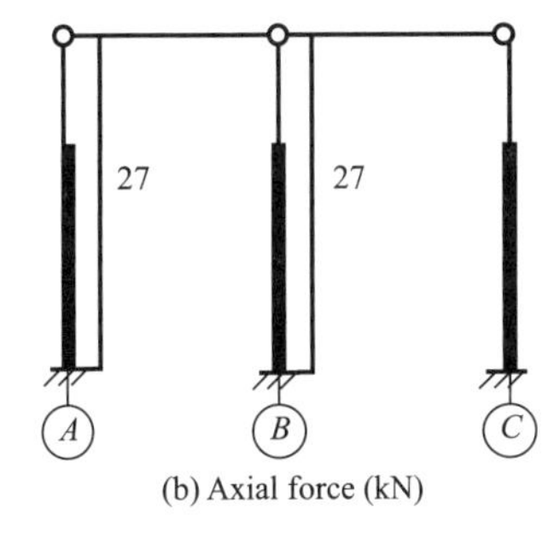

(b) Axial force (kN)

Fig. 2-81 Internal force diagram of bent frame under live load at span AB

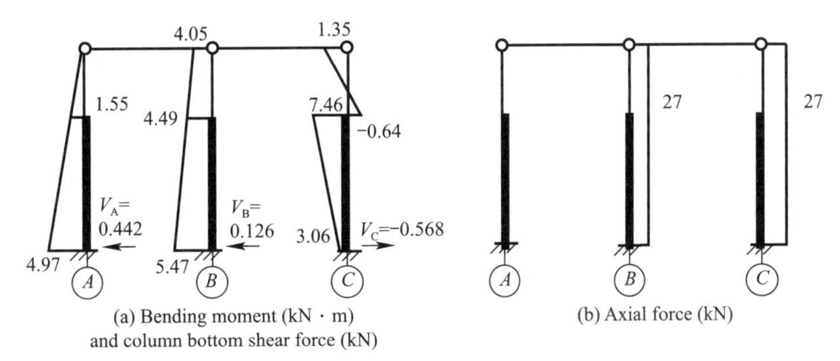

(a) Bending moment (kN · m) and column bottom shear force (kN)

(b) Axial force (kN)

Fig. 2-82 Internal force diagram of bent frame under live load at span BC

4. Internal force of bent under crane vertical load

1) Internal force of bent frame when there is a crane load at span AB, $D_{max,k}$ acts on the right of column A, and $D_{min,k}$ acts on the left of column B.

The calculation diagram is shown in Fig. 2-76(a).

The bearing reaction force of the fixed hinge support at the top of each column:

Column A:

$C_3 = 1.01$;

$$R_A = \frac{M_{max,k}}{H} C_3 = \frac{121.00}{11.25} \times 1.01 = 10.86 \text{ kN}(\leftarrow)$$

Column B:

$\lambda = H_u/H = 0.31$; $n = I_u/I_l = 0.173$

$$C_3 = 1.5 \frac{1-\lambda^2}{1+\lambda^3\left(\frac{1}{n}-1\right)} = 1.5 \times \frac{1-0.311^2}{1+0.311^3\left(\frac{1}{0.173}-1\right)} = 1.185$$

$$R_B = \frac{M_{min,k}}{H} C_3 = \frac{67.22}{11.25} \times 1.185 = 7.08 \text{ kN}(\rightarrow)$$

Column C:

$R_C = 0$; $R = R_A + R_B + R_C = -3.78$ kN$(\leftarrow)$

The shear force at the top of each column is:

$$V_A = R_A - \eta_A R = -10.86 + 0.26 \times 3.78 = -9.88 \text{ kN}(\leftarrow)$$

$$V_B = R_B - \eta_B R = 7.08 + 0.48 \times 3.78 = 8.89 \text{ kN}(\rightarrow)$$

$$V_C = R_C - \eta_C R = 0 + 0.26 \times 3.78$$

$=0.98\ \text{kN}(\rightarrow)$

The bending moment diagram, column bottom shear force and axial force diagram of bent-frame are shown in Fig. 2-83.

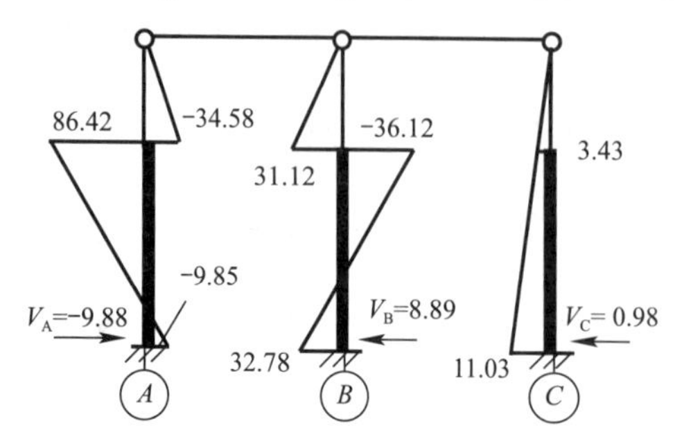

(a) Bending moment (kN · m) and column bottom shear force (kN)

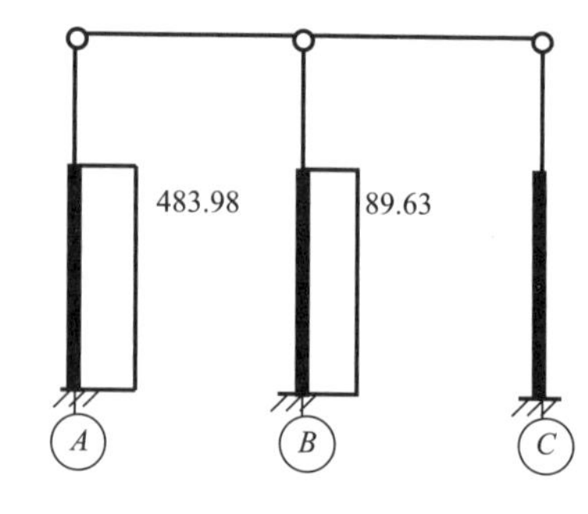

(b) Axial force (kN)

Fig. 2-83 Internal force diagram of bent frame under $D_{max,k}$ on the column A at span AB

2) Internal force of bent frame when there is a crane load on span AB, $D_{max,k}$ acts on the left of column B and $D_{min,k}$ acts on the right of column A.

The bending moment diagram, column bottom shear force and axial force diagram of bent frame are shown in Fig. 2-84.

3) Internal force of bent frame when there is a crane load on span BC, $D_{max,k}$ acts on the right of the column B, and $D_{min,k}$ acts on the left of the column C.

The bending moment diagram, column bottom shear force and axial force diagram of bent frame are shown in Fig. 2-85.

4) Internal force of bent frame when there is a crane load on BC span, $D_{max,k}$ acts on the left of the column C, and $D_{min,k}$ acts on the right of the Column B.

The bending moment diagram, column bottom shear force and axial force diagram of the bent are shown inFig. 2-86.

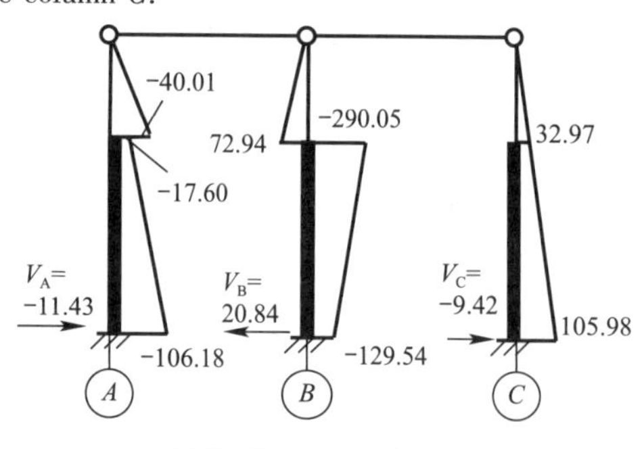

(a) Bending moment (kN · m) and column bottom shear force (kN)

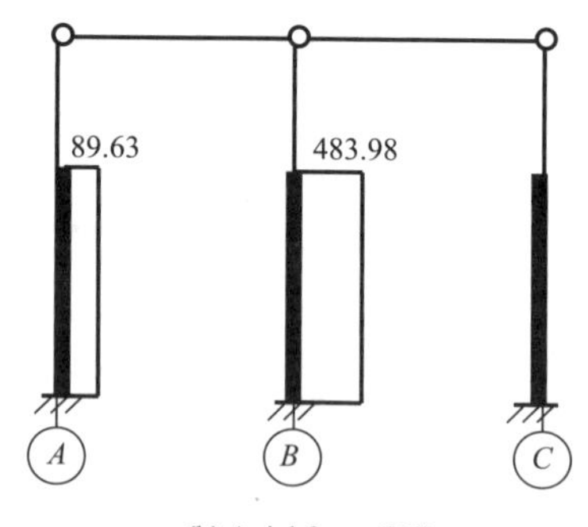

(b) Axial force (kN)

Fig. 2-84 Internal force diagram of bent frame under $D_{max,k}$ on the column B at span AB

5. Internal force of bent under crane transversal horizontal load

1) There is crane load at span AB, $T_{max,k}$ acts towards left

The distance from the point of action of $T_{max,k}$ to the column top $y=3.5-0.9=2.6$ m; $\alpha=y/H_u=0.743$. According to Appendix 2,

Column A, C: $\lambda=H_u/H=0.311$; $n=I_u/I_l=0.082$; $C_5=0.516$

$$R_A=C_5T_{max}=0.516\times18.73=9.66\ \text{kN}(\rightarrow)$$

Column B: $\lambda=H_u/H=0.311$; $n=I_u/I_l=0.173$; $C_5=0.588$

$$R_B=C_5T_{max}=0.588\times18.73=11.01\ \text{kN}(\rightarrow)$$

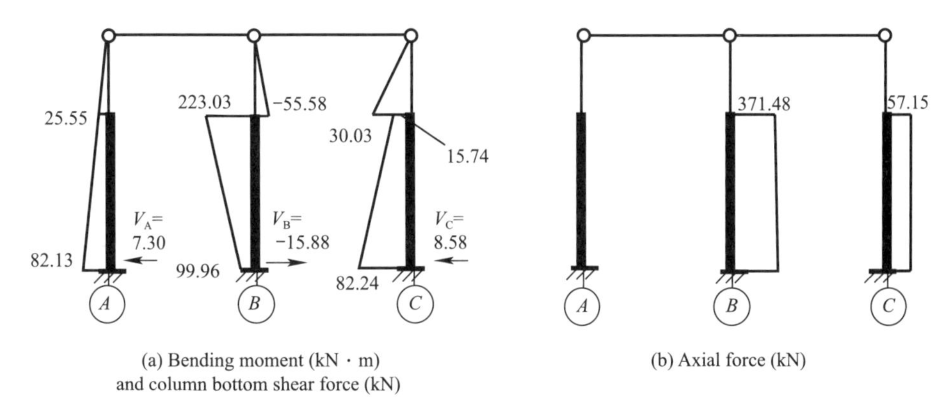

Fig. 2-85 Internal force diagram of bent frame under $D_{max,k}$ on the column B at span BC

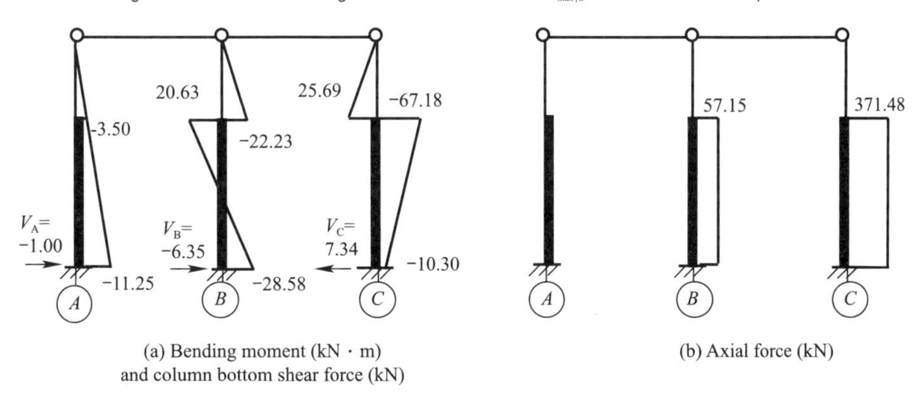

Fig. 2-86 Internal force diagram of bent frame under $D_{max,k}$ on the column C at span BC

Column C: $R_C=0$

The sum of the reaction forces of the fixed hinge support at the top of the column $R=R_A+R_B+R_C=20.67$ kN(→).

The actual shear force at the top of each column is:

$$V_A=R_A-\eta_A R=9.66-0.26\times20.67=4.29\ \text{kN}(\rightarrow)$$

$$V_B=R_B-\eta_B R=11.01-0.48\times20.67=1.09\ \text{kN}(\rightarrow)$$

$$V_C=R_C-\eta_C R=0-0.26\times20.67=-5.37\ \text{kN}(\leftarrow)$$

The bending moment and shear force at the bottom of the column are shown in Fig. 2-87, and the axial force of the column is zero.

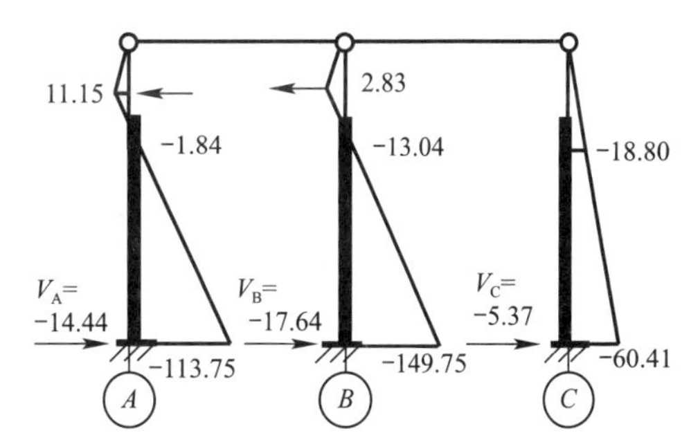

Fig. 2-87 Bending moment (kN · m) and column bottom shear force (kN) (span AB: $T_{max,k}$←)

2) There is crane load at span AB, $T_{max,k}$ acts towards right

Since the load is opposite to case 1, the values are equal and the opposite direction as shown in Fig. 2-88, and the axial force of the column is zero.

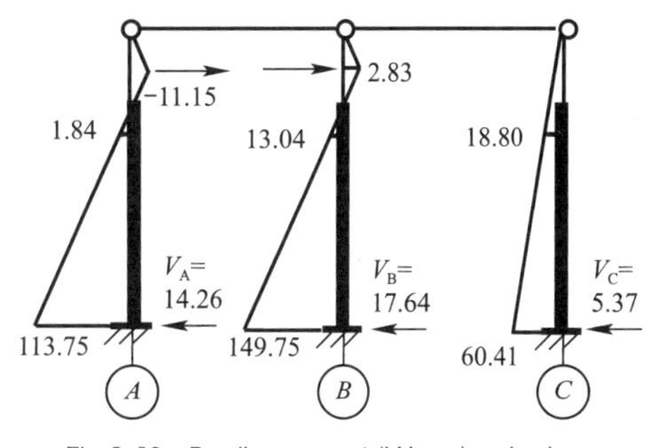

Fig. 2-88 Bending moment (kN · m) and column bottom shear force (kN) (span AB: $T_{max,k}$→)

3) There is crane load at span BC, $T_{max,k}$ acts towards left

The bending momentand column bottom shear force of bent frame are shown in Fig. 2-89. The axial force of the column is zero.

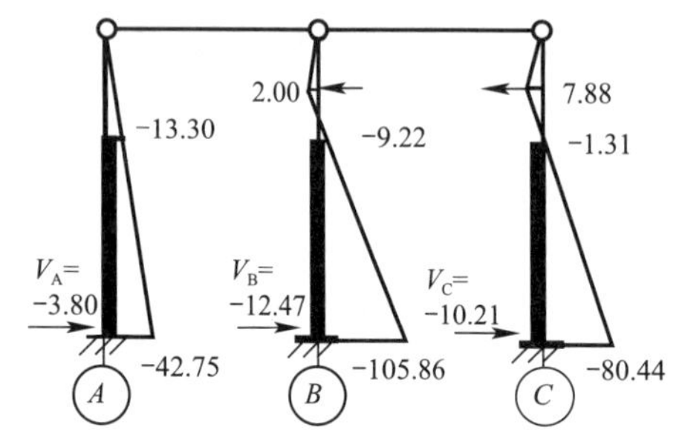

Fig. 2-89 Bending moment (kN · m) and column bottom shear force (kN) (span BC: $T_{max,k}$ ←)

4) There is crane load at span BC, $T_{max,k}$ acts towards right

The bending moment and the column bottom shear force are shown in Fig. 2-90.

6. Internal force of bent frame under wind load

The internal force diagram under wind load is shown in Fig. 2-91.

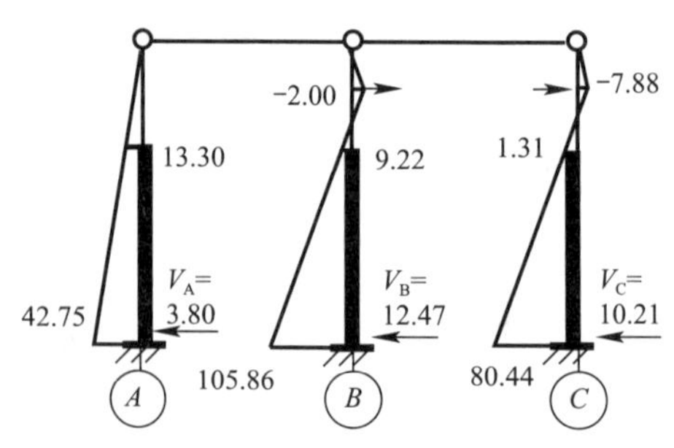

Fig. 2-90 Bending moment (kN · m) and column bottom shear force (kN) (span BC: $T_{max,k}$ →)

7. Internal force combination

According to the introduction in the fourth part of this chapter, the structural design service life is 50 years, $\gamma_L = 1.0$. For wind load, the combined value coefficient of variable load is $\psi_{ci} = 0.6$; for other loads, it is 0.7. When a combination expression uses 4 cranes to participate in the combination, the crane load effect should be multiplied by the conversion factor 0.8/0.9.

The internal force combination of column A as an example is shown in Table 2-11.

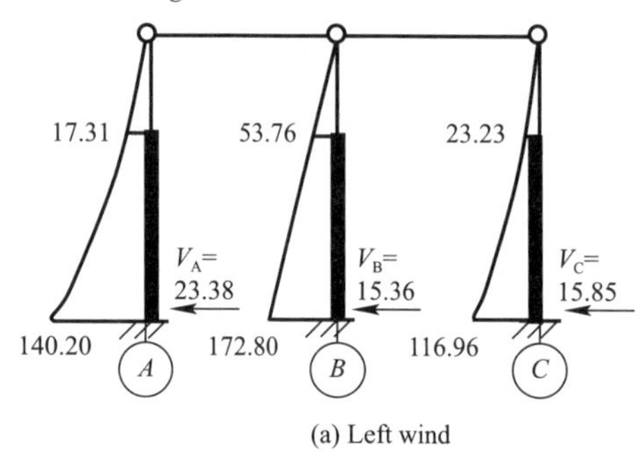

(a) Left wind

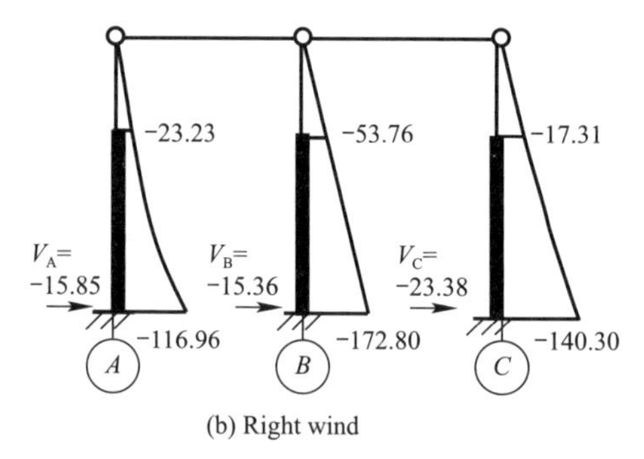

(b) Right wind

Fig. 2-91 Bending moment (kN · m) and column bottom shear force (kN) of bent frame under wind load

2.6.7 Design of bent column

Taking column A as an example, symmetrical reinforcement is adopted in design.

1. The Main Parameters

Concrete strength grade: C25 ($f_c = 11.9$ N/mm^2, $f_t = 1.27$ N/mm^2); steel bars: HRB335 ($f_y = f_y' = 300$ N/mm^2), $f_{yk} = 335$ N/mm^2.

The upper column is rectangular section (Fig. 2-92a). The lower column is I-shaped section, and the flange thickness of the section is taken as the average thickness (Fig. 2-92b).

$$a_s = a_s' = 35 + \frac{20}{2} = 45 \text{ mm}$$

Upper column: $h_0 = 400 - 45 = 355$ mm

$$N_b = \alpha_1 f_c b \xi_b h_0 = 1.0 \times 11.9 \times 400 \times 0.55 \times 355 = 92939 \text{ kN}$$

Lower column: $h_0 = 1000 - 45 = 955$ mm

$\xi_b h_0 = 0.55 \times 955 = 525.25$ mm$>h_f' = 160$ m.

Therefore, the boundary failure is the compression zone in the range of the web.

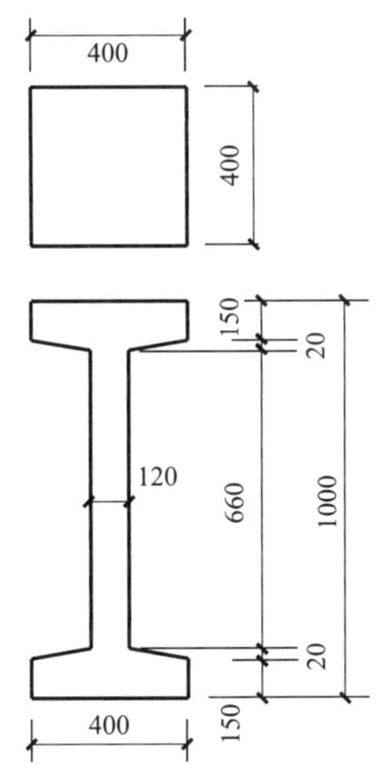

(a) Cross-section of column A

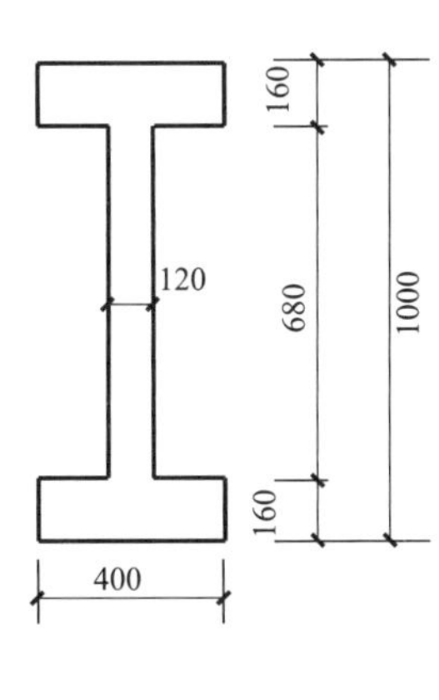

(b) Calculation diagram of lower column

Fig. 2-92 Cross-section of the column A

$N_b = \alpha_1 f_c [(b'_f - b)h'_f + b\xi_b h_0] = 1.0 \times 11.9 \times [(400-120)\times 120 + 120 \times 0.55 \times 955] = 1283.18$ kN

The upper column is 3.5 m high, the lower column is 7.75 m high, and the total height is 11.25 m.

In the bent direction, the calculated lengths of column when considering crane load are as follows.

Upper column: $l_u = 2.0H_u = 2.0 \times 3.5 = 7.0$ m

Lower column: $l_l = 1.0H_l = 7.75$ m

When the crane load is not considered,

Upper column: $l_u = 2.0H_u = 2.0 \times 3.5 = 7.0$ m

Lower column: $l_l = 1.25H = 1.25 \times 11.25 = 14.06$ m

In the perpendicular direction of the bent frame direction, when considering crane loads,

Upper column: $l_u = 1.5H_u = 1.5 \times 3.5 = 5.25$ m;

Lower column: $l_l = 1.0H_l = 7.75$ m

When the crane load is not considered,

Upper column: $l_u = 1.5H_u = 1.5 \times 3.5 = 5.25$ m

Lower column: $l_l = 1.2H = 1.2 \times 11.25 = 13.50$ m

2. Selection of the most unfavorable internal force of control section

Since symmetrical reinforcement is used in this design, only the absolute value of the bending moment needs to be considered when selecting the internal force. The longitudinal reinforcement of the column is calculated according to the M and N on the control section. The upper column reinforcement is calculated according to the unfavorable internal force of the section Ⅰ-Ⅰ. The lower column reinforcement is calculated according to the unfavorable internal forces of the sections Ⅱ-Ⅱ and Ⅲ-Ⅲ. The unfavorable internal force selection of the basic combination is shown in Table 2-12.

3. Reinforcement calculation of upper column

Rectangular section $b \times h = 400$ mm $\times$ 400 mm, $N_b = 929.39$ kN

$$l_{0x} = 7.0 \text{ m},\ l_{0y} = 5.25 \text{ m},\ a_s = a'_s = 45 \text{ mm},$$

$$h_0 = 400 - 45 = 355 \text{ mm}$$

1) According to the first group of unfavorable internal force

$$M_0 = M = 101.07 \text{ kN·m},\ N = 365.08 \text{ kN}$$

$$e_0 = M_0/N = 101.07/365.08 = 0.277 \text{ m} = 277 \text{ mm}$$

Take e_a as the larger value of $h/30$ and 20 mm, so take 20 mm.

$$e_i=e_0+e_a=277+20=297\ \text{mm}$$

Consider the influence of the second-order effect,

$$\zeta_c=\frac{0.5f_cA}{N}=\frac{0.5\times11.9\times160\times10^3}{365.08\times10^3}=2.61>1$$

Take $\zeta_c=1$,

$$\eta_s=1+\frac{1}{1500\dfrac{e_i}{h_0}}\left(\frac{l_0}{h}\right)^2\xi_c$$

$$=1+\frac{1}{1500\times\dfrac{297}{355}}\left(\frac{7000}{400}\right)^2\times1.0=1.24$$

$$M=\eta_sM_0=1.24\times101.07=125.33\ \text{kN}\cdot\text{m}$$

$$e_i=e_0+e_a=\frac{M}{N}+e_a=\frac{125.33\times10^6}{365.08\times10^3}+20$$

$$=363.29\ \text{mm}$$

$$e_i=e_0+e_a=363.29\ \text{mm}>0.3h_0=106.5\ \text{mm}$$

and $N<N_b$. Therefore, it belongs to the large eccentricity.

$$e=e_i+\frac{h}{2}-a_s=363.29+\frac{400}{2}-45$$

$$=518.29\ \text{mm}$$

$$x=\frac{N}{\alpha_1f_cb}=\frac{365.08\times10^3}{1.0\times11.9\times400}=77.70\ \text{mm}$$

Due to $x=77.70\ \text{mm}<\xi_bh_0=0.55\times355=195.25\ \text{mm}$ and $x<2a_s'=90$ mm

So take $x=2a_s'$,

$$e'=e_i-\frac{h}{2}+a_s'$$

$$=363.29-\frac{400}{2}+45=208.29\ \text{mm}$$

$$A_s=A_s'=\frac{Ne'}{f_y'(h_0-a_s')}$$

$$=\frac{365.08\times10^3\times208.29}{300\times(355-45)}$$

$$=817.66\ \text{mm}^2$$

The most unfavorable internal force of column *A* **Table 2-12**

Section	Serial number	Whether there are crane loads involved in the combination	M_0 (kN · m)	N (kN)	N_b (kN)	$e_0=M_0/N$ (mm)	e_a (mm)	$e_i=e_0+e_a$ (mm)	$0.3h_0$ (mm)	Preliminary discrimination of size eccentricity
Upper column I - I	1	Yes	101.07	365.08	929.39	276.8	20.0	296.8	106.5	Large eccentricity
	2	Yes	-79.23	328.63		241.1		261.1		Large eccentricity
	3	Yes	101.07	365.08		276.8		296.8		Large eccentricity
	4	Yes	100.20	328.63		304.9		324.9		Large eccentricity
Lower column Ⅲ - Ⅲ	1	Yes	457.94	1097.01	1283.18	417.4	33.3	450.7	286.5	Large eccentricity
	2	Yes	-437.2	601.09		727.5		760.8		Large eccentricity
	3	Yes	353.77	1133.46		312.1		345.4		Large eccentricity
	4	Yes	386.23	443.64		870.6		903.9		Large eccentricity

Reinforcement of section I-I can be calculated by the fourth group of unfavorable internal force as the above method. We can get $A_s = A'_s = 837.37\ \text{mm}^2$.

The reinforcement situation is shown in Fig. 2-93(a). Longitudinal reinforcement of upper column adopts 6⌽22, $A_s = A'_s = 1140\ \text{mm}^2$.

2) Checking calculation of reinforcement and reinforcement ratio

Check and calculate the minimum reinforcement ratio according to one side of the reinforced steel bar,

$$\frac{A_s}{A} = \frac{1140}{160 \times 10^3} = 0.71\% > 0.2\%$$

Check the minimum reinforcement ratio based on all longitudinally stressed steel bars,

$$\frac{\Sigma A_s}{A} = \frac{1140 \times 2}{160 \times 10^3} = 1.425\% > 0.6\%$$

3) Calculation of bearing capacityin the direction perpendicular to bent

According to the maximum axial force $N_{max} = 365.08$ kN, the compression bearing capacity is checked.

$$l_{0y}/b = 5.25/0.4 = 13.13;$$

$$\phi = \frac{0.92-0.95}{14-12}(13.13-12)+0.95 = 0.93$$

$$N_u = 0.9\phi(f_c A_c + f'_y A'_s) = 0.9 \times 0.93[11.9 \times (160,000-1140 \times 2)+300 \times 1140 \times 2]$$

$$= 2143.4\ \text{kN} > N_{max}$$

So, the compression bearing capacity in the direction perpendicular to bent frame meets the requirement.

4. Reinforcement calculation of lower column

I-shaped section: $b'_f \times h \times h'_f = 400\ \text{mm} \times 1000\ \text{mm} \times 160\ \text{mm}$,

$N_b = 1283.18$ kN, $l_{0x} = l_{0y} = 7.85$ m, $a_s = a'_s = 45$ mm, $h_0 = 1000-45 = 955$ mm

According to the second and fourth group of unfavorable internal forces, we can get the longitudinal reinforcement areas of lower column section, then 8⌽22 is adopted, $A_s = A'_s = 1520\ \text{mm}^2$. The reinforcement is shown in Fig. 2-93(b). Structural reinforcement is 10⌽12.

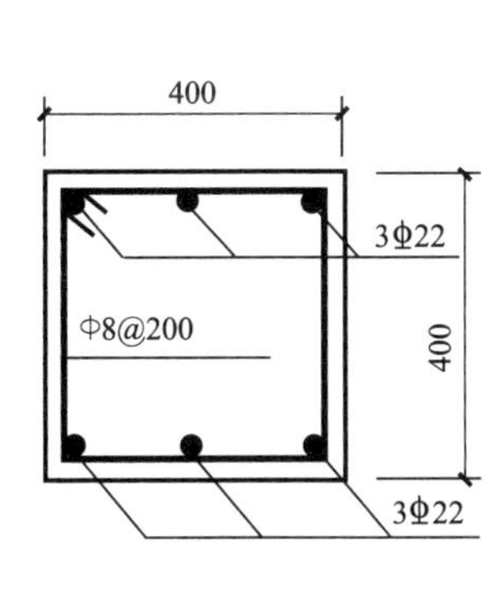

(a) Reinforcement of upper column

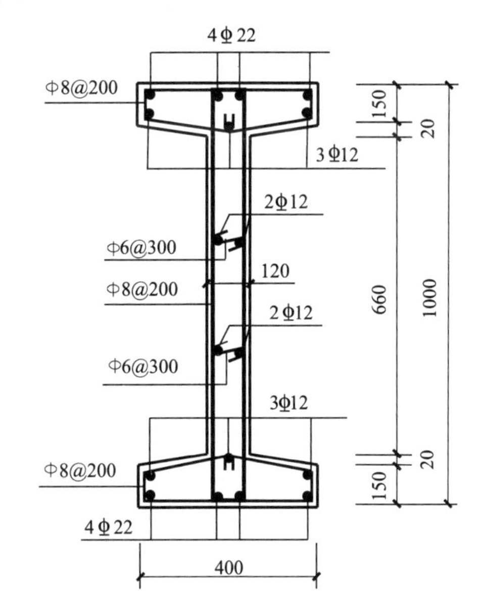

(b) Reinforcement of lower column

Fig. 2-93 Details of reinforcement of column A

According to the maximum axial force $N_{max} = 1133.46$ kN

$$l_{0y} = 7.75\ \text{m},\ i_y = 92.8\ \text{mm}$$

$$\frac{l_{0y}}{i_y} = \frac{7.75 \times 10^3}{92.8} = 83.51$$

Axial compression stability coefficient:

$$\phi=\frac{0.6-0.65}{90-83}(83.51-83)+0.65=0.65$$

$$N_u=0.9\times0.65[11.9\times(209.6\times10^3-1520\times2)+300\times1520\times2]=1971.5\ \text{kN}>N_{max}$$

So, the compression bearing capacity in the direction perpendicular to bent frame meets the requirement.

5. Crack width checking of column

Under the quasi-permanent combination of loads, when $e_0=\frac{M_q}{N_q}\geqslant0.55\ h_0$, the crack width need to be checked. For upper column, 0.55 h_0 = 195.25 mm. For lower column, 0.55 h_0 = 525.25 mm. The eccentricity of each section is as follows.

Section Ⅰ-Ⅰ: $e_0=\frac{M_q}{N_q}=\frac{17.92\ \text{kN}\cdot\text{m}}{262.24\ \text{kN}}=0.078\ \text{m}=78\ \text{mm}<195.25\ \text{mm}$

Section Ⅱ-Ⅱ: $e_0=\frac{M_q}{N_q}=\frac{50.35\ \text{kN}\cdot\text{m}}{306.04\ \text{kN}}=0.165\ \text{m}=165\ \text{mm}<525.25\ \text{mm}$

Section Ⅲ-Ⅲ: $e_0=\frac{M_q}{N_q}=\frac{18.40\ \text{kN}\cdot\text{m}}{341.26\ \text{kN}}=0.054\ \text{m}=54\ \text{mm}<525.25\ \text{mm}$

Therefore, crack width verification is not required for each section.

6. Column stirrup

The stirrups of single-storey factory building in non-seismic areas are generally controlled by structural requirements. According to the structural requirements, Φ8@200 stirrups are used for the upper and lower column.

2.6.8 Design of corbel

1. Section size of corbel

The preliminary size of the corbel is determined according to the structural requirements as shown in Fig. 2-94. The pressure F_{vk} is generated by the crane on the bent frame column and the weight of the crane beam, track and parts.

$$F_{vk}=D_{max,k}+G_{3,k}=483.98+43.8=527.78\ \text{kN}$$

$$h_0=\frac{0.5F_{vk}}{\beta\cdot f_{tk}b}=\frac{0.5\times527.78\times10^3}{0.65\times1.78\times400}=570.2\ \text{mm}$$

$h=h_0+a_s=570.2+40=610.2$ mm,

Take $h=650$ mm

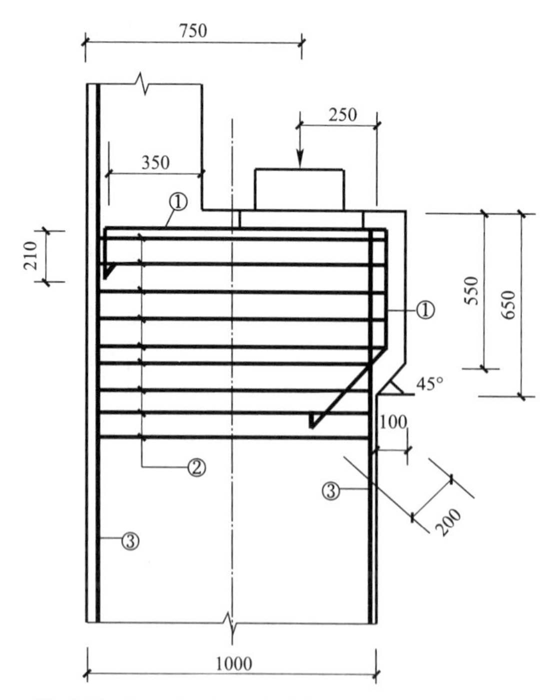

Fig. 2-94 Geometry size and reinforcement diagram of corbel

① The corbel longitudinal reinforcement 4Φ14; ② The corbel stirrup Φ8@100; ③ The column reinforcement

According to the control requirements of corbel crack,

$$F_{vk} \leqslant \beta\left(1-0.5\frac{F_{hk}}{F_{vk}}\right)\frac{f_{tk}bh_0}{0.5+\dfrac{a}{h_0}}$$

Where,

$F_{hk}=0$, $a=-250+20=-230$ mm, we take $a=0$, $\beta=0.65$, $f_{tk}=1.78$ N/mm^2. We can know h meets the requirement by calculation.

2. Reinforcement of corbel

The area of longitudinal reinforcement calculated by the minimum reinforcement ratio:

$$A_s \geqslant \rho_{min}=520 \text{ mm}$$

Take HRB335 grade steel, 4Φ14, A_s = 615 mm^2, meet the requirement.

According to the maximum reinforcement ratio: $A_s \leqslant \rho_{max}bh=1560$ mm^2.

Horizontal stirrup adopts Φ8@100. Due to $a=0$, no bent-up bars are provided. The reinforcement situation of the corbel is shown in Fig. 2-94.

2.6.9 Hoist checking of bent column

The calculation diagram of corbel hoisting is shown in Fig. 2-95.

The length of the column from the top of the foundation to the top of the column is 11.35 m. The depth of the column inserted into the cup opening is 900 mm according to Table 2-6, and the total length of the bent column is 12.25 m.

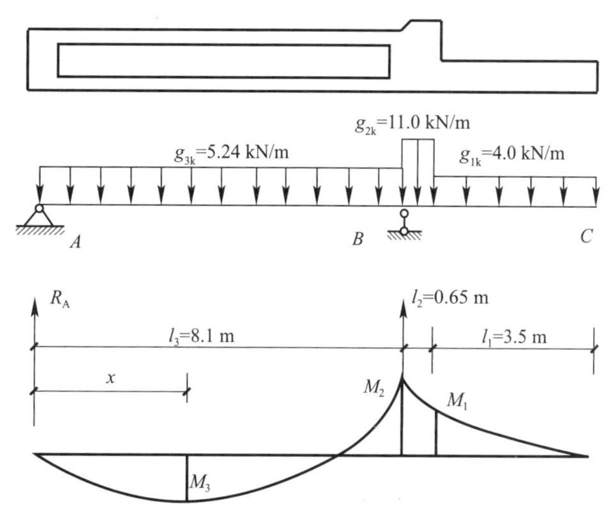

Fig. 2-95 Calculation diagram of column hoisting

After checking the bending bearing capacity and the crack width during hoisting, we can know that the hoisting check of column meets the requirements (the calculation process is omitted).

2.6.10 Design of foundation under column

Taking the foundation under column A as an example, the foundation adopts an independent cup-shaped foundation under the column. C20 is used for the foundation concrete, the steel bar is HPB300 grade, and the underlying layer of the foundation is C10 concrete. C30 fine stone concrete is filled between the precast column and the foundation.

1. Determine the height of the foundation

It is determined that the depth of the precast column inserted into the foundation is h_1 = 900 mm, a gap of 50 mm is left at the bottom of the column, and the gap between the column and the cup mouth is filled with C30 fine stone concrete. According to Table 2-7, the thickness of the bottom of the foundation cup a_1 = 200 mm, and the thickness of the cup wall $t=400$ mm (to meet the requirements of supporting the foundation beam). The height of the foundation is:

$h = h_1 + a_1 + 50 = 900 + 200 + 50 = 1150$ mm

The height of the cup wall h_2 is determined according to the punching resistance conditions of the foundation under the steps, and h_2 should be made as large as possible to reduce the amount of foundation concrete. It is preliminarily determined that $h_2 = 500$ mm, $t/h_2 = 0.8 > 0.75$, so the cup wall may not be reinforced. The size of the foundation bottom slab is determined according to the subgrade bearing capacity conditions. The basic structure of foundation is shown in Fig. 2-96.

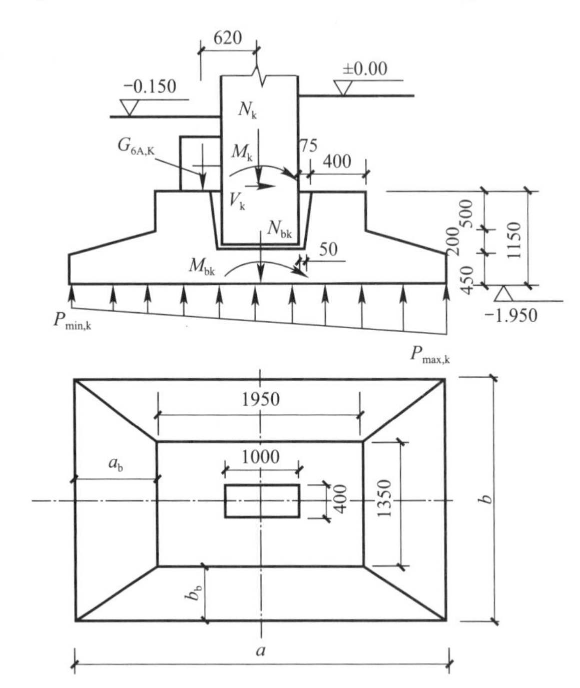

Fig. 2-96 Foundation under column

2. Determine loads on the top surface of the foundation

M, N and V of the top surface of the foundation (i.e., control section Ⅲ-Ⅲ) as shown in Table 2-13 are selected from Table 2-11. The standard combination of internal forces is used to determine the size of the foundation bottom slab. The basic combination of internal forces is used for the calculation of the anti-punching bearing capacity and the calculation of the reinforcement of the foundation.

Characteristic value of the top surface of the permanent load transmitted by the foundation beam is $G_{6A,K} = 276.5$ kN, and the eccentricity to the center line of the foundation is $e_6 = 620$ mm.

Unfavorable internal force on the top surface of foundation Table 2-13

Type of internal force	Basic combination				Standard combination			
	Group 1	Group 2	Group 3	Group 4	Group 1	Group 2	Group 3	Group 4
M(kN·m)	457.94	-437.27	353.27	386.23	284.94	-256.39	171.89	247.76
N(kN)	1097.01	601.09	1133.46	443.64	672.40	422.90	844.14	341.26
V(kN)	58.97	-44.75	49.89	57.95	40.50	-25.16	23.27	39.77

3. Determine the size of foundation bottom slab

1) Determination of characteristic value ofsubgrade bearing capacity

According to the designmaterial, the characteristic value of the bearing capacity of the subgrade bearing layer is $f_{ak}=240\ \text{kN/m}^2$. The corrected characteristic value of subgrade bearing capacity is:

$$f_a=f_{ak}+\eta_d\gamma_m(d-0.5)=240+1.6\times 17.5\times(1.8-0.5)=276.4\ \text{kN/m}^2$$

2) Characteristic value of internal forces of the foundation bottom

The characteristic value of bending moment M_{bk} and the characteristic value of axial force N_{bk} transmitted to the foundation bottom are shown in Table 2-14.

3) Determine the size of the foundation bottomslab

Average depth of foundation: $d=1.8+0.15/2=1.875$ m

Determine the area of the foundation bottom slab,

$$A_0=\frac{N_{bk}}{f_a-\gamma_G d}=\frac{1120.64}{265.4-20\times1.875}=4.92\ \text{m}^2$$

Characteristic values of internal forces of the foundation bottom Table 2-14

Type of internal force	Group 1	Group 2	Group 3	Group 4
$N_{bk}=N_k+G_{6A,K}$(kN)	918.9	699.40	1120.64	617.76
$M_{bk}=M_k+V_kh-G_{6A,K}e_6$(kN·m)	160.09	-456.75	27.22	122.07

Increase 25%, $A=1.25A_0=1.25\times4.92=6.15\ \text{m}^2$. So take $b=2.0$ m, $a=1.6b=3.2$ m. The above is a preliminary estimate of the size of the foundation bottom slab, and the subgrade bearing capacity must be checked.

$$W=\frac{1}{6}\times b\times a^2=\frac{1}{6}\times2\times3.2^2=3.41\ \text{m}^3$$

Average gravity of foundation and backfill: $G_K=\gamma_m dA=20\times1.875\times6.4=240$ kN

The calculation of subgrade bearing capacity shall meet the following requirements:

$$P_k=\frac{P_{max,k}+P_{min,k}}{2}\leqslant f_a=276.4\ \text{kN/m}^2$$

$$P_{max,k}=\frac{N_{bk}+G_K}{A}+\frac{M_{bk}}{W}\leqslant1.2f_a=331.68\ \text{kN/m}^2$$

$$P_{min,k}=\frac{N_{bk}+G_K}{A}-\frac{M_{bk}}{W}>0$$

We can know that the size of the foundation bottom slab meets the requirements for the bearing capacity of the subgrade.

4. Check the foundation height

The edge height of the foundation is 450 mm greater than 200 mm, the height of the conical foundation slope is 200 mm, the horizontal length of the slope is $a_b=725$ mm, and the slope is 1:3.6, which is less than the allowable slope of 1:3. The horizontal length from the cup wall to the edge of the foundation in the short-side direction is $b_b=325$ mm.

1) The basic combined design value of bending moment and axial force of foundation bottom

The loads and internal forces are shown in Fig. 2-97. The design value of bending moment M_b and the design value of axial force N_b of each group of internal forces transmitted to the foundation bottom are shown in Table 2-15.

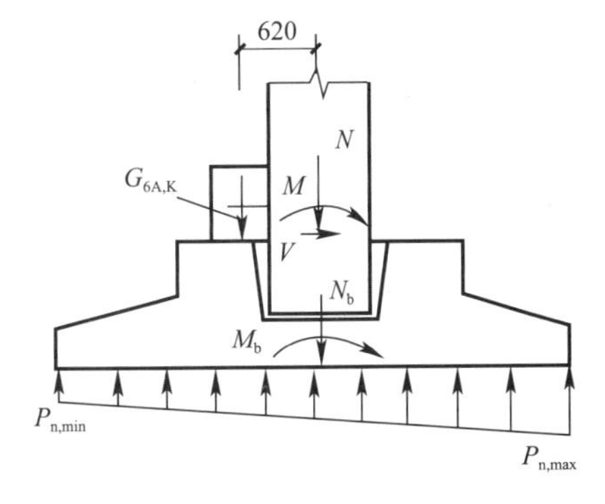

Fig. 2-97 Loads and internal forces on the foundation

Design values of internal force and net reaction force of foundation bottom

Table 2-15

	Group 1	Group 2	Group 3	Group 4
$N_b=N+1.2G_{6A,K}$(kN)	1428.81	932.89	1465.26	775.44
$M_b=M+Vh-1.2G_{6A,K}e_6$(kN·m)	320.04	-694.45	205.43	247.16
$P_{n,max}=\frac{N_b}{A}+\frac{M_b}{W}$(kN/m²)	317.10	349.42	289.19	193.64
$P_{n,min}=\frac{N_b}{A}-\frac{M_b}{W}$(kN/m²)	129.40	0.00	168.70	48.68

When $P_{n,min}<0$, $P_{n,max}$ should be recalculated (Fig. 2-98).

Eccentricity:

$$e_n=\frac{M_b}{N_b}=\frac{694.45}{932.89}=0.744\text{ m}$$

The distance from the resultant force to the maximum pressure side:

$$K=0.5a-e_n=0.5\times3.2-0.744=0.856\text{ m}$$

According to the force balance condition,

$$N_b=\frac{1}{2}P_{n,max}\times3Kb.$$

So,

$$P_{n,max}=\frac{2N_b}{3Kb}=\frac{2\times932.89}{3\times0.856\times2.0}$$

$$=363.27\text{ kN/m}^2;\ P_{n,min}=0$$

$$P_n=\frac{P_{n,max}+P_{n,min}}{2}=\frac{363.27+0}{2}=181.64\text{ kN/m}^2$$

2) Checking calculation of foundation anti-punching capacity

Anti-punching bearing capacity is checked according to the maximum net reaction force of the foundation under the second group of loads $P_{n,max}=363.27$ kN/m². The height of the cup wall $h_2=500$ mm (Fig. 2-98). 475 mm is less than the cup wall height $h_2=500$ mm, it means that the bottom of the upper step falls within the punching failure cone, so only the anti-punching bearing capacity below the step needs to be checked (Fig. 2-98).

The effective height of the punching failure cone:

$$h_0=1150-500-45=650-45=605\text{ mm}$$

The length of the upper side of the most unfavorable side of the punching failure cone:

$$a_t=400+2\times475=1350\text{ mm}$$

The length of the bottom side of the most unfavorable side of the punching failure cone:

$$a_b=1350+2\times605=2560\text{ mm}>2000\text{ mm}$$

So take $a_b=2000$ mm,

$$a_m=\frac{a_t+a_b}{2}=\frac{1350+2000}{2}=1675\text{ mm}$$

$$A_l=2.0\times\left(\frac{3.2}{2}-\frac{1.95}{2}-0.605\right)=0.04\text{ m}^2$$

Punching force: $F_l=P_{n,max}A_l=363.27\times0.04=14.53$ kN

$$0.7\beta_{hp}f_ta_mh_0=0.7\times1.0\times1.1\times1675\times605$$

$$=78030\text{ kN}>1453\text{ kN}$$

Therefore, the foundation height meets the requirement of anti-punching bearing capacity.

5. Calculation of reinforcement of foundation slab

The calculation of the reinforcement along the long side of the foundation is calculated according to the two control sections (section 1-1 and section 2-2). The calculation of the reinforcement along the short side of the foundation is calculated according to the control section 3-3 and control section 4-4 (Fig. 2-99).

1) Calculation of reinforcement along the long side of the foundation slab:

$$M_1=\frac{1}{24}\left(\frac{P_{n,max}+P_{n1}}{2}\right)(a-a_c)^2(2b+b_c)$$

$$=\frac{1}{24}\times289.3\times(3.2-1.0)^2(2\times2.0+0.4)$$

$$=256.77\text{ kN}\cdot\text{m}$$

$$A_{s1}=\frac{M_1}{0.9f_yh_{01}}=\frac{256.77\times10^6}{0.9\times270\times1105}=956.26\text{ mm}^2$$

$$M_2=\frac{1}{24}\left(\frac{P_{n,max}+P_{n2}}{2}\right)(a-a_1)^2(2b+b_1)=\frac{1}{24}\times309.84\times(3.2-1.95)^2(2\times2.0+1.35)$$

$$=131.82\text{ kN}\cdot\text{m}$$

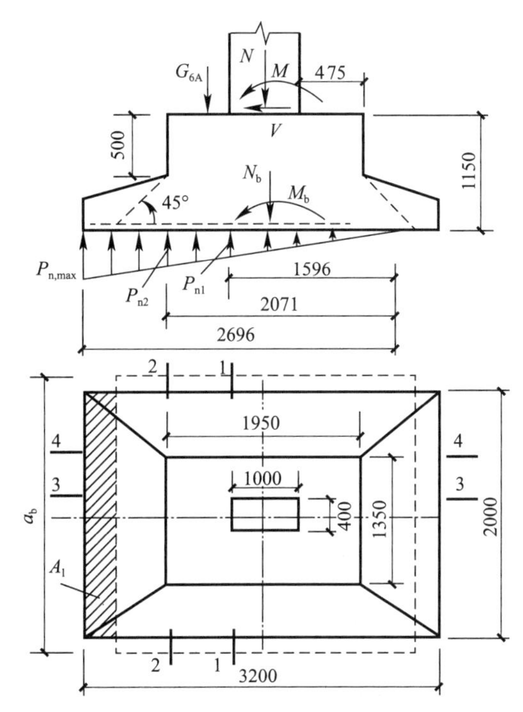

Fig. 2-98 Calculation diagram of anti-punching bearing capacity

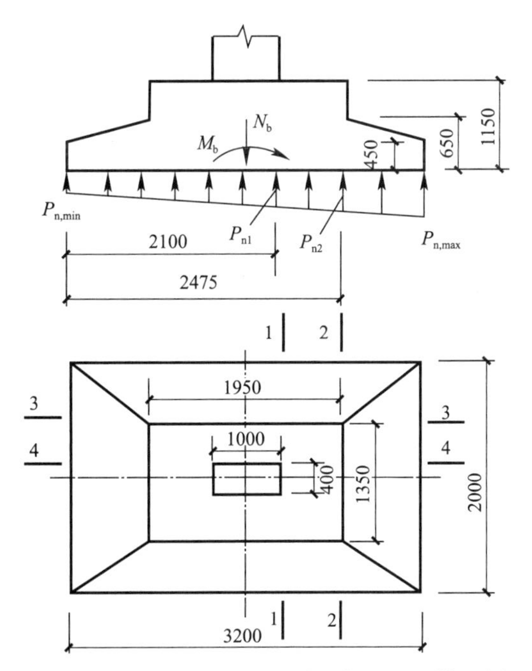

Fig. 2-99 Simplified calculation diagram of reinforcement of foundation slab

$$A_{s2}=\frac{M_2}{0.9f_y h_{02}}=\frac{131.82\times10^6}{0.9\times270\times605}=734.07\ \text{mm}^2$$

The area of steel bars is checked according to the minimum reinforcement ratio of 0.15%.

According to the foundation design specifications: the diameter of the steel bars is not less than 10 mm, and the spacing between the steel bars is not more than 200 mm and not less than

100 mm. The reinforcement of the foundation slab along the long side direction is Φ14@ 100. The number of bars is 20, which meets the design requirements. Since the long side of the foundation is greater than 2.5 m, the length of the steel bars in this direction is 0.9×3.2 m = 2.88 m = 2880 mm, and the steel bars are staggered.

2) Calculation of reinforcement along the short side of the foundation slab

The reinforcement of the foundation slab along the short side direction is Φ14 @ 110. This calculation process is the same as the above.

Exercises

2.1 What is the difference between the bent frame structure and the rigid frame structure?

2.2 Please list the main structure components and bracing types in the single-storey industrial building?

2.3 What are the basic steps of internal force analysis of bent frame structure?

2.4 How to determine crane vertical load D_{max}, D_{min} and crane transversal horizontal load T_{max}?

2.5 What is equal-height bent frame? Please describe three steps of internal force calculation of equal-height bent under any loads.

2.6 What is the spatial behavior of the single-storey industrial building?

2.7 How to make the internal force combination of the bent frame structure?

2.8 How to determine the cross-section height and reinforcement of the corbel?

2.9 What are the procedure and key points of designing the independent foundation under column?

2.10 What are the mechanical characteristics of the crane beam?

2.11 The calculation diagram of a bent frame structure is shown in Fig. 2-100. The shapes and cross-section sizes of columns A and B are equal. The moments of inertia of the upper and lower columns $I_1=2.13\times10^9$ mm^4, $I_2=1.438\times10^{10}$ mm^4, $I_3=7.2\times10^9$ mm and $I_4=1.85\times10^{10}$ mm^4, respectively. $H_u=3.3$ m and $H=11$ m. Please calculate the shear force at the column top according to the shear force distribution method under the action of $M_1=158$ kN · m and $M_2=75.9$ kN · m.

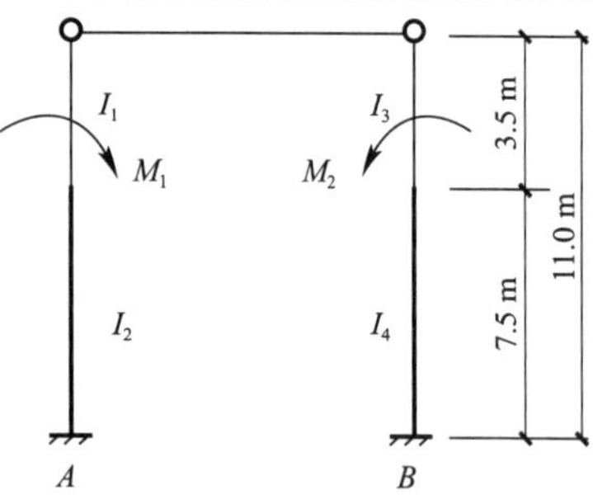

Fig. 2-100 Calculation diagram

2.12 The independent foundation under column of a single-storey workshop is known as follows. The standard combination values of loads on the top surface of foundation: axial force $N_k=210$ kN, moment $M_k=180$ kN · m and $V_k=19$ kN. The design values of loads on the top surface of foundation: axial force $N=301$ kN, moment $M=253$ kN · m and $V=26$ kN. The cross-section of column is $b\times h=400$ mm$\times$600 mm. The characteristic value of the bearing capacity of subgrade modified by correction factor of the width and depth of the foundation: $f_a=210$ kN/m^2. The embedded depth of foundation is 1.5 m. C25 concrete and HRB335 are used in the foundation. Please design the foundation and make the construction drawing.

Chapter 3
Multi-storey Frame Structure

Prologue

Main points

1. The structure composition and basic principle of structure layout of multi-storey frame structure.
2. Determination of calculation diagram of multi-storey frame structure.
3. Internal force calculation method of frame structure under the horizontal and the vertical load.
4. Reinforcement calculation and structural detailing of the frame beam and column.

Learning requirements

1. Know structural composition and layout of multi-storey frame structure.
2. Master the internal force calculation method of frame structure under the horizontal and the vertical load.
3. Master the internal force combination principle and practical calculation method of internal force of multi-storey frame structure.
4. Be familiar with horizontal displacement checking method of frame structure under the horizontal load.
5. Know reinforcement calculation and structural detailing of the frame beam and column.

3.1 Structure composition and layout of multi-storey frame structure

3.1.1 Structure composition and types of multi-storey frame structure

The frame structure is a structure formed by connecting a beam and a column, as shown in Fig. 3-1. The beam-column junctions shall be rigidly connected to form a bidirectional beam-column bearing structure to transmit the loads to the foundation. Generally, hinge joints shall not be used in the main structure except for individual parts and the bottom of the column shall be rigidly connected with the foundation. The axes of the beam and column shall be in the same vertical plane as far as possible, and the frame beam and column shall be aligned vertically and horizontally as far as possible. The self-weight of frame structure member is not large, the calculation theory is relatively mature, the cost is low in a certain height range, and the application is becoming more and more widely, but the pure frame structure system or frame structure house is not suitable for high-rise buildings.

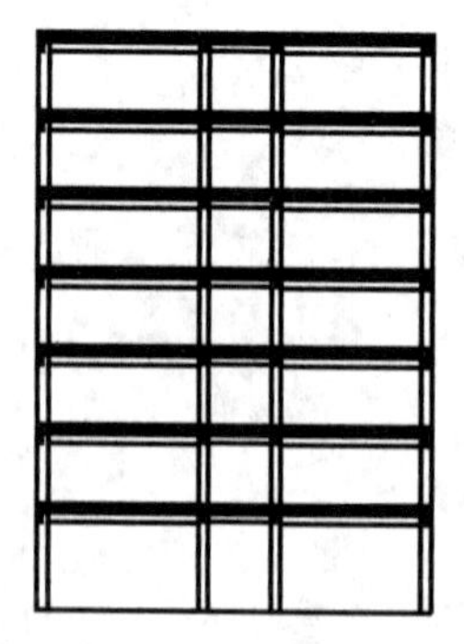

Fig. 3-1 Composition of multi-storey and multi-span frame

The frame structure may have the equal spans or unequal spans, and the layer height may be equal or not completely equal. Sometimes, due to the process requirements, the composite frame is formed by removing columns or beams on a certain floor, as shown in Fig. 3-2.

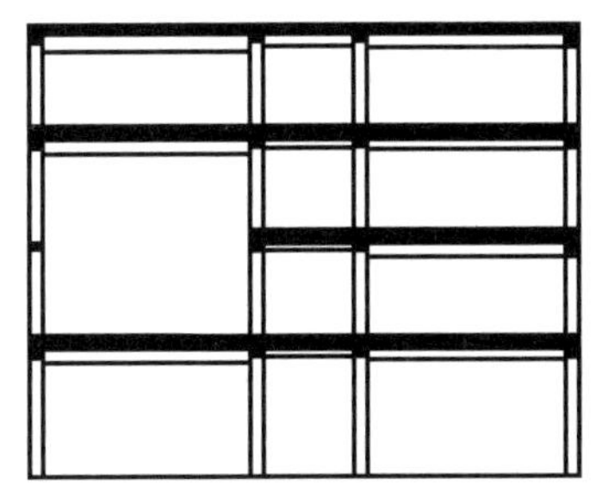

Fig. 3-2 Frame removing some columns and beams

The frame structure is a high-order statically indeterminate structure, which bears both vertical load and horizontal force (the wind load or earthquake action). In the frame structure, the non-load-bearing partition wall is often set up for functional needs. The position of partition wall is relatively fixed and masonry infill wall is often used. When considering the possible change of building function, lightweight partition wall can also be used for flexible separation. The masonry infill wall is built after the frame construction. The gap between the upper part of the masonry infill wall and the bottom of the frame beam must be filled with blocks. There are two kinds of connection modes between the wall and the frame column. One is to set a flexible connection between the column and the wall by the steel bar. The influence of the infill wall on the lateral stiffness of the frame is not considered in the calculation. The other is the rigid connection. The infill wall acts as a diagonal strut when the frame deforms laterally under the action of frequent horizontal earthquake, so as to improve the anti-lateral displacement ability of the frame. The infill wall can also play a positive role in preventing collapse.

According to the different construction methods, the reinforced concrete frame structure can be divided into cast-in-situ type, assembly type and assembly integral type (Fig. 3-3). In seismic areas, beams, columns and slabs are mostly cast-in-situ or beams and columns are cast-in-situ and slabs are prefabricated. In non-seismic areas, beams, columns and slabs can be prefabricated sometimes. Generally, the cast-in-situ reinforced concrete frame shall not exceed 70 m. When the seismic fortification intensity of cast-in-situ reinforced concrete frame is 6 degrees, 7 degrees, 8 degrees (0.2 g) and 8 degrees (0.3 g), its height generally does not exceed 60 m, 50 m, 40 m and 35 m. At present, cast-in-situ concrete frame structures are widely used.

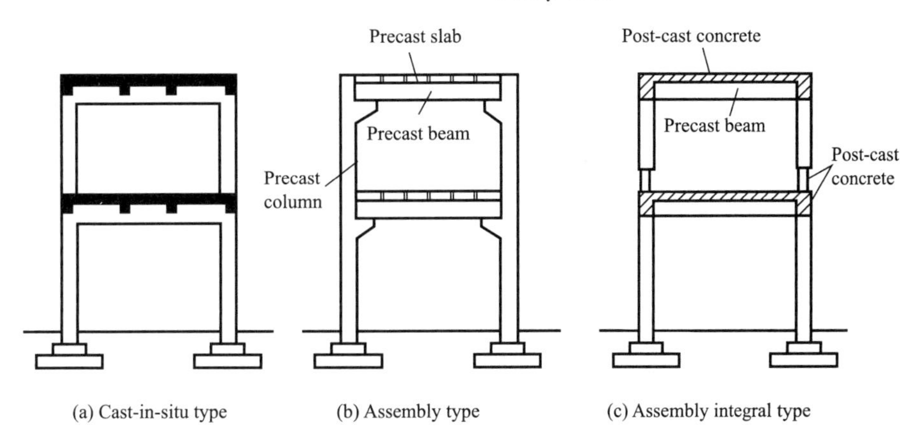

Fig. 3-3 Three types of frame structures

The design process of frame structure is as follows: structural layout → load calculation → structural analysis→internal force combination→ member reinforcement. This chapter mainly discusses cast-in-situ concrete frame structures.

3. 1. 2 Structure layout of multi-storey frame structure

Structural layout refers to the distribution mode, required quantity and mutual position relationship of various structural components which is a key link in structural design.

Basic principle of structural layout is "simple force transmission, clear force bearing, safe and reliable". The basic rules are as follows.

① The structural plane and elevation should be simple;

② The plane shall be as uniform and symmetrical as possible to make the centroid coincide with the center of mass and reduce the torsional effect of earthquake;

③ The vertical mass and stiffness shall be as uniform as possible to avoid stress concentration caused by sudden change of stiffness;

④ To control the length of structural units to reduce temperature stress, otherwise expansion joints shall be set;

⑤ To set settlement joints and seismic joints according to the requirement;

⑥ To control the inter-storey displacement to prevent the failure of non-structural members;

⑦ The height width ratio of frame column should be greater than 4 to prevent brittle shear failure;

⑧ The size and type of components shall be as few as possible to facilitate construction and save the formwork.

1. Layout of frame column grid

The frame columns are arranged in the vertical and horizontal directions of the plane, that is, the column grid is formed. The column grid layout of frame structure should not only meet the requirements of production technology, architectural function and architectural plane layout, but also be reasonable in bearing force and convenient in construction.

The column grid mainly depends on use requirements of the building , taking into account the economic rationality of the structure and construction conditions. There are three types of common column layout, i.e., internal corridor type, equal-span type and symmetrical unequal span type (Fig. 3-4). The column grid and storey height of civil buildings are determined according to the use function of buildings. The span of the side span of the internal corridor (Fig. 3-4a) is generally 6—8 m (genarally 6.0 m, 6.6 m, 7.2 m for the multi-storey workshop), and the middle span is generally 2—4 m (genarally 2.4 m, 2.7 m, 3.0 m, etc. for the multi-storey workshop). The span of the equal-span type (Fig. 3-4b) is generally 6—12 m (genarally 6.0 m, 7.5 m, 9.0 m, 10.5 m, 12.0 m, etc. for the multi-storey workshop). The storey height is generally 3.6—5.4 m.

At present, the column grid of residential buildings, hotels and office buildings can be divided into small column grid and large column grid. The small column grid refers to one bay with one column distance (Fig. 3-4b), and the column spacing is generally 3.3 m, 3.6 m, 4.0 m, etc. The large column grid refers to two bays with one column distance (Fig. 3-4c), and the column spacing is usually 6.0 m, 6.6 m, 7.2 m, 7.5 m, etc. The commonly used spans (depth of house) are 4.8 m, 5.4 m, 6.0 m, 6.6 m, 7.2 m, 7.5 m, etc.

2. Layout of load bearing frame

Frame is a spatial system composed of columns and longitudinal andtransversal beams, in which plane frame is the basic load-bearing structure. According to the different transfer paths of the vertical loads, the layout of load-bearing frame is divided into the transversal frame load-bearing scheme, the longitudinal frame load- bearing scheme and the vertical and horizontal frames load-bearing scheme.

1) Transversal frame load-bearing scheme

The main beams are arranged transversally, and the coupling beams are arranged longitudinally (Fig. 3-5a). In addition to bearing the self-weight and the weight of partition wall, the transversal frame beam also bears the load of building and roof. The load is relatively large and the section should be larger, which is called the frame main beam. The longitudinal frame beam only bears the self-weight and possible partition weight, and the section can be made relatively small. Its function is mainly to connect each transversal frame beam, so it is called the coupling beam. Because the vertical loads are mainly borne by the transversal frames, the cross-section height of the transversal frame beam is large which is conducive to increase the transversal stiffness of the building. The longitudinal beam is small, which is convenient for opening large window holes and daylight. But it may affect the clearance, which is not conducive to the laying of longitudinal pipes and the flexible layout of bays.

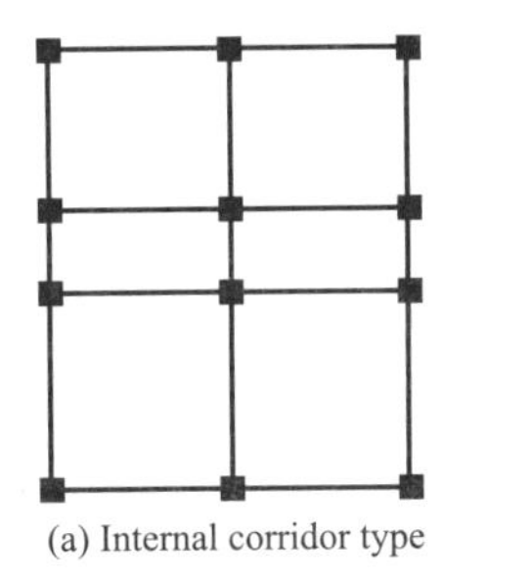
(a) Internal corridor type

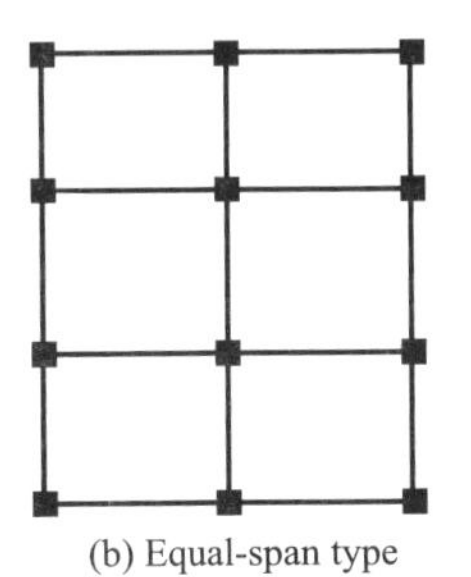
(b) Equal-span type

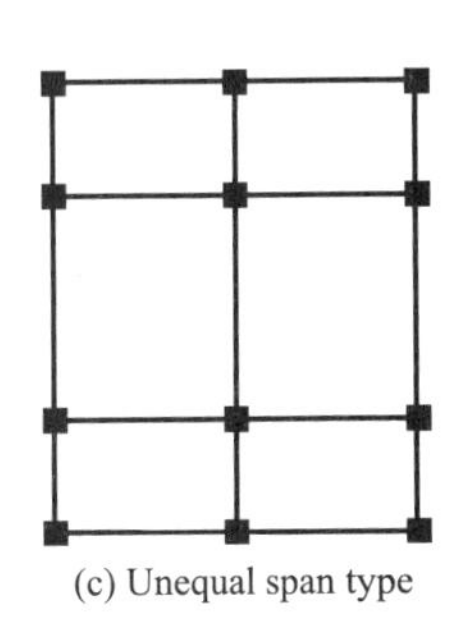
(c) Unequal span type

Fig. 3-4 Column grid layout

2) Longitudinal frame load-bearing scheme

The main beams are arranged longitudinally, and the coupling beam are arrangedtransversally (Fig. 3-5b). Because the slab load is transferred from the longitudinal beam to the column, the stiffness of the transversal beam is small which is conducive to the passage of equipment pipelines and obtain a higher indoor clear height. In addition, when the physical and mechanical properties of the local foundation soil are significantly different in the longitudinal direction of the building, the uneven settlement of the building can be adjusted by the stiffness of the longitudinal frame. However, its transversal stiffness is poor and it is rarely used in the practical projects.

3) Longitudinal andtransversal frames load-bearing scheme

Load-bearing frames are arranged in transversal and longitudinal directions (Fig. 3-5c). The slab often adopts two-way cast-in-situ slab. This load-bearing scheme has good overall working performance and is favorable for seismic performance of the frame structure.

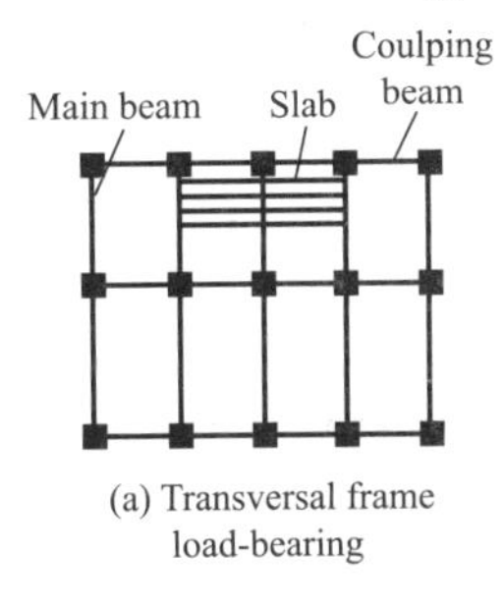

(a) Transversal frame load-bearing

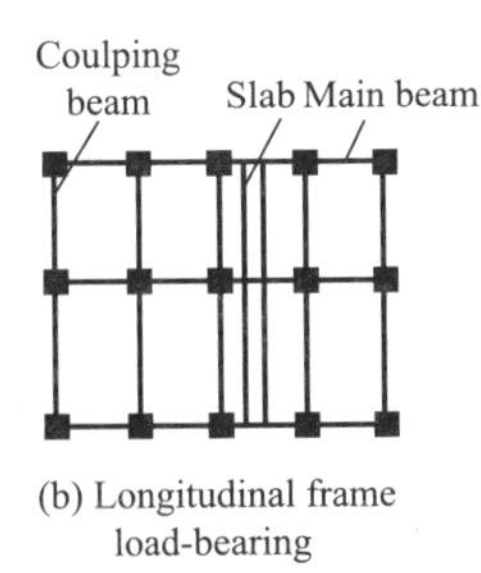

(b) Longitudinal frame load-bearing

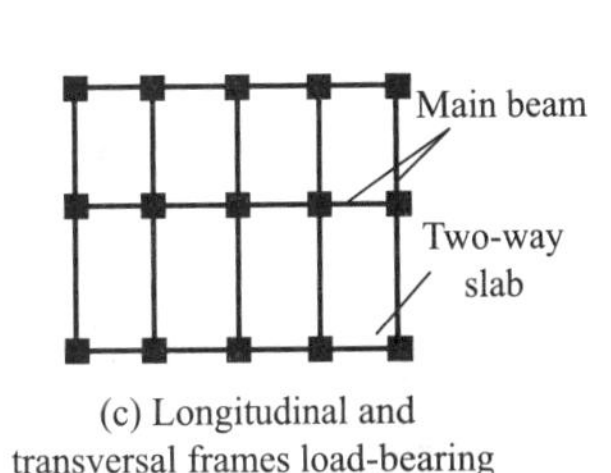

(c) Longitudinal and transversal frames load-bearing

Fig. 3-5 Layout of load-bearing frame

3. Deformation joints

Deformation joints include the expansion joint, settlement joint and seismic joint. In multi-storey structure, joints should be set as few as possible which can simplify the structure, facilitate construction, reduce the cost and enhance the structural integrity and spatial stiffness. Therefore, in the architectural design, we should adjust the plane, size, shape and take other measures. In the structural design, we should choose the joint connection mode, the structural reinforcement and the thermal insulation layer to prevent the damage of building caused by concrete shrinkage, uneven settlement, earthquake action and other factors. It is also necessary to set the deformation joints when the building plane is narrow and long or the stiffness, height and weight of each part are greatly different or the building shape is complex and asymmetric, and the above measures cannot solve these problems.

The setting of expansion joint is mainly related to the length of the structure. The ***Code for Design of Concrete Structures*** (GB 50010—2010) specifies the maximum spacing of expansion joints in reinforced concrete structure (Table 3-1). When the length of the structure exceeds the allowable value specified in the code, the temperature stress should be calculated and corresponding construction measures should be taken. The expansion joint width is generally 20—40 mm; when there are seismic requirements, it shall not be less than the width of seismic joint, generally ≥70 mm.

Maximum spacing of expansion joints in reinforced concrete frame structure(m)

Table 3-1

Construction method	Indoor or in soil	Outdoor
Assembly type or assembly integral type	75	50
Cast-in-situ type	55	35

The setting of settlement joint is mainly related to the loads acting on the foundation and the geological conditions of the site. When the loads difference is large or the physical and mechanical indexes of the subsoil differ greatly which may cause large uneven settlement, the settlement joint should be set.

The setting of seismic joint is mainly related to the plane shape, height difference, stiffness and mass distribution of the building. The setting of seismic joint should make each structural unit simple and regular, uniform distribution of stiffness and mass, which can avoid torsion effect under earthquake action. In order to avoid the structure collision between the units under the earthquake effect, the width of the seismic joint should not be less than 70 mm. Meanwhile, for the frame structure building whose height exceeds 15 m, the width of seismic joint should be widened 20 mm when the height is increased by 5 m, 4 m, 3 m and 2 m for seismic fortification intensity of 6 degrees, 7 degrees, 8 degrees and 9 degrees, respectively.

In non-earthquake area, the settlement joint can also be used as the expansion joint. In earthquake area, the expansion joint or settlement joint shall meet the requirements of seismic joint. Except that the settlement joint must separate the superstructure together with the foundation, the other two joints can start from the top surface of the foundation to completely break the superstructure, but the foundation does not have to be separated (Fig. 3-6). When only seismic joint is needed, the foundation cannot be separated, but the structure and connection of the foundation should be strengthened at the joint.

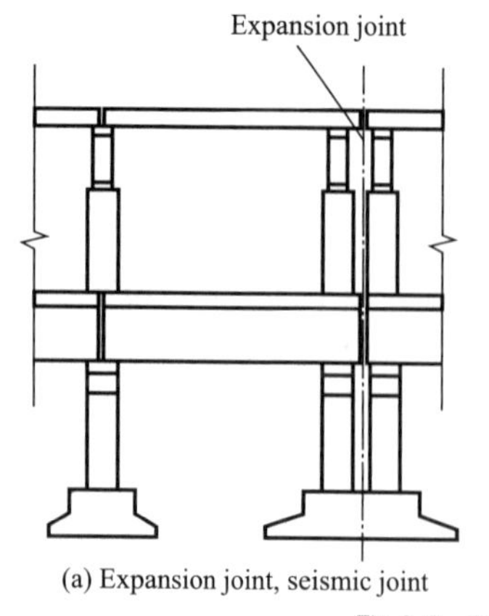

(a) Expansion joint, seismic joint

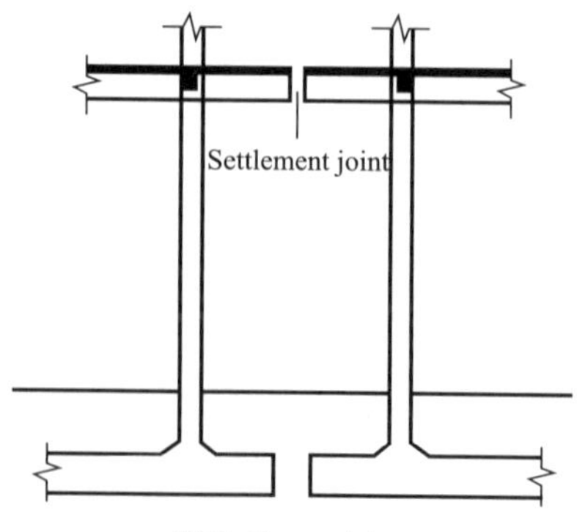

(b) Settlement joint

Fig. 3-6 Setting of deformation joints

3.2 Calculation diagram of multi-storey frame structure

3.2.1 Calculation diagram

1. Calculation unit

Multi-storey frame structure is a spatial structure system composed of longitudinal and transversal frames. Generally, the main frames are arranged at equal intervals, and their respective stiffness is basically the same. Under the action of vertical load, the mutual restraint, that is, the restraining effect between the frames is very small, and the influence of spatial stiffness on their stress can be ignored. Under the horizontal load, the spatial action will lead to the collaborative work of various frames, but the horizontal load of multi-storey frame house is mostly uniform, and the frames do not produce much binding force between each other. Generally, without considering the spatial effect of the house, the frame structure can be calculated according to the plane frame in the longitudinal and transversal directions, and each frame bears the external load separately according to its load area.

Generally, one or several representative calculation units in the whole frame are selected for internal force analysis and structural design as shown in Fig. 3-7.

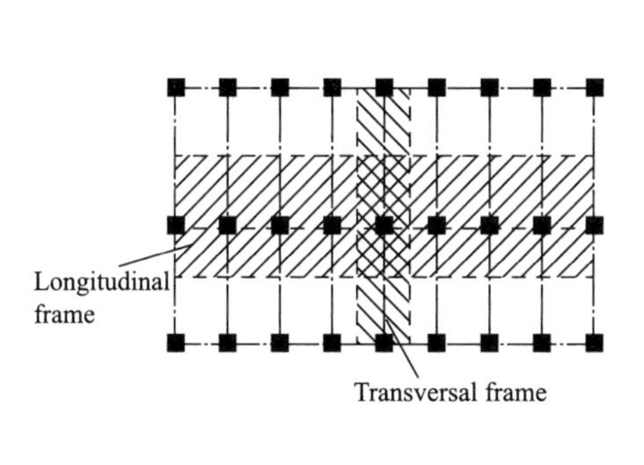

(a) Plane model

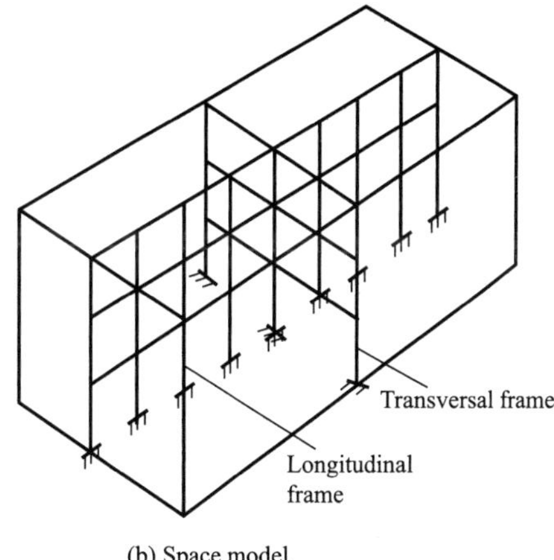

(b) Space model

Fig. 3-7 Calculation unit

2. Calculation diagram

Each member of the frame structure is represented by a single line in the calculation diagram, as shown in Fig. 3-8. Each single line represents the location of the centroid axis of each member. Therefore, the span of the beam is equal to the distance between the centroid axes of the left and right column sections of the span. The calculation height of column of the first floor can be calculated from the top surface of the foundation to the floor elevation, the column height of the middle floor can be calculated from the floor elevation to the upper floor elevation, and the top floor column height can be calculated from the top floor elevation to the roof elevation.

In engineering practice, the convenience of internal force calculation should be properly considered when determining the calculation diagram. Under the condition of ensuring the accuracy, the following calculation models are often simplified.

When the sections of upper and lower columns are different, the centroid axis of the column with smaller section is taken as the column element on the calculation diagram. When the internal force of the frame is calculated, the effect of load eccentricity should be considered when calculating the internal force of the member.

When the slope of frame beam $i \leqslant 1/8$, it can be calculated approximately as horizontal beam. When the span difference of each span is no more than 10%, it can be calculated approximately as a frame with equal span.

When tapered haunch is added at the end of the beam, and the ratio of the end section height to the mid span section height is less than 1.6, the influence of tapered haunch can be ignored and calculated as a beam with equal section.

The connection of beam and column is generally considered as a rigid joint, and the column bottom is rigidly connected on the top of the foundation. The calculation diagram is shown in Fig. 3-8.

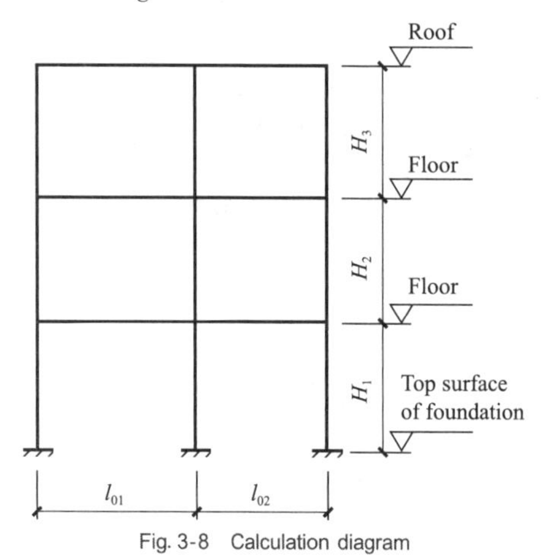

Fig. 3-8 Calculation diagram

3. Calculation of loads

There are two types of loads acting on the frame structure, i.e., the vertical load and the horizontal load. The vertical load includes the dead weight of the structure and the live load on the floor (roof), which is generally the distributed load and sometimes can be the concentrated load. The dead load is calculated and determined according to the design size of structural members and the self-weight per unit volume of materials according to the ***Load Code for the Design of Building Structures*** (GB 50009—2012).

The possibility of the characteristic value of all live loads fully acting on the floor is very small, and the probability of live loads acting on all the floors is smaller. Therefore, the reduction of floor live load can be considered in the structural design. The reduction factor is determined according to the dependent area of floor beam, number of floors and house type. For the floor beam subordinate area of multi-storey residence, dormitory, hotel and other houses > 25 m^2, the reduction factor is 0.9. In the design of column, wall and foundation, the reduction coefficient is taken as shown in Table 3-2.

The wind load can be calculated by the same method introduced in the chapter of single-storey workshop. However, the shape coefficient of wind load of multi-storey building is related to the plane shape of buildings, which can be taken according to ***Load Code for the Design of Building Structures*** (GB 50009—2012). When the building height is less than 40 m and the distribution of mass and stiffness are evenly distributed along the height, the bottom shear method can be used to calculate the horizontal seismic reaction. Wind load and horizontal earthquake reaction are generally simplified as horizontal concentrated forces acting on joints of the frame structures. The horizontal wind load is multiplied by the load width of the frame to obtain the line load distributed along the height; and then the line load is simplified as the node load.

3.2.2 Cross-section size and sectional moment of inertia of frame beam and column

The internal force and deformation of the frame structure are related to the structure form, the loads and the stiffness of the member, and the stiffness of the member is related to the cross-section size of the member. The cross-section size of the member is related to the loads and internal force, but is difficult to determine accurately the size of the cross-section before the internal force of the member is calculated. Therefore, the cross-section size of the member should be estimated firstly, then the internal force and deformation can be calculated. If the estimated section sizes meet the requirements, the estimated section sizes can be taken as the final section sizes of the members of the frame structure. If the required section sizes are significantly different from the estimated section sizes, it is necessary to make estimation and calculation again. In order to reduce the types of members, the shapes and sizes of multi-storey beams and columns are often the same, and the section reinforcement is only changed in the design.

1. Frame beam

For the frame beam bearing main vertical load, the section form is T-shaped (Fig. 3-9a) in the full cast-in-situ integral frame; in the assembled frame, it can be made into rectangle, T-shape, trapezoid and flower basket shape (Figs.3-9b-g).

Reduction factor of floor live load **Table 3-2**

Numberof floors above the calculated section of wall, column and foundation	1	2—3	4—5	6—8	9—20	>20
Reduction factor of the total live loads of each floor above the section	1.00(0.9)	0.85	0.70	0.65	0.60	0.55

Note:

When the calculated area of floor beam is over 25 m^2, the factor in the "( )" is adopted.

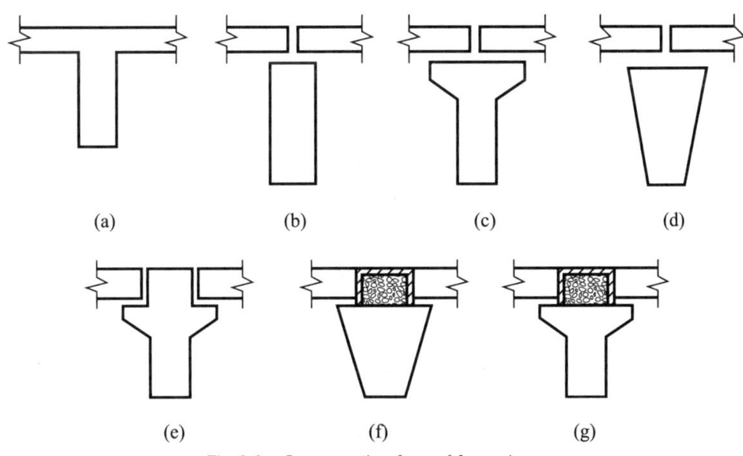

Fig. 3-9 Cross-section form of frame beam

For the coupling beam not bearing main vertical load, T-shaped, Γ-shaped, rectangle, ⊥-shaped, L-shaped, etc. are often used.

The cross-section size of the frame beam should be determined according to the beam span, the vertical load, the seismic fortification requirements, the strength of the selected concrete materials, etc. Generally, the section size of frame beam can be estimated according to the following three formulas.

For the main beam:

$$h_b=\left(\frac{1}{12}-\frac{1}{8}\right)l_0 \tag{3-1}$$

$$b_b=\left(\frac{1}{3}-\frac{1}{2}\right)h_b \tag{3-2}$$

For the coupling beam:

$$h_b = \left(\frac{1}{15} - \frac{1}{12}\right) l_0 \tag{3-3}$$

Where,

l_0—— the calculation span of beam(mm);

h_b—— the cross-section height of beam (mm);

b_b—— the cross-section width of beam (mm).

The ratio of clear span to cross-section height should not be less than 4 in order to prevent shear brittle failure of the beam and the cross-section width of beam should not be less than 200 mm. Generally, the section height and width of the beam shall be a multiple of 50 mm.

The slab can be made into assembly type, assembly integral type and cast-in-situ type. But the connection structure of three types of slab and frame beams are different. For the cast-in-situ slab, the reinforcement of slab and frame beam is put together and the concrete is poured at the same time, so the integrity is good. For the assembly integral slab, the prefabricated floor slab is supported on the frame beam, and then the reinforced concrete layer is made on the precast slab, which leads to weaker integrity. For the assembly floor, the floor slab is directly supported on the frame beam and the integrity is poorer.

When calculating the sectional moment of inertia of frame beam, it is necessary to consider the favorable influence of the floor slab on the sectional moment of inertia of beam. For the sake of simplification, it can be calculated according to the simplified formula in Table 3-3 and Fig. 3-10. The linear stiffness of the beam is $i_b = E_c I / l_0$.

Sectional moment of inertia of frame beam **Table 3-3**

Slab type	Edge frame beam	Middle frame beam
Cast-in-situ slab	$I=1.5I_0$	$I=2.0I_0$
Assembly integral slab	$I=1.2I_0$	$I=1.5I_0$
Assembly slab	$I=I_0$	$I=I_0$

Note:

I_0 is the sectional moment of inertia of the beam calculated according to the rectangular section, $I_0 = \frac{1}{12} b_b h_b^3$.

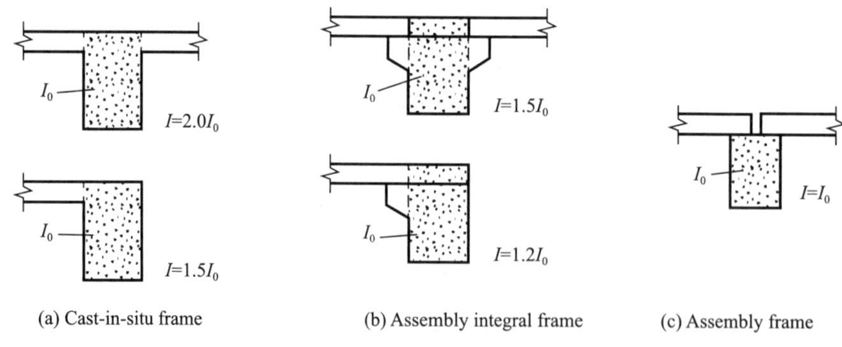

Fig. 3-10 Sectional moment of inertia of frame beam

2. Frame column

The cross-section of frame column is generally rectangular or square. In multi-storey frame building, the cross-section size of frame column can be estimated according to the following two formulas:

$$h_c = \left(\frac{1}{12} - \frac{1}{6}\right) H_i \tag{3-4}$$

$$b_c = \left(\frac{2}{3} - 1\right) H_i \tag{3-5}$$

Where,

H_i—— the height of the i-th floor(mm);

h_c—— the cross-section height of column (mm);

b_c—— the cross-section width of column (mm).

The size of the cross-section of the frame column should not be less than 250 mm, the diameter of the circular column should not be less than 350 mm, and the height width ratio of the cross-section should not be greater than 3. The section height of the column shall be a multiple of 100 mm, and the section width of the column shall be a multiple of 50 mm.

In order to reduce the type of components and simplify the construction, the cross-section of the middle column of a multi-storey building generally is not variable along the height of the building. When the column section changes along the height of the building, the middle column should make the axes of the upper and lower column coincide, and the side column and corner column should make the outer lines of the upper and lower column coincide.

The sectional moment of inertia of frame column is $I_0=\frac{1}{12}b_c h_c^3$. The linear stiffness of the frame column is $i_c=E_c I/H_i$.

3.3 Internal force and lateral displacement checking calculation

3.3.1 Internal force calculation of frame structure under vertical load

The approximate calculation methods of the frame structure under the vertical load include multi-layered method and secondary moment distribution method.

1. Multi-layered method

When the multi-layered method is used to calculate the internal force of frame structure under the vertical load, the following calculation assumptions are adopted.

① Ignore the lateral displacement of frame structure under the vertical load;

② The vertical load acting on a certain floor beam only produces the bending moment and shear force on the beams of this floor and the column connected with these beams of this floor, but does not produce the bending moment and shear force on the beams of the other floors and the columns connected with these beams.

The key points and procedure of multi-layered method are as follows:

1) According to linear superposition principle, an n-layer frame structure is divided into n frames whose loads act on each floor beam separately, and the internal force of each frame is calculated by moment distribution method, as shown in the Fig. 3-11 (a). According to the above assumption, when the vertical load is applied on the beams of each layer, the internal forces only occur in the beams of this layer and the columns connected with these beams with the solid lines shown in Fig. 3-11(b). The internal forces generated in the other members with the dotted lines shown in Fig. 3-11(b) can be neglected.

Except for the lower end of the ground floor column, the ends of all columns should be elastically restrained. For the convenience of calculation, they are treated as fixed ends. In this way, the bending deformations of the columns are reduced. In order to consider this effect, except the ground floor column, the linear stiffness of other columns in all layers can be multiplied by the correction factor of 0.9, and the moment transfer coefficients of other columns can be taken as 1/3. The moment transfer coefficient of the ground floor column is still 1/2.

2) Calculate the moments of fixed ends of each single-layer frame under the vertical loads

(as shown in Fig. 3-11c).

3) Calculate the linear stiffness and moment distribution coefficient of beams and columns, and calculate the moment of each frame by moment distribution method.

4) Superimpose the moments of each single-layer frame to get the moments of beam ends and column ends.

5) After superimposing the moment of n single-lager frames, the moments of beam ends calculated layer by layer are the final moments. However, each column belongs to two layers, so the final moments of each column ends need to plus the moment values calculated by the upper and lower layers. When the moment values of the upper and lower column ends are added, new unbalanced moments will occur. Therefore, further correction is needed. The unbalanced moments can be distributed again.

6) After calculate the moment, the shear force at the end of the beam and the moment in the middle of the beam can be calculated by using the static equilibrium condition of each beam; the axial force of the column can be obtained by superimposing the node concentrate force on the column layer by layer and the shear force at the beam ends connected with this node.

The multi-layered method is applicable to the internal force calculation of multi-storey frames with uniform structure and load distribution along the height under the action of vertical load. It is not suitable for multi-storey frames with large lateral displacement or irregular frames.

【Example 3-1】

Conditions: Fig. 3-12 is a frame structure with five floors and three spans (i is the linear stiffness of the beam and column).

Question: Use the multi-layered method to calculate the moment of the frame, and draw the moment diagram.

【Solution】

1) The frame is divided into 5single-layer frames (as shown in Fig. 3-13). The linear stiffness of each column is multiplied by a reduction coefficient of 0.9 except the ground floor column, the linear stiffness values are shown in the brackets in Fig. 3-12.

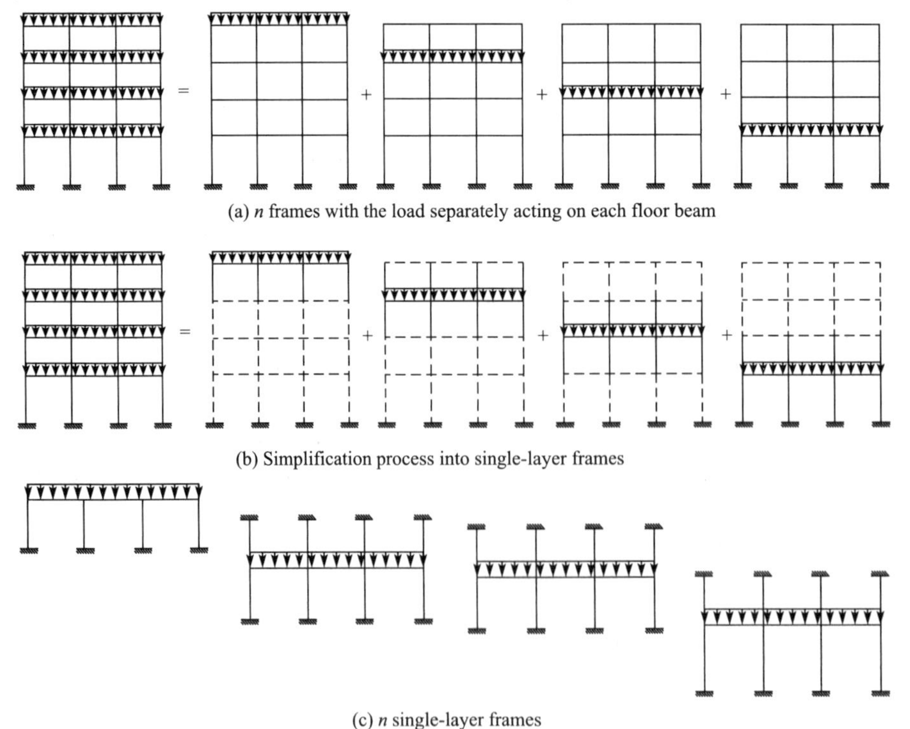

(a) n frames with the load separately acting on each floor beam

(b) Simplification process into single-layer frames

(c) n single-layer frames

Fig. 3-11 Simplified process of calculation diagram of multi-layered method

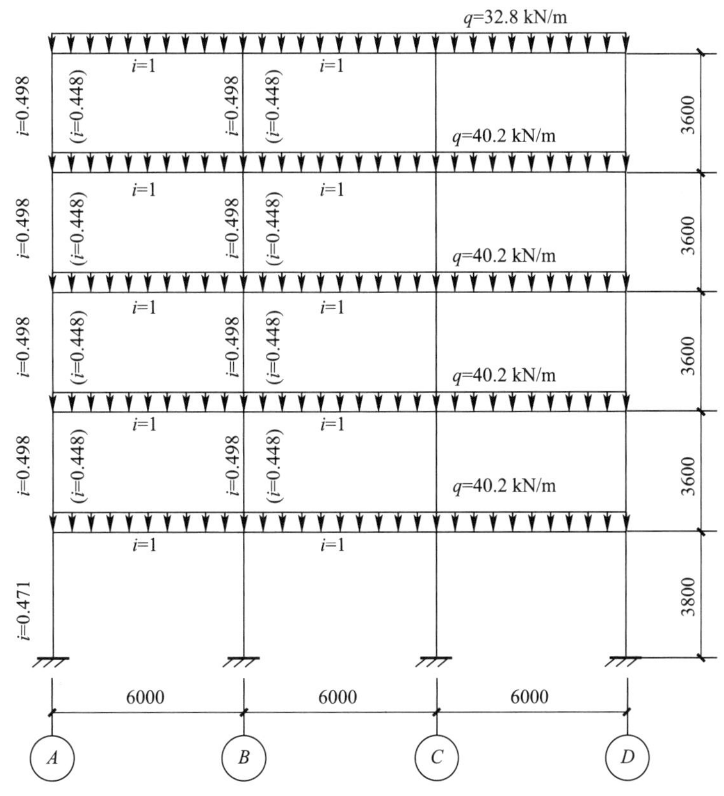

Fig. 3-12 Calculation diagram of the frame structure

2) The half span of the single-layer frame in Fig. 3-14 is taken as calculation model according to the symmetry principle. The moment distribution coefficient and the moments of each single-layer frame fixed ends can be calculated shown in Fig. 3-14.

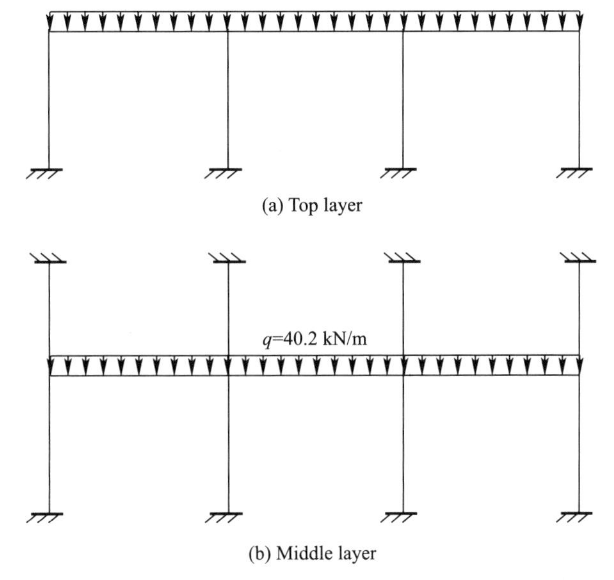

Fig. 3-13 Calculation diagram of each single-layer frame (One)

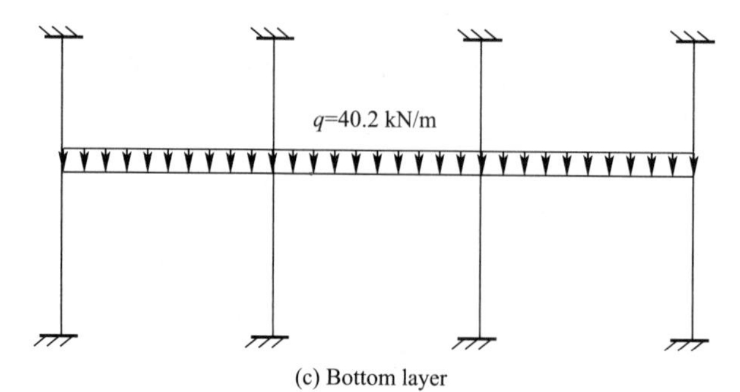

(c) Bottom layer

Fig. 3-13 Calculation diagram of each single-layer frame (Two)

Lower column	Upper column	Right beam
0.309	0	0.691
		−98.4
+30.4		+68.0
		−8.7
+2.7		+6.0
+33.1		−33.1
+11.0		

Left beam	Upper column	Lower column	Right beam	
0.513	0	0.230	0.257	
+98.4			−98.4	−49.2
+34.0				
−17.4		−7.8	−8.7	+8.7
+3.0				
−1.5		−0.7	−0.8	+0.8
+116.5		−8.5	−107.9	−39.7
		−2.8		

(a) Top layer

Lower column	Upper column	Right beam
	+10.1	
0.236	0.236	0.528
		−120.6
+28.5	+28.5	+63.7
		−6.7
+1.6	+1.6	+3.5
		−0.4
+0.1	+0.1	+0.2
+30.2	+30.2	−60.3
	+10.1	

Left beam	Upper column	Lower column	Right beam	
	−2.1			
0.417	0.187	0.187	0.209	
+120.6			−120.6	−60.3
+31.8				
−13.3	−5.9	−5.9	−6.6	+6.6
+1.8				
−0.8	−0.3	−0.3	−0.4	+0.4
+140.1	−6.2	−6.2	−127.6	−53.3
		−2.1		

(b) Middle layer

Fig. 3-14 Moment distribution of each single-layer frame (One)

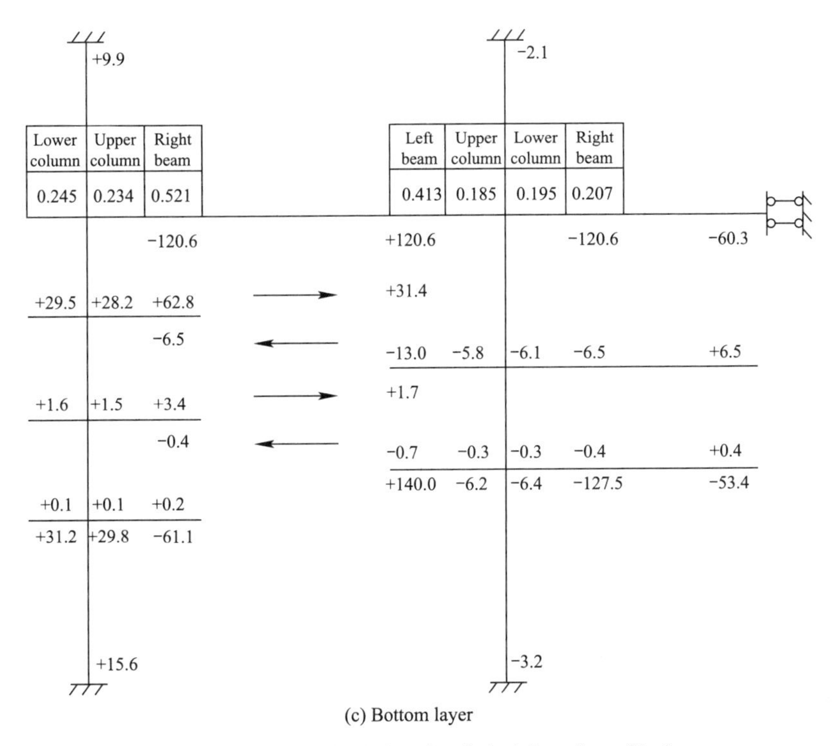

(c) Bottom layer

Fig. 3-14 Moment distribution of each single-layer frame (Two)

3) The moment distribution method is used to calculate the moment of each single-layer frame. The moment transfer coefficient of bottom column is 1/2, and the moment transfer coefficient of other columns is 1/3. The calculation process is shown in Fig. 3-14.

4) Superimpose the moments of each single-layer frame obtained by multi-layered method, and then distribute the unbalanced moment of each node and the final moments will not be transferred and distributed again. The final moment diagram of the frame is shown in Fig. 3-15. The values in the brackets are the final moments.

2. Secondary moment distribution method

The frame does not need to be decomposed into layers as the multi-layered method. The moments of all nodes are distributed and transferred to the far end at the same time, and then the unbalance moments of the nodes are redistributed. Under the condition that the engineering calculation accuracy is satisfied, the moment distribution is carried out twice. This method is suitable for manual calculation. The calculation steps of this method are as follows.

1) Calculate the moment distribution coefficient of each node according to the linear stiffness of each member, and calculate the moments of fixed ends of each beam under the vertical load.

2) The unbalanced moment of each node of the frame is calculated, and the first distribution of the unbalanced moments of all nodes is carried out after being reversed respectively.

3) The distributed moments of all members are transferred to the other ends of the members respectively (for example, the moment transfer coefficient is taken as 1/2 for the fixed end).

4) The new unbalanced moment caused by the transferred moment of each node is reversed and distributed for the second time, so that the moment of each node is in equilibrium. Here the whole process of moment distribution and transfer is finished.

5) The distributed moment and transferred moment of fixed ends are added to obtain the moments of each member end.

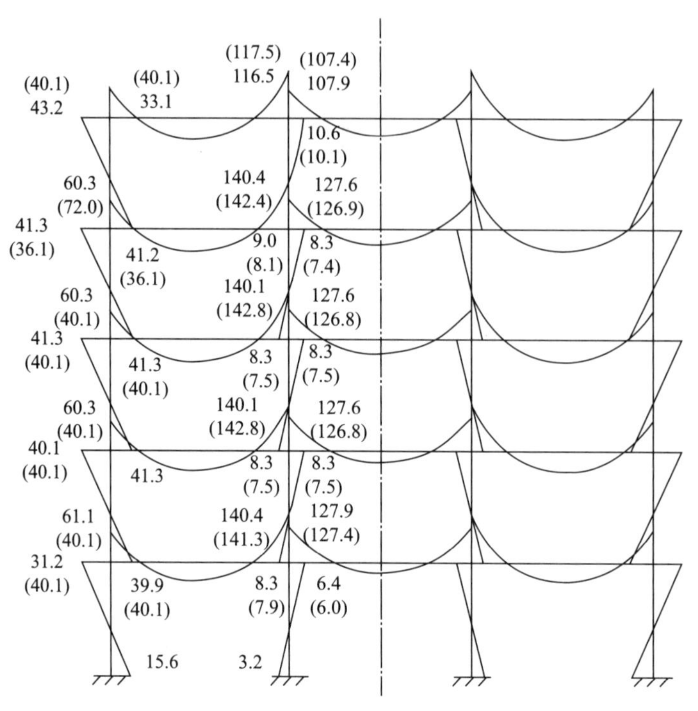

Fig. 3-15 Frame moment diagram

【Example 3-2】

Conditions: The frame is shown in Fig. 3-16. The values of the brackets are the relative linear stiffness of the beams and columns.

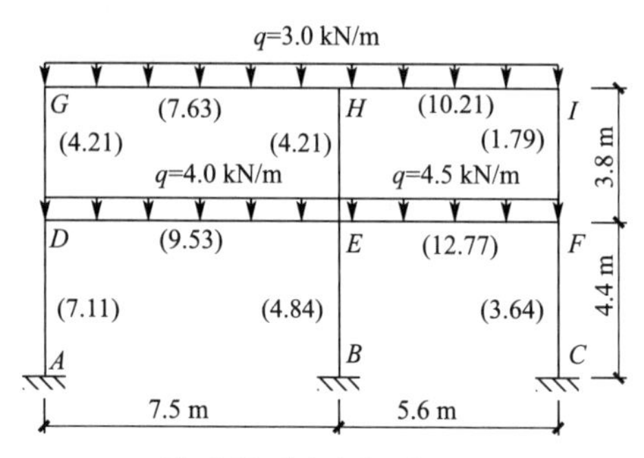

Fig. 3-16 Calculation diagram

Question: Calculate the moment of the frame by secondary moment distribution method.

【Solution】

1) Calculate the moment distribution coefficient of each node (Table 3-4)

2) Calculate the moments of fixed ends

$$M_{GH}=-M_{HG}=-\frac{1}{12}\times 3.0\times 7.5^2=-14.06\ \text{kN}\cdot\text{m}$$

$$M_{HI}=-M_{IH}=-\frac{1}{12}\times 3.0\times 5.6^2=-7.84\ \text{kN}\cdot\text{m}$$

$$M_{DE}=-M_{ED}=-\frac{1}{12}\times 4.0\times 7.5^2=-18.75\ \text{kN}\cdot\text{m}$$

$$M_{EF}=-M_{FE}=-\frac{1}{12}\times 4.5\times 5.6^2=-11.76\ \text{kN}\cdot\text{m}$$

3) Moment distribution and final moments

The calculation process of secondary moment distribution method is shoun in Fig. 3-17. Here, the moment diagram of the frame is omitted. The methods of shear force of the beam and the axial force of the column are the same as those of the multi-layered method.

3.3.2 Calculation of internal force and displacement of frame structure under horizontal load

The horizontal load mainly includes the wind load and the horizontal earthquake effect. In general, these loads can be simplified as horizontal concentrated forces acting on the frame nodes. The frame will produce lateral displacement and rotation angle under the horizontal load. The deformation diagram is shown in Fig. 3-18. It can be seen from Fig. 3-18 that there is no lateral

Relative linear stiffness and moment distribution coefficient **Table 3-4**

Layer	Note	Relative linear stiffness				Moment distribution coefficient			
		Left beam	Right beam	Upper column	Lower column	Left beam	Right beam	Upper column	Lower column
Top layer	*G*		7.63		3.79		0.668		0.332
	H	7.63	10.21		3.79	0.353	0.472		0.175
	I	10.21			1.61	0.864			0.136
Bottom layer	*D*		9.53	3.79	7.11		0.446	0.186	0.384
	E	9.53	12.77	3.79	4.84	0.308	0.413	0.123	0.156
	F	12.77		1.61	3.64	0.709		0.089	0.202

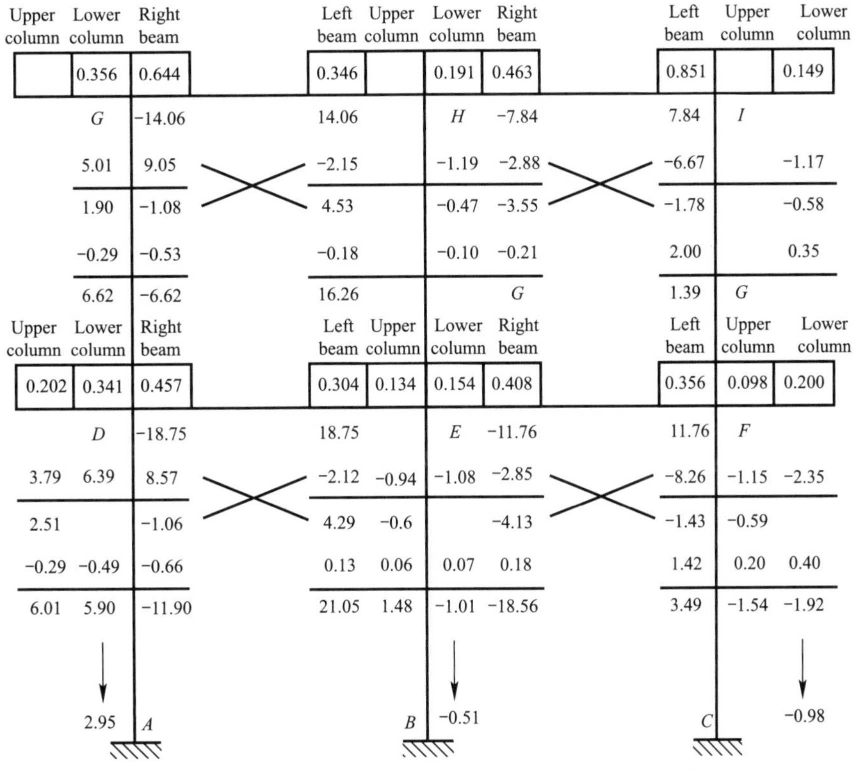

Upper column	Lower column	Right beam
	0.356	0.644
	G	-14.06
	5.01	9.05
	1.90	-1.08
	-0.29	-0.53
	6.62	-6.62

Left beam	Upper column	Lower column	Right beam
0.346		0.191	0.463
14.06		H	-7.84
-2.15		-1.19	-2.88
4.53		-0.47	-3.55
-0.18		-0.10	-0.21
16.26			G

Left beam	Upper column	Lower column
0.851		0.149
7.84	I	
-6.67		-1.17
-1.78		-0.58
2.00		0.35
1.39	G	

Upper column	Lower column	Right beam
0.202	0.341	0.457
	D	-18.75
3.79	6.39	8.57
2.51		-1.06
-0.29	-0.49	-0.66
6.01	5.90	-11.90
	2.95	A

Left beam	Upper column	Lower column	Right beam
0.304	0.134	0.154	0.408
18.75		E	-11.76
-2.12	-0.94	-1.08	-2.85
4.29	-0.6		-4.13
0.13	0.06	0.07	0.18
21.05	1.48	-1.01	-18.56
	B	-0.51	

Left beam	Upper column	Lower column
0.356	0.098	0.200
11.76	F	
-8.26	-1.15	-2.35
-1.43	-0.59	
1.42	0.20	0.40
3.49	-1.54	-1.92
C		-0.98

Fig. 3-17 Calculation process of secondary moment distribution method

displacement and rotation angle at the lower end of the frame column at the ground floor, and there is a lateral displacement and angle at other nodes. And the axial deformation of the beam can be ignored, so the nodes in the same floor have the same lateral displacement, and the columns in the same floor have the same inter-storey displacement.

As shown in Fig. 3-19, the moments of frame columns are all straight lines under the horizontal load, and generally have a zero bending moment in the columns which is called inflection point. If the position of inflection point and shear force of each column can be determined, the moments of the columns can be obtained, and then the moments of beams and other internal forces of the whole frame structure can be obtained by the equilibrium conditions.

The inflection point method and D-value method can be used for approximate calculation of multi-storey frame structure under the horizontal load.

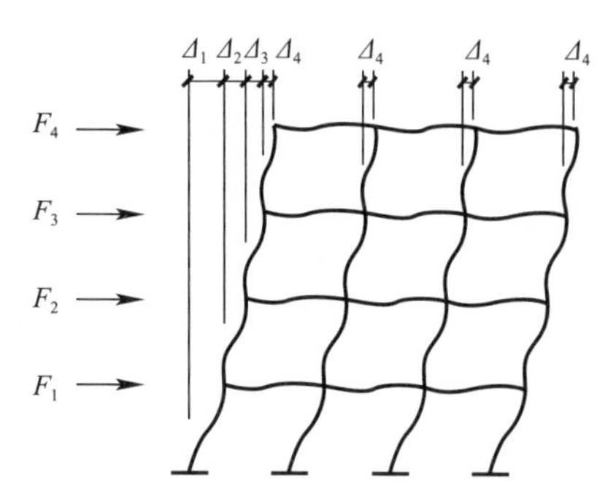

Fig. 3-18 Deformation diagram of frame under horizontal load

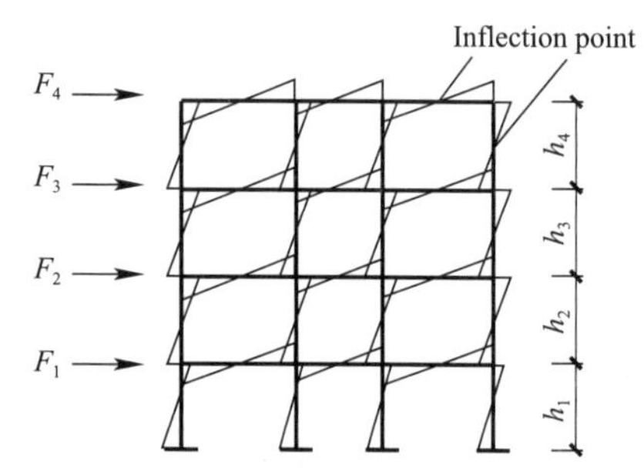

Fig. 3-19 Moment diagram of frame under horizontal load

1. Inflection point method

For the multi-storey frame building, the axial force of the column is not big, and the cross-section of the beam is larger than the cross-section of the column. Therefore, the linear stiffness of frame beam is much larger than that of column, and the rotation angle of frame node is very small. When the linear stiffness ratio of beam to that of column is greater than 3, the bending deformation of beam under the horizontal load is very small, so the stiffness of beam can be regarded as infinite and the rotation angle of note is considered as zero. The calculation assumptions of the inflection point method are as follows.

1) When calculating the shear force of each column, it is assumed that there is no angular displacement between the upper end and lower end of each column, and that the ratio of the linear stiffness of the beam to that of the column is infinite.

2) The axial deformation of the beam is ignored.

3) When the linear stiffness of the beam is assumed to be infinite, there is no rotation angle at the end of the column, and the moments at both ends of the floor column are equal. That is to say, except for the ground floor column, rotation angles of upper node and lower node of the columns of other floors are the same. The inflection point is at the middle of the column, and the inflection point height $y_h = h/2$. For the ground floor column, because the bottom end is fixed and there is a rotation angle at the upper end, the inflection point moves upward. It is usually assumed that the inflection point of the ground floor column is at 2/3 of the ground floor column height, that is, the inflection point height is $y_h = 2h_1/3$.

The shear force at the column end can be obtained by assumption 1). The frame structure has n storeys and m columns in each layer (Fig. 3-20a). If the frame is cut along the inflection points of columns in the j-th floor and replaced by shear force and axial force (Fig. 3-20b), the equilibrium condition of horizontal force is as follows:

$$V_j = \sum_{i=j}^{n} F_i \tag{3-6}$$

$$V_j = V_{j1} + \cdots + V_{jk} + \cdots + V_{jm} = \sum_{k=1}^{m} V_{jk} \tag{3-7}$$

Where,

F_i —— the horizontal force acting on the i-th floor;

V_j —— the total shear force of frame structure in the j-th floor;

V_{jk} —— the shear force of the k-th column in the j-th floor; the sum of the shear force of each column in each floor is equal to the sum of the horizontal loads above the floor;

m —— the number of columns in the j-th floor;

n —— the number of floor.

According to the assumption 1), the deformation of the k-th column of the j-th floor under the horizontal force is shown in Fig. 3-21. According to the structural mechanics, the shear force of the frame column is:

$$V_{jk}=D'_{jk}\Delta u_j,\ D'_{jk}=\frac{12i_{jk}}{h_j^2} \quad (3\text{-}8)$$

Where,

i_{jk}—— the linear stiffness of the k-th column in the j-th floor;

h_j—— the column height of the j-th floor;

Δu_j—— the inter-storey lateral displacement of the j-th floor of the frame;

D'_{jk}—— the lateral stiffness of the k-th column in the j-th floor.

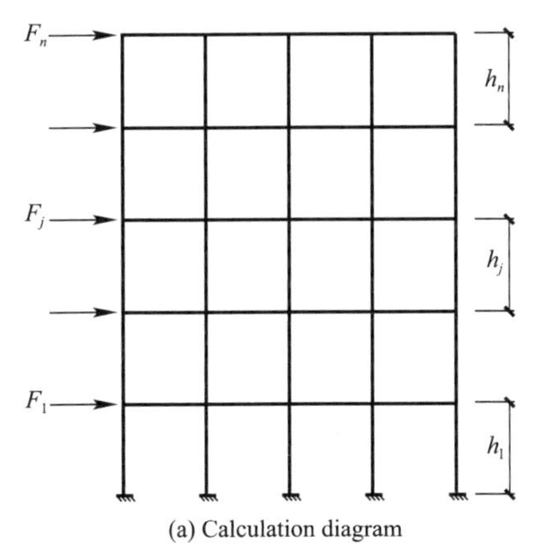

(a) Calculation diagram

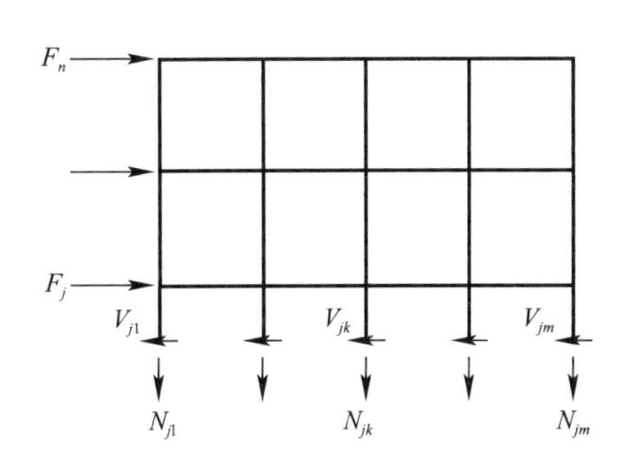

(b) Shear force and axial force of frame columns

Fig. 3-20 Schematic diagram of inflection point method

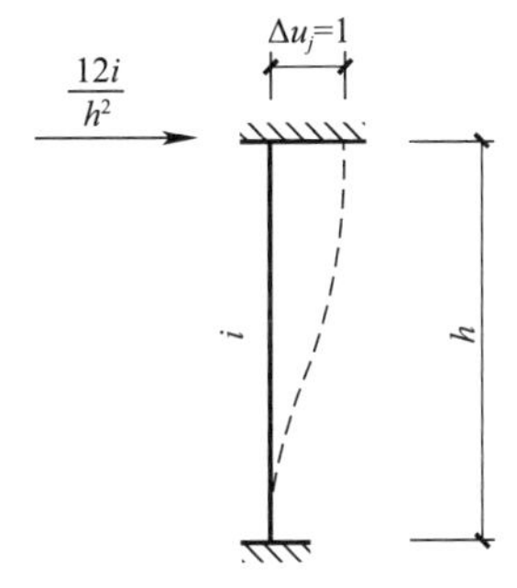

Fig. 3-21 Lateral stiffness of constant sectional column with two fixed ends

For the column shown in Fig. 3-21, D'_{jk} is called the lateral stiffness of the k-th column fixed at both ends. It represents the horizontal force to be applied at the top of the column to make the unit relative horizontal displacement ($\Delta u_j=1$) between the upper end and lower end of the column with fixed ends. Because the axial deformation of the beam is neglected, the columns in the j-th floor have the same inter-storey displacement Δu_j. Substituting formula (3-8) into formula (3-7), we can get Δu_j as follows:

$$\Delta u_j=\frac{V_j}{\sum_{k=1}^{m} D'_{jk}}=\frac{V_j}{\sum_{k=1}^{m}\frac{12i_{jk}}{h_j^2}} \quad (3\text{-}9)$$

If the column height in the samefloor is the same, the shear force of the k-th column in the j-th floor is:

$$V_{jk}=\frac{i_{jk}}{\sum_{k=1}^{m} i_{jk}}V_j \quad (3\text{-}10)$$

After the shear force of each column is obtained, the bending moment at the end of each column can be obtained by assumption 2). For the bottom column:

$$\left.\begin{aligned}M_{c1k}^{t}&=V_{1k}\cdot\frac{h_1}{3}\\ M_{c1k}^{b}&=V_{1k}\cdot\frac{2h_1}{3}\end{aligned}\right\} \quad (3\text{-}11)$$

For the k-th column in the j-th floor of the other floor:

$$M_{cjk}^{t}=M_{cjk}^{b}=V_{jk}\cdot\frac{h_j}{2} \quad (3\text{-}12)$$

t and b denote the top and bottom of the column respectively.

The moments of beam ends can be calculated from the equilibrium condition of the node moment, and then distributed according to the linear stiffness of the beams as shown in Fig. 3-22.

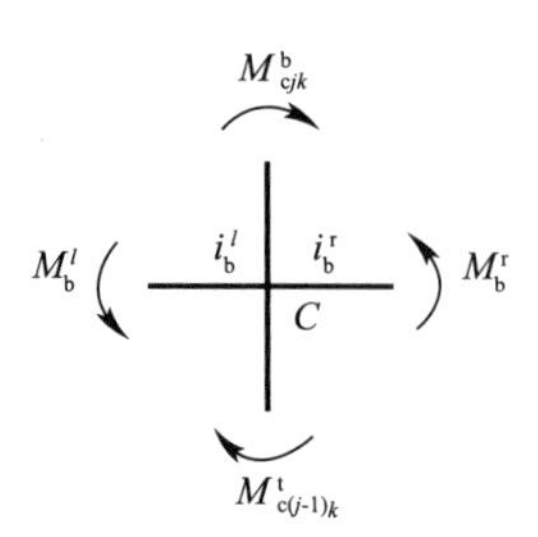

Fig. 3-22 Node equilibrium conditions

$$M_{b}^{l}=\frac{i_{b}^{l}}{i_{b}^{l}+i_{b}^{r}}(M_{cjk}^{b}+M_{c(j-1)k}^{t})$$
$$M_{b}^{r}=\frac{i_{b}^{r}}{i_{b}^{l}+i_{b}^{r}}(M_{cjk}^{b}+M_{c(j-1)k}^{t}) \tag{3-13}$$

Where,

M_b^l, M_b^r—— the moment of beam end on the left side and right side of the node;

M_{cjk}^b, $M_{c(j-1)k}^t$—— the moment of column end at the top and bottom of the node;

i_b^l, i_b^r—— the linear stiffness of the left beam and right beam of the node.

The shear force of the beam can be obtained by dividing the sum of the bending moments at the left end and right end of the beam by the span length of the beam. The axial force in the column can be obtained by adding the shear forces at the beam ends of the node layer by layer.

The inflection point method is applicable to: ① regular frames or approximate regular frames (i.e., the height, span and linear stiffness of each layer change little); ② the ratio of linear stiffness of beams and columns connected at the same frame node $i_b/i_C \geqslant 3$; ③ house height width ratio $H/b<4$.

2. *D*-value method

The inflection point method assumes that the ratio of linear stiffness between beams and columns is infinite, and theinflection point height is a certain value. In this way, the internal force calculation of frame structure under the horizontal load is greatly simplified, but at the same time, it also brings some errors. When the height of the upper and lower floors changes and the linear stiffness of beams and columns is relatively small, the internal force error calculated by the inflection point method is large. Therefore, it is necessary to modify the lateral stiffness and the inflection point height of the column. The modified method is called "*D*-value method". The modified column lateral stiffness is expressed by D, so it is called "*D*-value method". After determining the inflection point and lateral stiffness, its internal force calculation is the same as that of the inflection point method, so it is also called "improved inflection point method".

1) Modified lateral stiffness of column

The lateral stiffness of the column is the shear force that the column bears when the unit relative horizontal displacement occurs at the upper end and lower end of the column. For the k-th column in the j-th floor of the frame structure, the lateral stiffness of the column after considering the elastic restraint effect of the upper node and lower node of the column is as follows:

$$D_{jk}=\frac{V_{jk}}{\Delta u_{j}} \tag{3-14}$$

Taking the k-th column in the j-th floor of the frame structure shown in Fig. 3-23, the calculation formula of D_{jk} is derived.

Assumptions are as below:

(1) The linear stiffness of column AB and its adjacent columns is i_c;

(2) The inter-storey displacement of column AB and its adjacent columns is Δu_j;

(3) The rotation angle of the joints at both ends of column AB and the adjacent joints is θ;

(4) The linear stiffness of the beams intersecting column AB are i_1, i_2, i_3, i_4, respectively.

After the frame structure is subjected to the load, the deformations of column AB and its adjacent members are shown in Fig. 3-23 (b). Where, θ is the rotation angle of joint and φ is the shear angle along the height direction of the frame. $\varphi=\frac{\Delta u_j}{h_j}$. According to the moment equilibrium conditions of node A and node B, the following formulas can be obtained respectively:

$$4(i_3+i_4+i_c+i_c)\theta+2(i_3+i_4+i_c+i_c)\theta-6(i_c\varphi+i_c\varphi)=0$$
$$4(i_1+i_2+i_c+i_c)\theta+2(i_1+i_2+i_c+i_c)\theta-6(i_c\varphi+i_c\varphi)=0$$

By adding the above two formulas, we can get the following result:

$$\theta=\frac{2}{2+\frac{\sum i}{2i_c}}\varphi=\frac{2\varphi}{2+K} \quad (3\text{-}15)$$

Where,

$$\sum i=i_1+i_2+i_3+i_4;$$

K —— the ratio of linear stiffness of beam to that of column.

$$K=\frac{\sum i}{2i_c} \quad (3\text{-}16)$$

Column AB is restrained by the lateral displacement Δu_j and the joint angle θ. The shear force in the column is:

$$V_{jk}=-\frac{M_{AB}+M_{BA}}{h_j} \quad (3\text{-}17)$$

So,
$$V_{jk}=\frac{12i_c}{h_j}(\varphi-\text{i}) \quad (3\text{-}18)$$

Replace formula (3-15) with formula (3-18):

$$V_{jk}=\frac{K}{2+K}\frac{12i_c}{h_j}\varphi=\frac{K}{2+K}\frac{12i_c}{h_j^2}\Delta u_j=\alpha_c\frac{12i_c}{h_j^2}\Delta u_j \quad (3\text{-}19)$$

Where,

$$\alpha_c=\frac{K}{2+K}$$

Therefore, the lateral stiffness of the k-th column in the j-th floor is obtained as follows:

$$D_{jk}=\frac{V_{jk}}{\Delta u_j}=\alpha_c\frac{12i_c}{h_j^2} \quad (3\text{-}20)$$

The value of α_c reflects the influence coefficient of the ratio of linear stiffness of beam to that of column on the lateral stiffness of column, which is also called the reduction coefficient of lateral stiffness of frame column. When the linear stiffness of the frame is infinite, i.e., $k=\infty$, $\alpha_c=1$, then D_{jk} is equal to the lateral stiffness D'_{jk} of the columns fixed at both ends. The reduction coefficient of lateral stiffness α_c of the ground floor column can be derived in the same way. Table 3-5 lists the calculation formula of α_c and corresponding K under various conditions.

After obtain the modified lateral stiffness D_{jk}, the inter-layer shear force V_{jk} can be distributed to the columns of the same floor according to the condition that the relative displacement of each column in the same floor is equal:

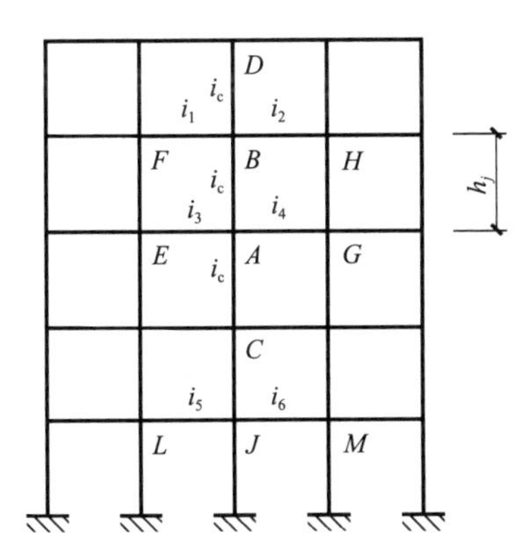

(a) Linear stiffness

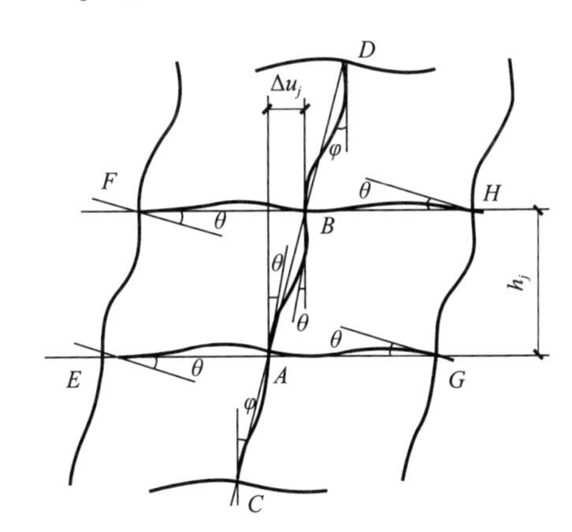

(b) Deformation diagram

Fig. 3-23 D-value method

Correction coefficient of lateral stiffness the column **Table 3-5**

Floor	Diagram	K	α_c
Other floor	i_1, i_c, i_4; i_1, i_2, i_c, i_3, i_4	$K=\frac{i_1+i_2+i_3+i_4}{2i_c}$	$\alpha_c=\frac{K}{2+K}$

Continued

Floor	Diagram	K	α_c
Ground floor	i_2 i_1 i_2 i_c i_c	$K=\frac{i_1+i_2}{i_c}$	$\alpha_c=\frac{0.5+K}{2+K}$

Note: In the case of side column, i_1 and i_3 are taken as 0.

$$V_{jk}=\frac{D_{jk}}{\sum_{k=1}^{m} D_{jk}}V_j \tag{3-21}$$

Where,

V_{jk}—— the hear force of the k-th column in the j-th floor;

D_{jk}—— the lateral stiffness of the k-th column in the j-th floor;

m —— the number of frame columns in the j-th floor;

V_j—— the total shear force produced by the external load in the j-th floor.

2) Modified inflection point height of column

The position of the inflection point of the column depends on the rotational stiffness of the upper end and lower end of the column. If the rotation angles of upper end and lower end of the column are the same, the inflection point is in the center of the column height. If the rotation angles of upper end and lower end are different, the inflection point is inclined to move to the section with larger rotation angle, that is, to the end with smaller constraint stiffness. The factors that affect the inflection point of the column include the form of horizontal load, the total number of storeys and the floor level of the column, the ratio of linear stiffness of beam to that of column, the ratio of the linear stiffness of upper beams to that of lower beams, the change of the upper floor height and the lower floor height. In order to analyze the influence of the above factors on the inflection point height, it can be assumed that under the horizontal force, the rotation angles of the joints on the same floor are equal, that is to say, the inflection points of all beams on the same floor are in the middle of the span of each beam, and there is no vertical displacement at this point. In this way, a multi-storey and multi-span frame can be simplified to the calculation diagram shown in Fig. 3-24. When the above factors change one by one, the distance from the bottom of the column to the inflection point of the column (the inflection point height) can be calculated respectively, and the corresponding table about the inflection point height can be made for reference.

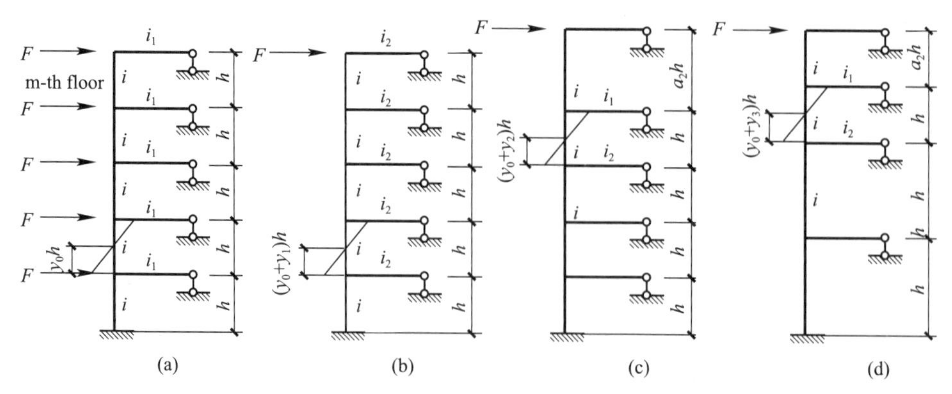

Fig. 3-24 Inflection point height of column

(1) Influence of the ratio of linear stiffness of beam to that of column, number of storeys and floor level on the inflection point height

Assuming that the linear stiffness of the frame beam, the linear stiffness of the frame column and the storey height remain unchanged along the frame height, then the inflection point height $y_0 h$ of the column at each floor can be calculated according to Fig. 3-24 (a); y_0 is called the ratio of standard inflection point height, and its value is related to the total number of storeys n, the floor level j of the column, the ratio K of linear stiffness of beam and that of column, and the form of horizontal load, which can be found in appendix Table 3-1 and Table 3-2. The value of K in the appendix Table can be calculated according to the formula.

(2) Influence of the ratio of linear stiffness of upper beams to that of lower beams on the inflection point height

If the linear stiffness of the upper beams and lower beams connected with a certain column are different, the position of the inflection point of the column will shift to the side with less beam stiffness, so the standard inflection point height must be modified, and this modified-value is the $y_1 h$ (Fig. 3-24b).

y_1 can be found from appendix Table 3-3 according to the I and K. When $(i_1+i_2)<(i_3+i_4)$, the inflection point moves up, and the y_1 can be obtained by checking the appendix Table 3-3 from $I=\frac{i_1+i_2}{i_3+i_4}$. When $(i_1+i_2)>(i_3+i_4)$, the inflection point moves down. $I=\frac{i_3+i_4}{i_1+i_2}$ should be adopted when looking up this table, and y_1 should be marked with a negative sign. For the bottom column, the modified-value y_1 is not considered, that is, $y_1=0$.

(3) Influence of floor height change on inflection point height

If the column height of the floor is different from that of the adjacent floor, the position of the inflection point of the column is different from that of the standard inflection point and needs to be modified. As shown in Fig. 3-24 (c), when the upper floor height changes, the upward displacement increment of the inflection point height is $y_2 h$. As shown in Fig. 3-24 (d), when the lower layer height changes, the upward displacement increment of the inflection point height is $y_3 h$. y_2 and y_3 can be found in appendix Table 3-4. For the top column, the modified-value y_2 is not considered, that is, $y_2=0$. For the bottom column, the modified-value y_3 is not considered, that is, $y_3=0$.

The height yh from the column bottom to the inflection point can be calculated by the following formula after considering various correction values:

$$yh=(y_0+y_1+y_2+y_3)h \qquad (3\text{-}22)$$

After the lateral stiffness of the frame column is obtained according to formula (3-20), the shear force of each column is calculated according to formula (3-21), and the inflection point height y_h of each column is calculated according to formula (3-22), the bending moments of the ends of each column can be calculated. After calculating the bending moments of the ends of the column, the bending moments of the beam ends can be obtained according to formula (3-13), and then the shear force at each beam end and the axial force of each column can be calculated according to the same method as the inflection point method.

3.3.3 Inter-storey displacement of frame structure under horizontal load

1. Inter-storey horizontal displacement Δu_j

From formula (3-20) and formula (3-21), the relationship between the inter-storey horizontal displacement Δu_j and the shear force V_j of the j-th floor is as follows:

$$\Delta u_j=\frac{V_j}{\sum_{k=1}^{m} D_{jk}} \qquad (3\text{-}23)$$

Where,

D_{jk}—— the lateral stiffness of the k-th column in the j-th floor;

m —— the number of frame columns in the j-th floor.

In this way, the horizontal displacement of each floor can be obtained floor by floor. The total horizontal displacement of the frame structure should be the sum of the inter-storey displacements, namely.

$$u = \sum_{j=1}^{n} \Delta u_j \tag{3-24}$$

Where,

n—— the total number of floors of the frame structure.

It should be pointed out that the horizontal displacement of the frame structure obtained by the above method is only the deformation caused by the bending deformation of the beam and column, and the structural lateral displacement caused by the axial deformation of the beam and column and the section shear deformation are not considered.

It can be seen from formula (3-23) that the frame inter-storey displacement Δu_j is proportional to the shear force V_j generated by the horizontal loads on the floor. Because the lateral stiffness of frame columns generally does not change much along the height, and the inter-storey shear force V_j is accumulated floor by floor from the top floor to the ground floor, the inter-storey horizontal displacement Δu_j increases from the top floor to the ground floor. The displacement curve of the frame is shown in Fig. 3-25(a). This kind of displacement curve is called a shear type, which is similar to the shear deformation curve of a cantilever column under the uniform horizontal load caused by the shear force in the section, as shown in Fig. 3-25 (b). The deformation curve of the cantilever column caused by the bending moment is shown in Fig. 3-25 (c). For the multi-storey frame structure with few floors, the lateral displacement caused by the axial deformation of the column is very small and can be ignored. In the approximate calculation, only the lateral displacement caused by the bending deformation of beam and column, that is, shear deformation needs to be calculated.

2. Allowable value of inter-storey displacement angle of frame structure

The inter-storey horizontal displacement Δu_j calculated by the elastic method is divided by the floor height h, then the tangent of the displacement angle θ_e between the floors is obtained. Since θ_e is small, $\theta_e = \Delta u/h$ can be taken approximately. If the elastic inter-storey displacement angle θ_e of the frame is too large, the cracks of the non-load-bearing filling members such as partition walls in the frame will occur. The ***Code for Seismic Design of Buildings***

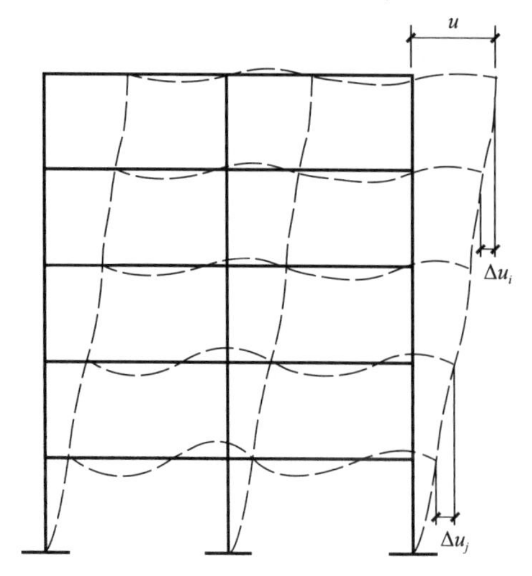

(a) Deformation of the frame under horizontal load (shear type)

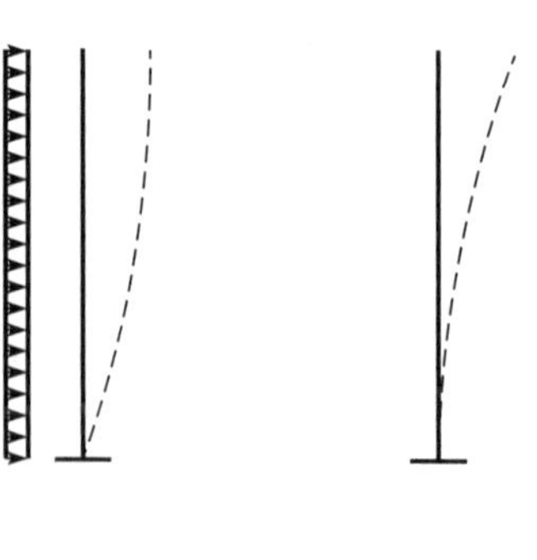

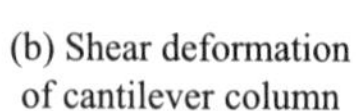

(b) Shear deformation of cantilever column

(c) Bending deformation of cantilever column

Fig. 3-25 Horizontal displacement curve of structure

(GB 50011—2010) stipulates that the ratio of the maximum elastic inter-storey displacement Δu to the floor height of the frame cannot exceed its limit, that is:

$$\frac{\Delta u}{h} \leqslant [\theta_e] \tag{3-25}$$

Where,

Δu —— the inter-storey horizontal displacement calculated by elastic method;

h —— the floor height;

$[\theta_e]$ —— the allowable value of elastic inter-storey displacement angle. For the reinforced concrete frame structure, $[\theta_e]$ is 1/550.

3.4 Second-order effect of frame structure

The lateral horizontal displacement of frame structure under the action of horizontal force and gravity load will produce additional internal force in the structure, namely, the so-called P-Δ effect, also known as the second-order effect of gravity or the second-order effect of lateral displacement. When calculating the P-Δ effect of frame structure, the method of increasing coefficient from the ***Code for Design of Concrete Structures*** (GB 50010—2010) can be used. The bending moment at the end of frame column, bending moment at beam end and inter-storey horizontal displacement which are calculated in the first-order elastic analysis can be increased as follows:

$$M = M_{ns} + \eta_s M_s \tag{3-26}$$

$$\Delta u_j = \eta_s \Delta u_{js} \tag{3-27}$$

Where,

M —— the design values of bending moments of column end and beam end considering P–Δeffect;

M_{ns}—— according to the first-order elastic analysis, the design values of bending moments of column ends and beam ends are obtained from the loads which do not cause lateral displacement of the frame (i. e., the vertical loads);

M_s—— according to the first-order elastic analysis, the design values of bending moments of column ends and beam ends are obtained from the loads or actions which cause lateral displacement of the frame (i.e., the horizontal loads);

Δu_j—— the inter-storey horizontal displacement of the j-th floor considering the P-Δ effect;

Δu_{js}—— the inter-storey horizontal displacement value of the j-th floor according to the first-order elastic analysis;

η_s—— the increasing coefficient of P-Δ effect.

In the frame structure, the same increasing coefficient of P-Δ effect is adopted for all columns in the same floor. The increasing coefficient of P-Δ effect of the j-th floor is as follows:

$$\eta_{s,j} = \frac{1}{1 - \dfrac{\sum_{k=1}^{m} N_{jk}}{\sum_{k=1}^{m} D_{jk} h_j}} \tag{3-28}$$

Where,

$\sum_{k=1}^{m} D_{jk}$—— the sum of lateral stiffness of all columns in the j-th floor. When calculating the moment increasing coefficient of P-Δ effect, it is better to multiply the cross-sectional elastic bending stiffness (EI) of the column and beam by the reduction coefficient. The reduction coefficient is 0.4 for the beam and the reduction coefficient is 0.6 for the column. There is no stiffness reduction when calculating the displacement increasing coefficient of P-Δ effect;

$\sum_{k=1}^{m} N_{jk}$——the sum of axial forces of all columns in the j-th floor;

h_j——the floor height of the j-th floor.

The increasing coefficient of P-Δ effect of the beam ends is the average value of the increasing coefficient of the upper end and lower end of the column of the corresponding node, that is, the increasing coefficient of P-Δ effect of the frame beam end is $\eta_s=\frac{1}{2}(\eta_{s,j}+\eta_{s,j}+1)$.

3.5 Internal force combination of multi-storey frame structure

3.5.1 Control sections

For frame columns, the bending moment, axial force and shear force vary linearly along the column height, so the top and bottom end sections of each column can be taken as the control sections. For the multi-storey frame structure, the section size, concrete strength and reinforcement of the whole column are the same, so we can only take the top section of the column at the top of the frame and the bottom section of the column at the bottom of the frame as the control sections. For the frame with larger height or more storeys, the whole column should be divided into several parts for reinforcement. The top and bottom end sections of each part are taken as the control sections, and each part generally takes 2—3 floors.

For the frame beam, under the action of the horizontal load and vertical load, the shear force changes linearly along the beam axis, while the bending moment changes non-linearly (parabola change under the vertical load and linear change under the horizontal load). Therefore, in addition to taking both ends of the beam as the control sections, the section with the maximum positive bending moment (generally mid-span section) should be taken as the control section.

Control sections of some column and some beam are shown in Fig. 3-26.

In addition, the internal force of the end section of the beam (i.e., the column edge) should be used in the reinforcement calculation of the beam, rather than the internal force at the axis, as shown in Fig. 3-27. The design value of bending moment and shear force at beam ends shall be calculated according to the following formula:

$$V_b=V-(g+q)\frac{b}{2} \tag{3-29}$$

$$M_b=M-V_b\frac{b}{2} \tag{3-30}$$

Where,

V_b, M_b—— the shear force design value and bending moment design value of the column side section;

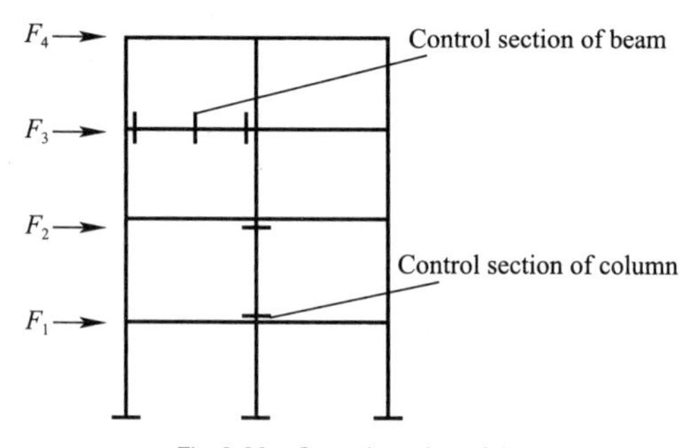

Fig. 3-26 Control section of the frame structure

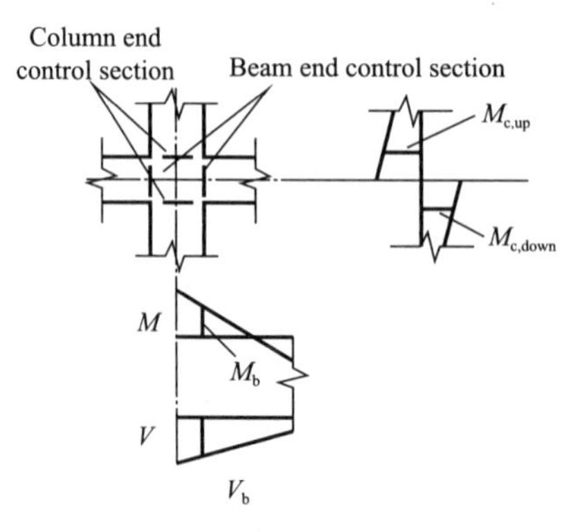

Fig. 3-27 Bending moment and shear force of control sections at beam ends

V, M —— the shear force design value and bending moment design value at the column axis obtained by internal force analysis;

g, q —— the design values of the vertical uniformly distributed dead load and live load acting on the beam;

b —— the column width.

The control sections of column should be the beam edge sections. The internal force of the column axis is taken as the internal force of the control section of column while designing in order to simplify the calculation.

3.5.2 Internal force combination

1. Load effects combination

According to ***Load Code for the Design of Building Structures*** (GB 50009—2012), simplifid load effects combination can be adopted for the bent frame and frame structure.

1) "dead load" + any "live load"

$$s=\gamma_G s_{GK}+\gamma_{Q1} s_{Q1K} \tag{3-31}$$

2) "dead load" + 0.9 any two or more "live loads"

$$s=\gamma_G s_{GK}+0.9\sum_{i=1}^{n}\gamma_{Qi} s_{QiK} \tag{3-32}$$

2. The most unfavorable combination of internal forces

The most unfavorable internal force combinations of frame beams are as follows.

Beam end section: $+M_{max}$, $-M_{max}$, V_{max};

Mid-span section of beam: $+M_{max}$.

The most unfavorable internal force combinations of cross-section frame column ends are as follows:

$|M|_{max}$ and corresponding N, V;

N_{max} and corresponding M;

N_{min} and corresponding M.

In some cases, the above internal force combinations may not be the most unfavorable. For example, for the large biased column, the larger the eccentricity (i.e., larger M, smaller N), the more section reinforcement is. However, although the bending moment is not the maximum value but slightly smaller than the maximum value, and its corresponding axial force decreases a lot, the sectional reinforcement calculated according to this group of internal force combination will be larger. Therefore, when calculating $|M|_{max}$, the corresponding N shall be as small as possible. When calculating N_{max} or N_{min}, the corresponding M shall be as large as possible.

3.5.3 Unfavorable distribution of vertical live load

There are four methods to consider the most unfavorable distribution of live load, i.e., floor and span cross combination calculation method, the most unfavorable load location method, multi-layered combination method and full load distribution method.

1. Cross combination calculation method by floor and span

In this method, the live load is applied to the structure floor by floor and span by span separately (as shown in Fig. 3-28), and the internal forces of the whole structure are calculated respectively. According to different components, different sections and different types of internal forces, the most unfavorable internal forces are combined. Therefore, for a multi-storey and multi-span frame, there are (span×storey) kinds of live load distributions, and the internal force of the structure also needs to be calculated by (span×storey) times. In other words, the calculation workload is very heavy. However, after these internal forces are obtained, the maximum internal forces of any section can be obtained, and the process is relatively simple. This method is often used in combination of internal forces by the computer.

2. The most unfavorable load location method

In order to find out the most unfavorable internal force of a given section, the live load-distribution which produces the most unfavorable internal force can be directly determined according to the influence line method. The frame with the four floors and four spans in Fig. 3-29(a) is taken as an example. In order to find the most unfavorable arrangement of live load of the maximum positive bending moment M_c of the section C in the middle of a certain span beam AB, imposing it with the positive restraint force to make the unit virtual displacement $\theta_c = 1$ produced along the positive direction of the binding force. Thus, the virtual displacement diagram of the whole structure can be obtained as shown in Fig. 3-29(b).

According to the principle of virtual displacement, in order to find the maximum positive bending moment in the span of beam AB, the live load must be arranged in all the spans that produce positive virtual displacement in Fig. 3-29(b). That is to say, the span AB must be arranged with live load, and other spans should be arranged alternately. The loads are arranged alternately in the vertical direction to form a checkerboard layout as shown in Fig. 3-29(c). It can be seen that the most unfavorable distribution of the live load when the span AB has the maximum bending moment in the middle of the span can also makes the bending moments of other live load spans reach the maximum values. Therefore, as long as the second kind of chessboard live load distribution is carried out, the maximum positive bending moments of all beams of the whole frame can be obtained.

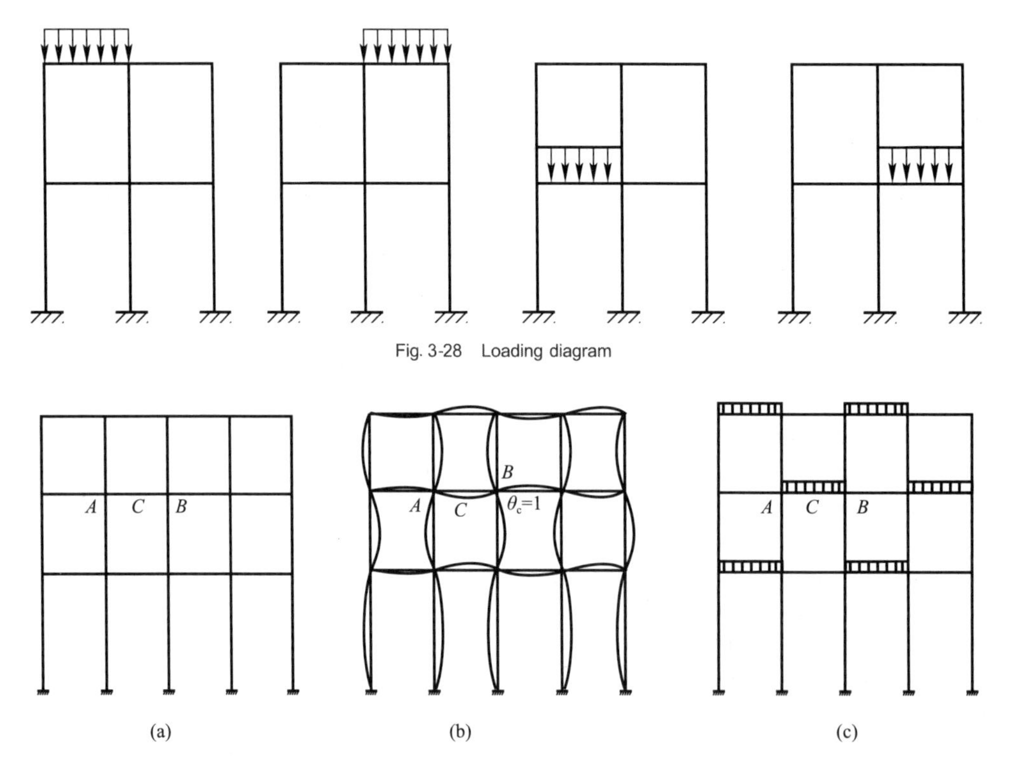

Fig. 3-28 Loading diagram

Fig. 3-29 Arrangement of the most unfavorable load

Maximum negative bending moments at beam ends or the maximum bending moments at the ends of the column can also be obtained by the above method. However, it is very difficult to accurately determine the influence line of multi-storey and multi-span frame structure with different linear stiffness of the member of each span and each floor. For the frame joints far away from the calculation section, it is difficult to accurately determine the direction of the virtual displacement (rotation angle). Fortunately, the load far away from the calculation section has little influence on the internal force of the calculated section, which can be ignored in practice.

Obviously, the most disadvantageous distribution of the live load for the maximum axial force of the column is that the beam spans adjacent to the column are full of live loads in each floor above this column.

3. Multi-layered combination method

It is very difficult to calculate the internal force of the structurewith the most unfavorable distribution of live load either by floor and span cross combination or by most unfavorable load position method. The multi-layered combination method is based on the multi-layered method of the frame under the vertical load and is relatively simple. The most unfavorable arrangement of live load is simplified as follows.

1) For the beam, only the unfavorable arrangement of live load in this layer is considered, and the influence of other layers is not considered. Therefore, the layout method of live load is the same as that of the most unfavorable live load of continuous beam.

2) For the moments at the ends of the column, only the influence of the live load of the adjacent upper and lower layers of the column is considered, and the influence of the live load of other layers is not considered.

3) For the maximum axial force of the column, only the case of full distribution of live loads on the beams adjacent to the column in all layers above the floor is considered. The influence of the live load of other layers on the bending moment is not considered.

4. Full load distribution method

When the internal force generated by the live load is far less than that generated by the dead load and horizontal force, the most unfavorable arrangement of live load can be ignored, and the live loads can be applied to all frame beams at the same time. In this way, the internal force obtained at the support is very close to that obtained by the most unfavorable load location method, so the internal force combination can be carried out directly. However, the bending moment at the mid-span is smaller than that calculated by the most unfavorable load location method. Therefore, the bending moment at the mid-span should be multiplied by the coefficient of 1.1—1.2. For the multi-storey workshop and public building when the floor live load is larger than 5.0 kN/m^2, full load method cannot be adopted.

3.5.4 Moment modulation at beam ends

According to the reasonable failure mode of the frame structure, the plastic hinges at the beam ends are allowed. Also, it is often hoped that the negative reinforcement of the beam at the joint should be placed less in order to facilitate the concrete pouring. And the actual bending moment of the beam ends are less than its elastic calculation values because the joint is not absolutely rigid for the prefabricated or assembled integral frame structure. Therefore, the bending moments of beam ends are generally adjusted in the design of frame structure, that means the negative bending moments at the beam ends would be reduced artificially to reduce the reinforcement amount on the top of the beam near the joint.

Supposing that the maximum negative bending moments at the ends of a frame beam AB under the vertical load are M_{A0} and M_{B0}, respectively, and the maximum positive bending moment in the beam span is M_{C0}, then the bending moments of beam ends after amplitude modulation can be taken as:

$$M_A = \beta M_{A0} \tag{3-33}$$

$$M_B = \beta M_{B0} \tag{3-34}$$

Where,

β —— the moment modulation coefficient.

For the cast-in-situ frame, β can be taken as 0.8—0.9. For prefabricated concrete frame structure, because the joint welding is not firm or the concrete pouring in the joint area is not dense, the joint is easy to occur deformation and cannot reach the absolute rigidity, the actual bending moment at the beam end of the frame is smaller than the elastic calculation value. The bending moment modulation coefficient for prefabricated concrete frame structure is allowed to be lower, generally 0.7—0.8.

It must be pointed out that the moment modulation is only applied to the internal force under the vertical load, that is, the bending moment generated under the horizontal load does not participate in the moment modulation. Therefore, the bending moment modulation should be carried out before the internal force combination.

After the bending moments of beam ends are adjusted, the bending moments at the mid-span will increase under the corresponding loads as shown in Fig. 3-30. At this time, the static equilibrium condition of the beam should be checked, that is, the sum of M_{C0} and the average value of M_A and M_B after adjustment should be greater than or equal to the bending moment M_0 at the mid-span calculated for the simply supported beam. That is:

$$\frac{M_A+M_B}{2}+M_{C0}\geqslant M_0 \tag{3-35}$$

In the section design, the design value of the positive bending moment of the mid-span section of the frame beam should not be less than 50% of the design value calculated according to the simply supported beam under the vertical load.

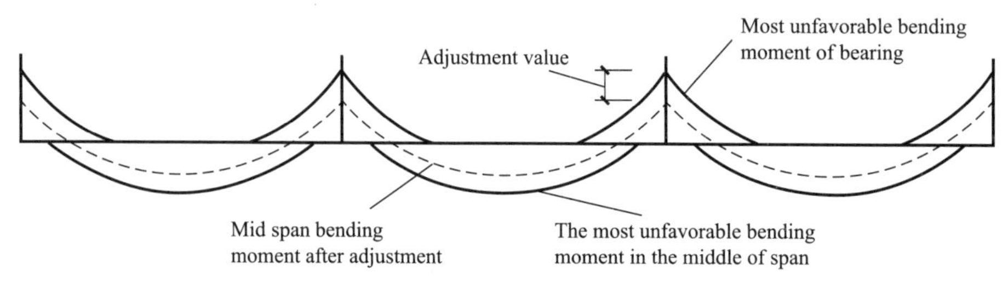

Fig. 3-30 Moment adjustment at beam ends

3.6 Members design of frame structure without seismic fortification requirements

3.6.1 General construction requirements of frame structure

The concrete strength grade of reinforced concrete frame is not lower than C20. The longitudinal reinforcement can be HRB400 or HRB500 and the stirrup is generally HRB400 or HRB335.

Concrete cover should be determined according to the environmental category of the buildings. For example, when the environment category is class 1, the thickness of concrete cover for longitudinal stressed reinforcement of frame beam and frame column shall not be less than 20 mm. If the concrete strength grade is not greater than C25, the protective layer thickness shall not be less than 25 mm.

The frame beams and columns should meet the structural requirements of flexural members and compression members respectively. And the frame in seismic area should also meet the seismic design requirements.

Frame column generally adopts symmetrical reinforcement, the reinforcement ratio of all longitudinal reinforcement in the column should not be greater than 5% , and the minimum reinforcement ratio is 0.6%. Generally, the bending reinforcement is not used for shear resistance of frame beam.

3.6.2 Effective length of column

The effective length of reinforced concrete frame column with rigid connection between beam and column should be determined according to different lateral restraint conditions and load conditions of the frame, and the influence of the second-order effect of column (additional bending moment caused by axial force and horizontal displacement) on column section design shall be considered.

Generally, the effective length of columns on each floor(l_0) can be taken from Table 3-6.

3.6.3 Structural requirements of frame members and joints

1. Frame beam

1) Longitudinal reinforcement of beam

In order to prevent brittle failure of beam and control crack width, the quantity of longitudinal tensile reinforcement should be determined according to calculation. The minimum reinforcement ratio ρ_{min}(%) of longitudinal tensile bars shall not be less than the greater value of 0.2 and $0.45f_t/f_y$. At the same time, in order to prevent over-reinforced beam, the maximum reinforcement ratio of longitudinal tensile reinforcement shall not exceed $\rho_{max}=\frac{\xi_b\alpha_1 f_c}{f_y}$.

At least two longitudinal steel bars shall be provided on the top and bottom surface of the whole length of the beam and the diameter of the steel bars shall not be less than 12 mm. The longitudinal bars of frame beam shall not be welded with stirrups, tie bars and embedded parts. If there are at least two longitudinal bars extending from the upper part of the beam to the lower part of the frame, the length of the upper bending point shall not be less than $10d$.

2) Stirrup of beam

Stirrups should be set along the whole length of frame beam. The spacing should meet the requirements of ***Code for Design of Concrete Structures***(GB 50010—2010). When the beam section $h>800$ mm, the stirrup diameter should be $\geqslant 8$ mm; when $h\leqslant 800$ mm, the stirrup diameter should be $\geqslant 6$ mm.

2. Frame column

1) Longitudinal reinforcement of column

For the frame column, symmetrical reinforcement is usually used. In non-seismic fortification, all longitudinal reinforcement ratio of columns shall meet the following requirements.

The minimum reinforcement ratio of longitudinal reinforcement of middle column, side column and corner column shall not be less than 0.5%. The minimum reinforcement ratio of longitudinal reinforcement of frame column shall not be less than 0.7%. At the same time, the reinforcement ratio of each side of the column section should not be less than 0.2%, and the reinforcement ratio of all reinforcement of the column should not be greater than 5%. When the section height of column is greater than or equal to 600 mm, 10—16 mm longitudinal constructional bars and corresponding composite stirrups or tie bars should be set on the side of the column.

Effective length of column **Table 3-6**

Slab type	Types of columns	l_0
Cast-in-situ slab	Ground floor column	$1.0H$
	Columns of other floors	$1.25H$
Prefabricated slab	Ground floor column	$1.25H$
	Columns of other floors	$1.5H$

Note:

H is the height from the top of the foundation to the top of the first floor for the ground floor column. For the rest of the columns, H is the height between the top surfaces of the upper and lower floors.

2) Stirrup of column

For non-seismic fortification design, the stirrup around the section of frame column should be closed stirrup. The spacing between stirrups shall not be greater than 400 mm, and shall not be greater than 15 times of the short side size of the member section and the minimum longitudinal reinforcement diameter. The diameter of stirrup should not be less than 1/4 of the maximum diameter of longitudinal bars, and should not be less than 6 mm. When the reinforcement ratio of all longitudinal reinforcement in the column is more than 3%, the diameter of stirrup should not be less than 8 mm, and the stirrup spacing shall not be greater than 10 times of the minimum longitudinal reinforcement diameter or 200 mm. The end of stirrup shall be made into 135 degrees hook, and the length of straight part at the end of hook shall not be less than 10 times of longitudinal reinforcement diameter. When the short side size of the column is more than 400 mm and the number of longitudinal bars on each side is more than or equal to 3, or the short side size is less than or equal to 400 mm and the number of longitudinal bars on each side is more than or equal to 4, compound stirrups should be set.

The common reinforcement forms of columns are shown in the Fig. 3-31.

3. Frame joint

1) Section size of joint

If the joint size is too small and the number of reinforcement of the beam and the column is too high, the concrete in the joint area will be crushed obliquely due to the excessive pressure in the core area diagonal bar mechanism. Therefore, the relative quantity of negative reinforcement in beams and columns should be limited. According to ***Code for Design of Concrete Structures***(GB 50010—2010), the cross-sectional area of longitudinal reinforcement on the top of beam at the top floor of frame structure is A_s shall meet the requirements of the following formula:

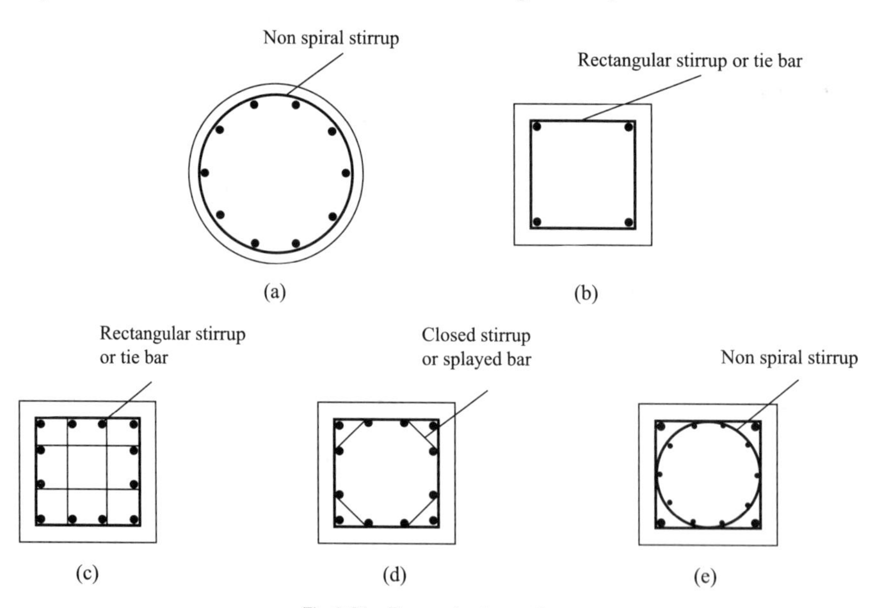

Fig. 3-31 Forms of column stirrup

$$A_s \leqslant \frac{0.35\beta_c f_c b_b h_0}{f_y} \tag{3-36}$$

Where,

β_c—— the influence coefficient of concrete strength;

b_b—— the web width of beam;

h_0—— the section height of beam.

2) Stirrup design

The beam-column connections of cast-in-situ frames are all made of rigid joints. The beam-column joint is in the state of composite stress state of shear and compression. In order to ensure that the joint has enough shear bearing capacity and prevent the joint from shear brittle failure, it is necessary to set enough horizontal stirrups in the joint. In addition to meet the structural requirements of the stirrup of the frame column, the stirrup spacing should not be greater than 250 mm. For the joint with beams on four sides connected with it, rectangular stirrup can only be set around the joint.

3) Longitudinal reinforcement of frame beam

As shown in Fig. 3-32(a), the anchorage length of the upper longitudinal reinforcement extending into the end of the beam shall not be less than l_a when anchoring in a straight line. The length extending through the center line of the column should not be less than 5 times of the diameter of the upper longitudinal reinforcement of the beam. When the section size of the column is insufficient, the longitudinal reinforcement at the upper part of the beam should extend to the opposite side of the joint and bend downward. The horizontal projection length of the anchorage section before bending should not be less than $0.4l_{ab}$ (l_{ab} is the basic anchorage length of tensile bar). The vertical projection length after bending should be 15 times of the beam longitudinal reinforcement diameter. As shown in Fig. 3-32(b), the upper longitudinal reinforcement of beam can also be anchored by adding mechanical anchor head at the end of reinforcement, and it should extend to the inner edge of longitudinal reinforcement outside the column. The horizontal projection anchorage length including mechanical anchor head shall be no less than $0.4l_{ab}$, as shown in Fig. 3-32(b).

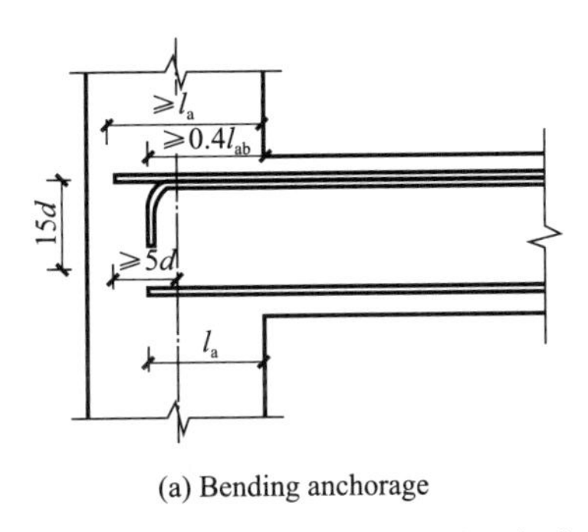

(a) Bending anchorage

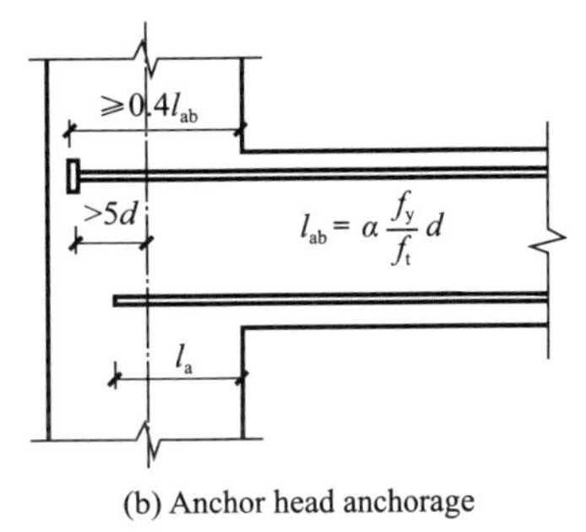

(b) Anchor head anchorage

Fig. 3-32 Anchorage length of upper longitudinal reinforcement of beam

The anchorage of the longitudinal reinforcement at the lower part of the frame beam extending in to the end node is shown in Fig. 3-33. When the strength of the longitudinal reinforcement at the lower part of the beam is not used in the calculation, the anchorage length extended into the joint shall be no less than 12 times of the longitudinal reinforcement diameter of the beam. When the tensile strength of the reinforcement at the lower part of the beam is fully utilized in the calculation, the lower longitudinal reinforcement of the beam can be bent 90° upward or anchored by anchor head. The anchorage length in the straight-line anchorage shall not be less than l_a. For bending anchorage, the horizontal projection length of anchorage section shall not be less than $0.4l_{ab}$, the vertical projection length should be 15 times of the longitudinal reinforcement diameter of the beam.

When the reinforcement of the middle joint (or the middle support of continuous beam) in the middle floor of frame beam is anchored or overlapped, the upper longitudinal bar shall pass through the middle joint or support. When it is necessary to anchor, the following anchoring requirements shall be met.

When the strength of the reinforcement is not fully utilized in the calculation, the anchorage length extending into the joint or support must be greater than or equal to 12d for ribbed reinforcement, 15d for smooth reinforcement (d

is the maximum diameter of reinforcement). When the calculation makes full use of the compressive strength of the steel bar, the steel bar should be anchored in the middle node or middle support as a compression bar. And its linear anchorage length is greater than or equal to 0.7 l_a. When the tensile strength of reinforcement is fully utilized, the reinforcement can be anchored in the middle node or middle support in a straight line, and the anchorage length is greater than or equal to l_a as shown in Fig. 3-34(a). The lap reinforcement can be set out of the joint or the support where the bending moment is small, and the distance between the starting point of the lap reinforcement and the edge of the joint or support is greater than or equal to $1.5h_0$ (h_0 is the effective height), as shown in Fig. 3-34(b).

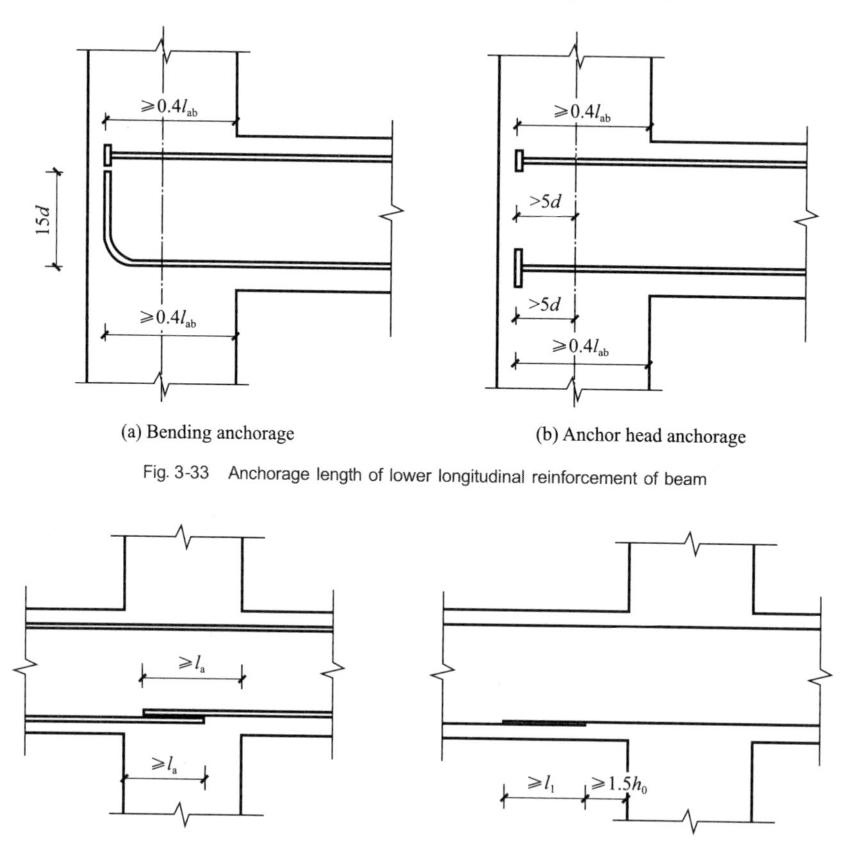

Fig. 3-33 Anchorage length of lower longitudinal reinforcement of beam

(a) In the joint

(b) Out of the joint

Fig. 3-34 Anchorage of support reinforcement of continuous beam

In addition, the longitudinal tensile reinforcement on the upper part of the beam should extend to the mid-span as (1/4—1/3) l_n (l_n is the clear span of the beam), and it is overlapped with the erection bar (no less than 2 Φ12) in the middle of the span, and the lap length can be 150 mm.

4) Longitudinal reinforcement of the frame column

In the middle joints or end joints of the middle floors, the longitudinal reinforcement of the column should continuously pass through the middle floor joints and the end joints. In the middle joint of the top floor, the longitudinal reinforcement of the column should be extended to the top of the column, and the anchorage length from the bottom of the beam should be greater than or equal to the anchorage length l_a. If the section height does not meet the requirements of straight-line anchorage, it should extend to the top of the column and bend inward horizontally. When the tensile strength of longitudinal reinforcement is fully utilized, the vertical projection length of reinforcement including bending arc is greater than or equal to $0.5l_{ab}$, the length of the horizontal projection including the curved arc in the bending plane should be greater than or equal to $12d$, as shown in Fig. 3-35(a).

When there are cast-in-situ floor slab on the top of the column, the slab thickness is greater than or equal to 100 mm, and the concrete strength is greater than or equal to C20, the longitudinal reinforcement of the column can also be bent outward, and the horizontal projection length after bending should be greater than or equal to $12d$. When the section height of the column is insufficient, mechanical anchorage with anchor head can also be used. At this time, the vertical anchorage length including anchor head is greater than or equal to $0.5l_{ab}$ as shown in Fig. 3-35(b).

The longitudinal bar outside the column at the top joint can be bent into the beam as the upper longitudinal bar of the beam. Also, the upper longitudinal bar of the beam and the outer longitudinal bar of the column can be overlapped at the top joint and its adjacent parts. There are two kinds of lapping modes.

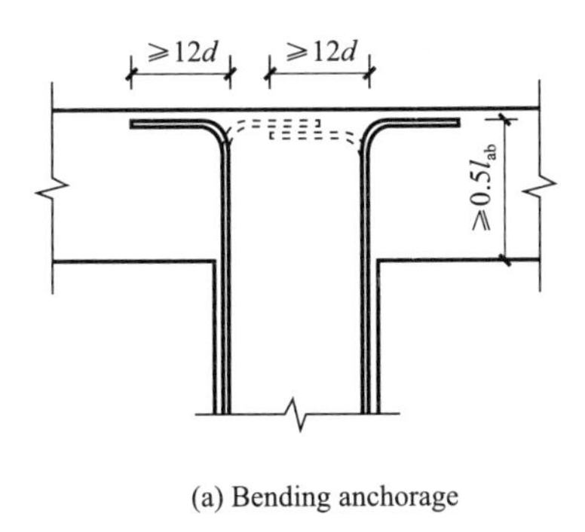

(a) Bending anchorage

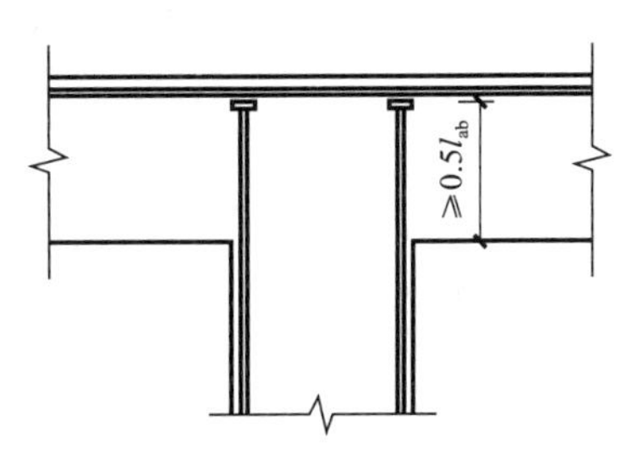

(b) Anchor head anchorage

Fig. 3-35 Anchorage structure requirements of column longitudinal reinforcement

The lapping reinforcement can be arranged along the outer side of the top joint and the top of the beam end, and the lapping length is greater than or equal to $1.5l_{ab}$ as shown in Fig. 3-36 (a). The longitudinal reinforcement of the column extending into the beam should be greater than or equal to 62% of the total longitudinal reinforcement of the outer column. The longitudinal reinforcement of the beam should extend into the column along the top of the joint. When the longitudinal reinforcement of the column is located in the first layer of the top of the column, it can be bent down to the inner edge of the column more than $8d$ (d is the diameter of the longitudinal bar outside the column) and then cut off. When there is the cast-in-situ slab and the slab thickness is greater than or equal to 100 mm and the concrete strength is greater than or equal to C20, the longitudinal reinforcement of the beam width can extend into the cast-in-situ slab, and its length is the same as that of the column longitudinal bar extending into the beam. The upper longitudinal reinforcement of the beam should extend to the outside of the joint and be bent down to the elevation of the lower edge of the beam and then cut off.

The lapping reinforcement joint can also be arranged in a straight line along the outside of the column top, as shown in Fig. 3-36(b), and the lapping length is greater than or equal to $1.7l_{ab}$. When the reinforcement ratio of the upper longitudinal reinforcement of beam is greater than 1.2%, the lapping length of the longitudinal reinforcement bent into the outer side of the column is greater than or equal to $1.7l_{ab}$ and it should be cut off in two batches. The distance between the cut-off points should be greater than or equal to $20d$. The anchoring of the longitudinal reinforcement of the column shall comply with the relevant anchoring regulations of the joint of the top floor.

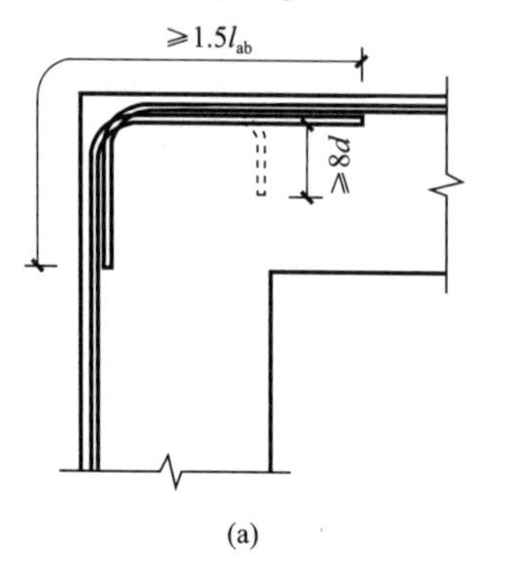

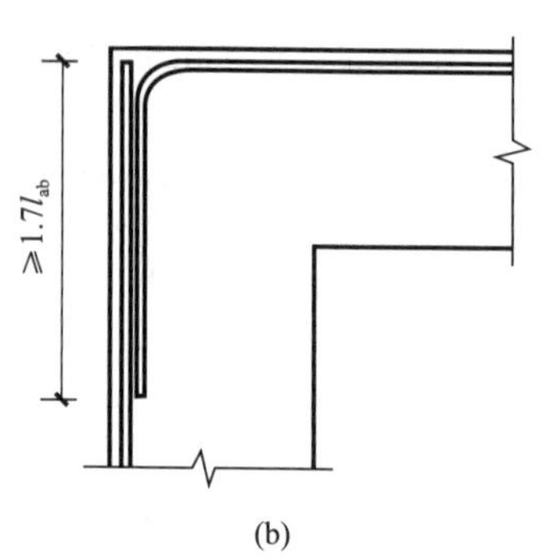

Fig. 3-36 Lap joint structure of column longitudinal bar at top end node

3.7 Foundation of multi-storey frame structure

3.7.1 Selection of foundation

In the design and construction of buildings, the soil and foundation occupy a very important position, which has a great impact on the safe use of buildings and project cost. When choosing the type of foundation, two factors are considered. The first is the nature of the building (including the function, importance, structural type, loads of the building, sensitivity of superstructure to even settlement of foundation). The second is the engineering geological properties and hydrogeological conditions of the soil (including the distribution of rock, subsoil layers, the properties of rock and soil and groundwater, etc.).

The foundation can be divided into the shallow foundation and deep foundation.

1. Shallow foundation

The individual foundation and strip foundation (Fig. 3-37a) are often used in single-storey industrial plants or multi-storey building. The strip foundation is arranged in strip shape, which can be arranged longitudinally or transversally.

When the soil is weak and the load is large, and the strip foundation cannot meet the design requirements, the raft foundation can be used (Fig. 3-37b). The raft foundation has a large foundation, which can greatly reduce the foundation pressure and increase the overall rigidity of the foundation.

The box foundation is often used in multi-storey or high-rise building. The box foundation is a spatial structure composed of reinforced concrete bottom plate, top plate and enough crisscross internal and external walls (Fig. 3-37c). The box foundation is like a huge hollow thick plate, which makes it have much larger spatial stiffness than the raft foundation. It can be used to resist uneven settlement to avoid excessive secondary stress of the superstructure. In addition, the box foundation has good seismic performance, and it can also be used as the basement. However, the use of reinforcement and concrete in the box foundation is large, and the construction technology is complex.

2. Deep foundation

The buried depth of deep foundation is large, and the lower solid soil or rock layer is taken as the bearing stratum. The deep foundation transfers the load to the deep layer of the soil.

Therefore, when the shallow soil cannot meet the requirements of bearing capacity and deformation, and is not suitable for the use of soil treatment measures, it is necessary to consider the use of deep foundation.

Deep foundation includes pile foundation (Fig. 3-37d), pier foundation, diaphragm wall and caisson.

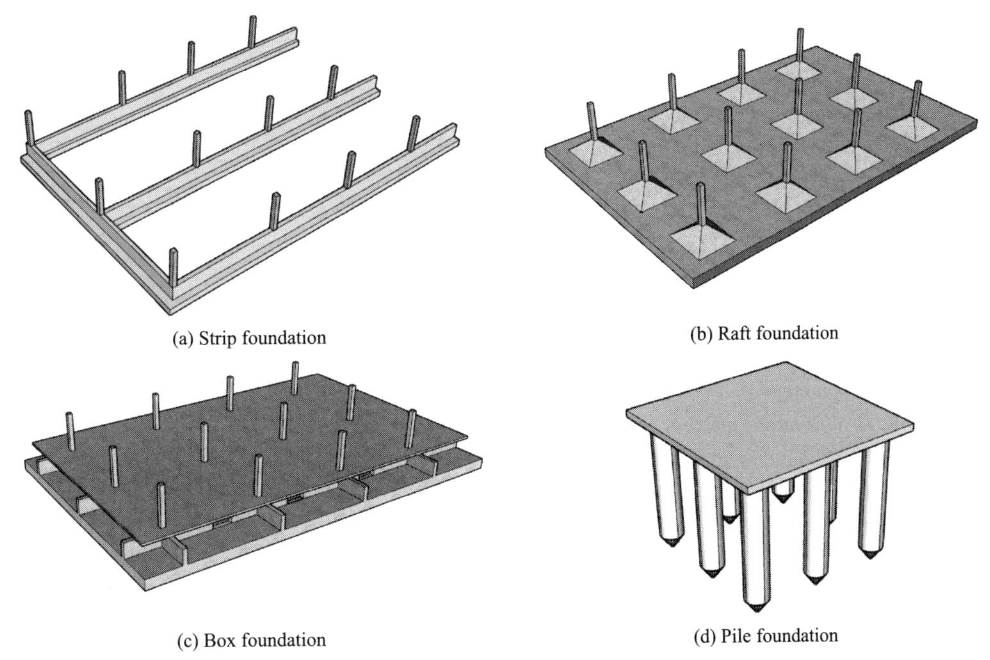

(a) Strip foundation (b) Raft foundation

(c) Box foundation (d) Pile foundation

Fig. 3-37 Types of foundation

The design method of independent foundation under column has been introduced in the last chapter. In this chapter the design of strip foundation will be introduced.

3.7.2 Internal force calculation of strip foundation

The reinforced concrete strip foundation under the column can be regarded as a flexural member in the design. Generally speaking, there are two methods to calculate the internal force of reinforced concrete strip foundation under column, that is simplified calculation method and elastic foundation beam method.

1. Simplified calculation method

This method assumes that the foundation reaction is a linear distribution which is suitable for the calculation of reinforced concrete strip foundation with sufficient relative stiffness. According to the stiffness of the superstructure, the simplified calculation method can be divided into static analysis method (static beam method) and inverted beam method.

1) Static analysis method

Basic assumptions and applicable conditions of static analysis method are as follows.

(1) The stiffness of the upper structure is very small;

(2) The stiffness of the foundation itself is large;

(3) The deformation restraint of the superstructure to the foundation is very small;

(4) And, it still assumed that the foundation reaction is linear distribution.

Firstly the net base reaction force can be calculated according to the assumption of linear distribution.

$$\begin{matrix} p_{max} \\ p_{min} \end{matrix} = \frac{\sum N}{BL} \pm \frac{6\sum M}{BL^2} \quad (3\text{-}37)$$

Where,

$\sum N$ —— the sum of all vertical loads

(kN);

$\sum M$ —— the sum of eccentricity of external load to foundation centroid (kN/m);

B, L —— the width and length of the foundation (m).

At this point all forces on the foundation beam have been determined, and the internal forces of any section can be calculated according to the static equilibrium condition by treating the foundation beam as a statically beam without any excess restraint, as shown in Fig. 3-38. Since the static analysis assumes that the superstructure is flexible, and does not consider the influence of the stiffness of the superstructure, the foundation beam will bend as a whole under the loads. Compared with other methods, the absolute value of bending moment on the unfavorable section of foundation calculated by this method may be larger.

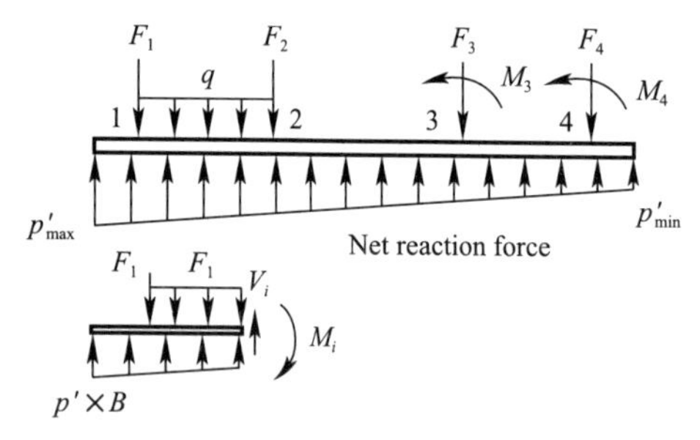

Fig. 3-38 Internal force calculation of strip foundation by static equilibration condition

2) Inverted beam method

Basic assumptions of inverted beam method are as follows.

(1) The foundation beam is absolutely rigid compared with the foundation soil;

(2) The bending deflection of the foundation does not change the foundation pressure, and the foundation pressure is linear distribution;

(3) The superstructure has enough rigidity.

The inverted beam method considers that the superstructure is rigid and there is no settlement difference between the columns, so the columns can be regarded as the hinged supports of the reinforced concrete strip foundation under the columns, and there is no uneven settlement between the supports. In this way, in the calculation, only the local bending between the columns on the strip foundation can be considered, and the influence of the overall bending on the foundation cannot be considered. Calculation diagram is shown in Fig. 3-39.

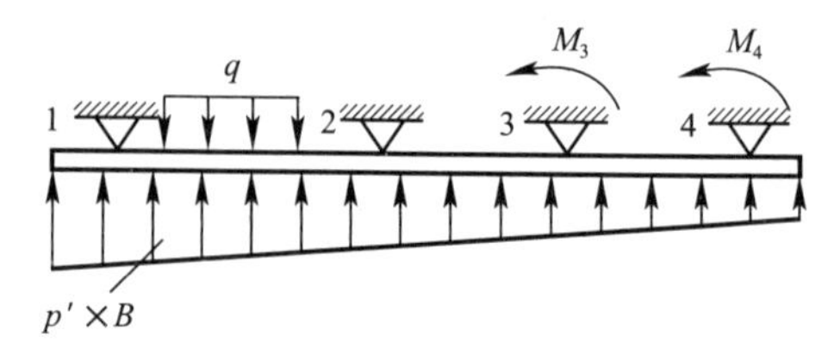

Fig. 3-39 Calculation diagram of inverted beam method

The reaction force of the support obtained by the inverted beam method may not be equal to the vertical load of the column that originally acted on the foundation. At this time, the foundation reaction can be adjusted locally, that is, difference between supper reaction and column axial force is evenly distributed in the 1/3 span range of both sides of the support, and then it is superimposed with the internal force calculated by the inverted beam method. After repeated for 1-2 times, the support reaction is nearly same as the column axial force.

Applicable conditions of the inverted beam method are as follows. The ground is relatively uniform. The stiffness of the superstructure is better, and the load distribution and column spacing distribution are relatively uniform (the difference is not more than 20%). When the height of the strip foundation beam under the column is not less than 1/6 of the column spacing, the base reaction can be in a straight line distribution, and the internal force of the foundation beam can be calculated by the inverted beam method.

According to the loads from the columns, the netbase reaction is calculated according to the assumption of linear distribution. Then, the columns are regarded as the fixed hinge suppers, and the net base reaction is taken as the load on the foundation beam. The internal force of the beam is calculated according to the multi-span continuous beam, and the moment distribution method can be used for the internal force calculation of foundation beam.

2. Elastic foundation beam method

When the reinforced concrete strip foundation under column does not meet the applicable conditions of simplified calculation, the internal force of foundation should be calculated by elastic foundation beam method.

Elastic foundation beam theory is simply assumed that the foundation is elastic body, which is placed on the elastic beam. The foundation and ground are studied as a whole which is separated from the superstructure. The superstructure is only as a load on the foundation. The foundation bottom surface and ground surface are always attached to the platform in the process of deformation under load, that is, they not only meet the static equilibrium conditions, but also meet the deformation coordination conditions. Then using a variety of geometric and physical simplification, the internal force and deformation of foundation and ground are solved by mathematical and mechanical methods. The specific calculation method can be referred to the textbooks.

3.7.3 Construction requirements

In order to make the distribution of base pressure uniform and make the bending moment under each column close to that in the middle of the span, the end of strip foundation should be extended longitudinally to the side columns at both ends, and the extension length should be 1/4 span of the side span. If the load is unsymmetrical, the extension length of the two ends may not be equal, so that the foundation centroid coincides with the action point of the resultant force of load. However, the extension length should not be too large in order to avoid the greater bending moment of the foundation under the column.

Strip foundation under column generally adopts inverted T-section, which is composed of the rib beam and wing plate. In order to resist uneven settlement, the height of rib beam should not be too small, and it should meet the requirements of shear capacity. It generally equals to 1/8—1/4 of column spacing. When the column load is large, the height of rib beam can be increased locally on both sides of the column. Generally, the rib beam is of equal section along the longitudinal direction, and each side of the beam is at least 50 mm wider than the column.

The thickness of wing plate should not be less than 200 mm. When the thickness of the wing plate is 200—250 mm, the equal thickness wing plate should be used. When the thickness of the wing plate is greater than 250 mm, the variable thickness plate should be used, and the gradient should be less than or equal to 1:3.

The reinforcement of rib beam shall be arranged according to bending moment diagram and shear force diagram. The longitudinal reinforcement at the column position is arranged at the bottom of the rib beam. The overlapping position of the bottom longitudinal reinforcement should be arranged in the middle of the span, and that of the top longitudinal reinforcement should be in the column. The overlapping length should meet the requirements of the current code for design of concrete structure. When the diameter of longitudinal stressed reinforcement is greater than 22 mm, non-welded lapping joint is not suitable.

When the web height of the foundation beam is greater than or equal to 450 mm, longitudinal structural reinforcement shall be arranged along the height on both sides of the beam. The area of structural reinforcement on each side shall not be less than 0.1% of the sectional area of web, and the spacing shall not be greater than 200 mm. The longitudinal structural steel bars on both sides of the beam should be connected with the tie bars. The diameter of the tie bars is the same as that of the stirrups, and the spacing is 500—700 mm, which is generally two times of the spacing of the stirrups.

Stirrups should be closed type, its diameter is generally 6—12 mm. When the beam height is greater than 800 mm, the stirrup diameter shall not be less than 8 mm. Stirrup spacing should be determined according to relevant regulations. When the beam width is less than or equal to 350 mm, double-leg stirrups are used. When the beam width is 350—800 mm, four-limb stirrups are used. When the beam width is more than 800 mm, six-limb hoops are used.

The reinforcement of flange plate shall be determined by calculation, but the diameter shall not be less than 10 mm and the spacing shall be 100—200 mm. The diameter of longitudinal distribution reinforcement of non-rib part should be 8—10 mm, and the spacing should not be more than 300 mm. The construction requirements can refer to the relevant provisions of reinforced concrete foundation. The example of the reinforcement diagram of strip foundation is shown in Fig. 3-40.

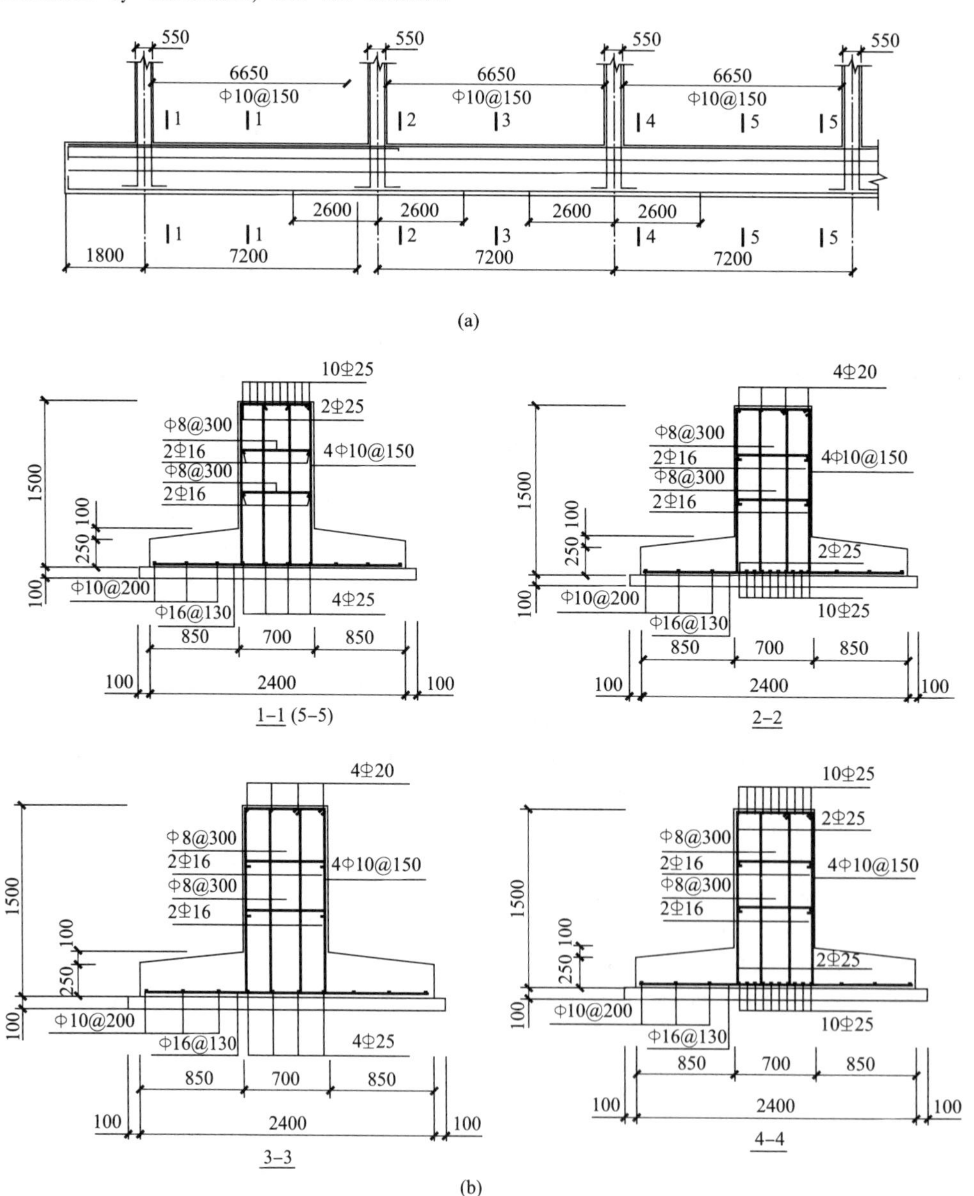

Fig. 3-40 Reinforcement diagram of strip foundation

The concrete strength grade of strip foundation should not be lower than C20.

Frame structural design mainly includes frame structure layout, preliminary determination of beam and column cross-section dimensions, structural calculation diagrams, loads calculation, and then internal force analysis, internal force combination and reinforcement calculation etc. Among them, the approximate analysis of the internal force of the frame under the vertical load generally adopts the multi-layered method or the secondary moment distribution method, and the approximate analysis of the internal force of the frame under the horizontal load generally adopts the inflection point method and the D-value method.

For multi-storey frame structure, the horizontal displacement of the frame structure is caused by the bending deformation of the beams and columns. The structural horizontal displacement caused by the axial deformation of the beams and columns and the section shear deformation is not considered.

The longitudinal reinforcement and stirrups shall not only meet the requirements for the calculation of the bearing capacity of the bending members and the compression members, but also meet the construction requirements of the diameter, spacing, number of reinforcements, anchorage length, lapping length and joints construction. The component connection joint is an important part of the frame structural design.

3.8 Design example of multi-storey frame structure

A three-storey office building is the cast-in-situ reinforced concrete frame structure. The floor plan of the office building is shown in Fig. 3-41. Please make the structure design of the office building (there is no requirement of seismic fortification for this building).

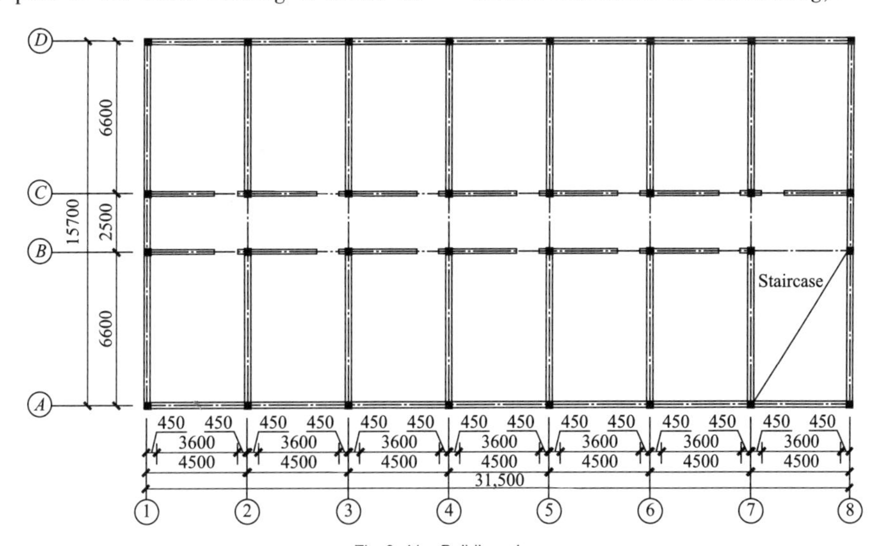

Fig. 3-41 Building plan

3.8.1 Design data

1) Design elevation: the indoor design elevation is ±0.000. The difference between indoor elevation and outdoor elevation is 60 mm.

2) Wall surface: the wall body adopts ordinary machine-made bricks and M5 mixed mor-

tar. The internal painting is mixed mortar bottom, and then paper reinforced mortar surface (20 mm). The external painting is 1∶3 cement mortar bottom (20 mm) and then ceramic brocade brick veneer.

3) Floor slab surface: the top surface of the floor slab is leveled with 20 mm thick cement mortar, 5 mm thick 1∶2 cement mortar and "108" glue coloring powder layer. The bottom surface of the floor slab is 15 mm thick paper reinforcement surface lime plastering and coating twice.

4) Roof slab surface: the cast-in-situ floor slab is paved with expanded perlite insulation layer (the cornice is 100 mm thick, 2% slope from the cornices on both sides to the middle), 20 mm thick 1∶2 cement mortar leveling layer, and then APP modified asphalt waterproof layer.

5) Geological data: it belongs to class Ⅲ construction site, the rest is omitted.

6) Basic wind pressure is 0.55 kN/m^2 (the ground roughness belongs to class B).

7) Live load: roof live load is 1.5 kN/m^2. Office floor live load is 2.0 kN/m^2. Corridor floor live load is 2.5 kN/m^2.

3.8.2 Structural layout and structural calculation diagram

For simplification, planar frame of axis ③ in Fig. 3-42 is taken for calculation as shown in Fig. 3-43.

The cross-section sizes of the beam and the column are determined as follows.

Side span (span *AB* or *CD*) beam:

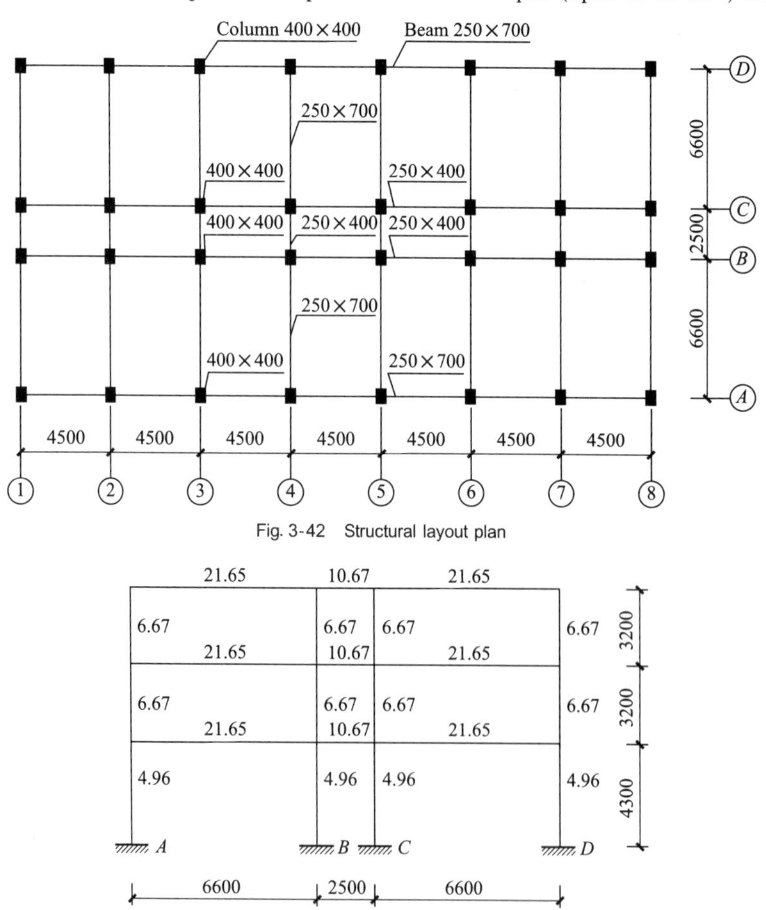

Fig. 3-42 Structural layout plan

Fig. 3-43 Calculation diagram

Note: The data in the Fig. 3-43 is the linear stiffness, unit is $\times 10^{-4} E_c \mathrm{m}^3$.

$$h=\left(\frac{1}{12}l—\frac{1}{8}l\right)=\frac{1}{12}\times6600—\frac{1}{8}\times6600$$
$$=550—825\ \text{mm},\ \text{take}\ h=700\ \text{mm}.$$

$$b=\left(\frac{1}{3}h—\frac{1}{2}h\right)=\left(\frac{1}{3}—\frac{1}{2}\right)\times700$$
$$=233—350\ \text{mm},\ \text{take}\ b=250\ \text{mm}.$$

Middle span (span BC) beam: $h=400$ mm, $b=250$ mm.

Main beam of axis A or axis D:

$$h=\left(\frac{1}{12}l—\frac{1}{8}l\right)=\frac{1}{12}\times4500—\frac{1}{8}\times4500$$
$$=375—563\ \text{mm},\ \text{take}\ h=500\ \text{mm}.$$

$$b=\left(\frac{1}{3}h—\frac{1}{2}h\right)=\left(\frac{1}{3}—\frac{1}{2}\right)\times500$$
$$=167—250\ \text{mm},\ \text{take}\ b=250\ \text{mm}.$$

For the main beam of middle column (axis B, axis C), take $b\times h=250$ mm×500 mm.

The thickness of continuous two-way slab is generally not less than 1/35 of its short span, so the slab thickness is taken as 120 mm.

Loading area of side column:

$$4.5\times3.3=14.85\ \text{m}^2$$

Loading area of middle column:

$$(6.6/2+2.5/2)\times4.5=20.48\ \text{m}^2$$

According to the axial compression ratio, the cross-sectional area of the column:

Side column:

$$A_c\geqslant\frac{N}{\lambda f_c}=\frac{\beta F g_E n}{\lambda f_c}=\frac{1.4\times14.85\times14\times3\times10^3}{0.9\times14.3}$$
$$=67{,}846\ \text{mm}^2$$

Middle column:

$$A_c\geqslant\frac{N}{\lambda f_c}=\frac{\beta F g_E n}{\lambda f_c}=\frac{1.4\times20.48\times14\times3\times10^3}{0.9\times14.3}$$
$$=93{,}568\ \text{mm}^2$$

The column section is taken as square, $b\times h=400$ mm×400 mm.

The structural calculation diagram is shown in Fig. 3-43. According to the geological data, it is determined that the top surface of the foundation is 500 mm away from the outdoor ground, so the height of the first floor is 4.3 m. The linear stiffness of the beams and columns are listed in Fig. 3-43 according to the calculation. When calculating the sectional moment of inertia of beam, considering the effect of cast-in-situ floor slab, take $I=2I_0$ (I_0 is the moment of inertia of beam section without considering the action of slab).

Span AB or CD beam:

$$i=2E_c\times\frac{1}{12}\times0.25\times0.70^3/6.6$$
$$=21.65\times10^{-4}E_c\ \text{m}^3$$

Span BC beam:

$$i=2E_c\times\frac{1}{12}\times0.25\times0.40^3/2.5$$
$$=10.67\times10^{-4}E_c\ \text{m}^3$$

Bottom floor column:

$$i=E_c\times\frac{1}{12}\times0.40\times0.40^3/4.3$$
$$=4.96\times10^{-4}E_c\ \text{m}^3$$

Other columns:

$$i=E_c\times\frac{1}{12}\times0.40\times0.40^3/3.2$$
$$=6.67\times10^{-4}E_c\ \text{m}^3$$

3.8.3 Load calculation

1. Dead load

1) Characteristic values of linear loads on frame beams of roof

20 mm thick 1 : 2 cement mortar leveling layer $\qquad 0.02\times20=0.4\ \text{kN/m}^2$

APP modified asphalt waterproof layer $\qquad 0.3\ \text{kN/m}^2$

100-140 mm thick (2% slope making) expanded perlite $\qquad (0.10+0.14)/2\times7=0.84\ \text{kN/m}^2$

120 mm thick cast-in-situ reinforced concrete floor slab $\qquad 0.12\times25=3.0\ \text{kN/m}^2$

15 mm thick paper reinforced lime plastering bottom $\qquad 0.015\times16=0.24\ \text{kN/m}^2$

Roof dead load $\qquad 4.78\ \text{kN/m}^2$

Dead load of span AB or CD beam

$$0.25\times(0.7-0.1)\times25=3.75\ \text{kN/m}$$

Beam side painting

$$2\times(0.7-0.1)\times0.02\times17=0.41\ \text{kN/m}$$

Dead load of span BC beam

$$0.25\times(0.4-0.1)\times25=1.88\ \text{kN/m}$$

Beam side painting

$$2\times(0.4-0.1)\times0.02\times17=0.2\ \text{kN/m}$$

Therefore, the linear loads acting on the

top frame beams are as follows. (Note: subscript 3 here indicates the third floor, that is, frame beam of roof floor.)

$g_{3AB1} = g_{3CD1} = 3.75 + 0.41 = 4.16$ kN/m

$g_{3BC1} = 1.88 + 0.2 = 2.08$ kN/m

$g_{3AB2} = g_{3CD2} = 4.78 \times 4.5 = 21.51$ kN/m

$g_{3BC2} = 4.78 \times 2.5 = 11.95$ kN/m

2) Characteristic values of linear loads of frame beams of floor

25 mm thick cement mortar surface layer

$0.025 \times 20 = 0.50$ kN/m^2

120 mm thick cast-in-situ reinforced concrete floor slab $0.10 \times 25 = 3.0$ kN/m^2

15 mm thick paper reinforced lime plastering bottom $0.015 \times 16 = 0.24$ kN/m^2

Floor dead load 3.74 kN/m^2

Dead load of span *AB* or *CD* beam and beam side painting 4.16 kN/m

Self-weight of infilled wall

$0.24 \times (3.2 - 0.7) \times 19 = 11.4$ kN/m

Wall painting

$(3.2 - 0.7) \times 0.02 \times 2 \times 17 = 1.7$ kN/m

Dead load of span *BC* beam and beam side painting 2.08 kN/m

Therefore, the linear load acting on the middle floor frame beam is:

$g_{AB1} = g_{CD1} = 4.16 + 11.4 + 4.7 = 17.26$ kN/m

$g_{BC1} = 2.08$ kN/m

$g_{AB2} = g_{CD2} = 3.74 \times 4.5 = 16.83$ kN/m

$g_{BC2} = 3.74 \times 2.5 = 9.35$ kN/m

3) Characteristic values of concentrated loads on frame joints of roof

Dead load of frame beam of axis *A* or axis *D* $0.25 \times 0.5 \times 4.5 \times 25 = 14.06$ kN

Painting

$0.02 \times (0.50 - 0.10) \times 2 \times 4.5 \times 17 = 1.22$ kN

Self-weight of 1m parapet

$1 \times 4.5 \times 0.24 \times 19 = 20.52$ kN

Painting

$1 \times 0.02 \times 2 \times 4.5 \times 17 = 3.06$ kN

Roof self-weight from frame beam

$1/2 \times 4.5 \times 1/2 \times 4.5 \times 4.78 = 24.2$ kN

Concentrated load on roof frame joints of axis *A* or axis *D* $G_{3A} = G_{3D} = 63.06$ kN

Dead weight of frame beam of axis *B* or axis *C* $0.25 \times 0.5 \times 4.5 \times 25 = 14.06$ kN

Painting

$0.02 \times (0.40 - 0.10) \times 2 \times 4.5 \times 17 = 0.92$ kN

Roof self-weight from frame beam

$1/2 \times (4.5 + 4.5 - 2.5) \times 1.25 \times 4.78 = 19.42$ kN

Concentrated load on roof frame joints of axis *B* or axis *C* $G_{3B} = G_{3C} = 58.6$ kN

4) Characteristic values of concentrated loads on frame joints of floor

Dead load of frame beam of axis *A* or axis *D* 14.06 kN

Painting 1.22 kN

Dead load of column

$0.40 \times 0.40 \times 3.2 \times 25 = 12.8$ kN

Painting

$0.40 \times 4 \times 0.02 \times (3.2 - 0.7) \times 17 = 1.36$ kN

Floor self-weight from coupling beam

$$1 \times 4.5 \times \frac{1}{2} \times 4.5 \times 3.74 = 18.93 \text{ kN}$$

Concentrated load on floor frame joints of axis *A* or axis *D* $G_A = G_D = 48.37$ kN

Dead weight of frame beam of axis *B* or axis *C* 14.06 kN

Painting 0.92 kN

Self-weight of inner wall

$4.2 \times (3.2 - 0.5) \times 0.24 \times 19 = 51.71$ kN

Painting

$4.2 \times (3.2 - 0.5) \times 2 \times 0.02 \times 17 = 7.71$ kN

Deduct weight of door opening and door

$-2.1 \times 1.0 \times (19 \times 0.3 - 0.2) = -11.55$ kN

Painting 0.85 kN

Floor self-weight from frame beam

$1/2 \times (4.5 + 4.5 - 2.5) \times 1.25 \times 3.74 = 15.19$ kN

Concentrated load on floor frame joints of axis *B* or axis *C* $G_B = G_C = 79.86$ kN

5) Structural calculation diagram under dead loads

Structural calculation diagram under dead loads is shown as Fig. 3-44.

2. Live load

Structural calculation diagram under live loads is shown as Fig. 3-45.

The load-values in the figure are calculated as follows.

$P_{3AB} = P_{3CD} = 1.5 \times 4.5 = 6.75$ kN/m

$P_{3BC}=1.5\times2.5=3.75\ \text{kN/m}$

$P_{3A}=P_{3D}=\frac{1}{2}\times4.5\times\frac{1}{2}\times4.5\times1.5$
$=7.59\ \text{kN}$

$P_{3B}=P_{3C}=\frac{1}{2}\times(4.5+4.5-2.5)\times1.25\times$
$1.5+\frac{1}{4}\times4.5\times4.5\times1.5=13.69\ \text{kN}$

$P_{AB}=P_{CD}=2.0\times4.5=9.00\ \text{kN/m}$

$P_{BC}=2.5\times2.5=6.25\ \text{kN/m}$

$P_{A}=P_{D}=\frac{1}{4}\times4.5\times4.5\times2.0=10.13\ \text{kN}$

$P_{B}=P_{C}=10.13+\frac{1}{2}\times(4.5+4.5-2.5)\times$
$\frac{2.5}{2}\times2.5=20.29\ \text{kN}$

3. Wind loads

The wind loads are converted into the concentrated loads on the nodes of each floor of the frame. The calculation diagram is shown in Fig. 3-46. The calculation results are shown in Table 3-7. In the table, Z is the height from the frame node to the outdoor ground, and A is the wind-receiving area of node on each floor of the frame.

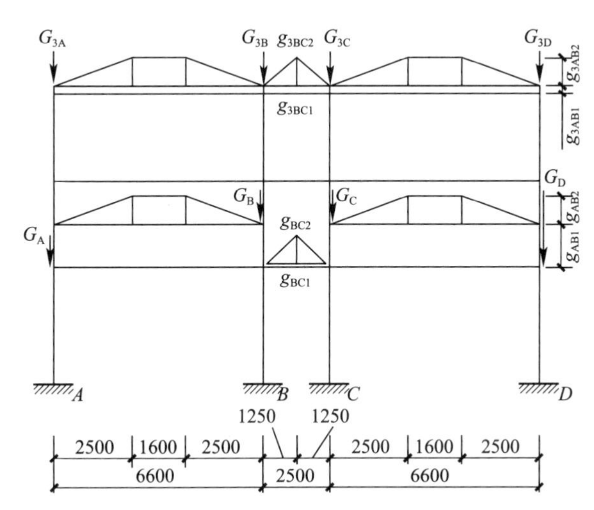

Fig. 3-44 Calculation diagram under dead loads

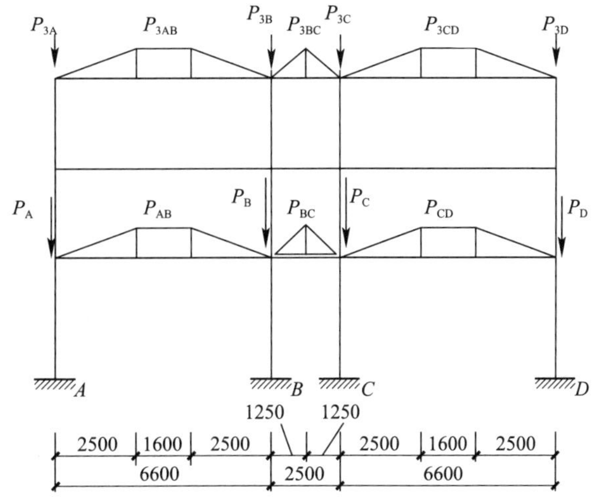

Fig. 3-45 Calculation diagram under live loads

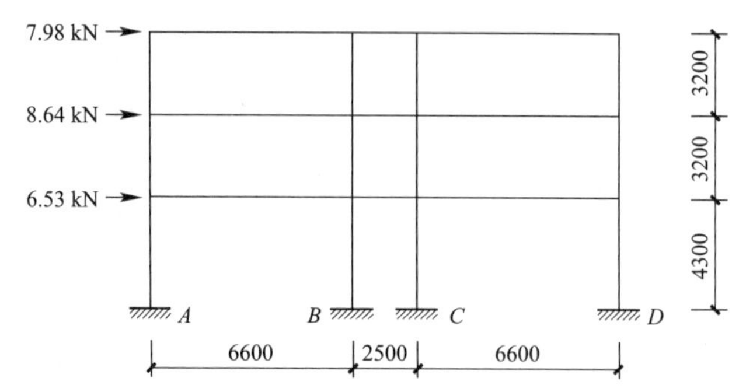

Fig. 3-46 Calculation diagram under wind loads

Wind loads calculation **Table 3-7**

Floor	β_z	μ_s	Z(m)	μ_z	ω_0(kN/m^2)	A(m^2)	P_w(kN)
3	1.0	1.3	10.2	1.006	0.55	11.1	7.98
2	1.0	1.3	7.0	0.880	0.55	13.73	8.64
1	1.0	1.3	3.8	0.608	0.55	15.02	6.53

3.8.4 Internal force calculation

1. Calculation of internal force under dead loads

The internal force under the dead loads (vertical load) is calculated by multi-layered method. Here, the middle layer is taken as an example to illustrate the calculation process of the multi-layered method. The structural calculation diagram of middle layer is shown as an example in Fig. 3-47(a).

In Fig. 3-47(a), the distributed loads on the beam consist of rectangular and trapezoidal distributed loads. The bending moments at the fixed ends can be calculated directly according to the loads shown in the Fig. 3-47(a), or the bending moments at the fixed ends can be calculated according to the principle of equal bending moment of the fixed end. The trapezoidal distributed load and triangular distributed load can be transformed into equivalent uniform load (as shown in Fig. 3-47b). The calculation formula of equivalent uniform load is shown in Fig. 3-48.

Trapezoidal load is transformed into equivalent uniform load.

$$\alpha_1 = 0.5\times\frac{4.5}{6.6} = 0.341$$

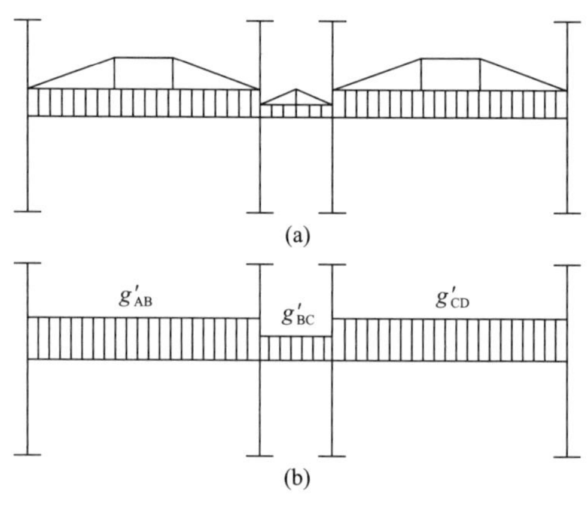

Fig. 3-47 Calculation diagram of multi-layered method

$$g'_{AB} = g_{AB1} + (1-2\alpha_1^2+\alpha_1^3)g_{AB2}$$
$$= 17.26 + (1-2\times0.341^2+0.341^3)\times16.83 = 30.84\ \text{kN/m}$$

$$g'_{BC} = g_{BC1} + \frac{5}{8}g_{BC2}$$
$$= 2.08 + \frac{5}{8}\times9.35$$
$$= 7.92\ \text{kN/m}$$

The internal forces of the structure shown in Fig. 3-44 and Fig. 3-45 can be calculated by the moment distribution method and one-half of the structure can be calculated due to the structural symmetry.

The bending moments at the fixed ends of each beam:

$$M_{AB}=\frac{1}{12}g'_{AB}l_{AB}^2=\frac{1}{12}\times30.84\times6.6^2=111.95\ \text{kN}\cdot\text{m}$$

$$M_{BC}=\frac{1}{3}g'_{BC}l_{BC}^2=\frac{1}{3}\times7.92\times1.25^2=4.13\ \text{kN}\cdot\text{m}$$

$$M_{CB}=\frac{1}{6}g'_{CB}{}^2l_{CB}^2=\frac{1}{6}\times7.92\times1.25^2=2.06\ \text{kN}\cdot\text{m}$$

The calculation process of the moment distribution method of middle layer is shown in Fig. 3-49. The bending moments of the top layer and the bottom layer can also be obtained by the multi-layered method.

The bending moment diagrams of the whole frame structure under dead loads and live loads can be obtained by superimposing the bending moment diagrams of each layer obtained by the multi-layered method, and then the unbalanced bending moments of the nodes are redistributed to obtain the final bending moments of the frame structure. The bending moment diagrams of the frame under dead loads and live loads are shown in Fig. 3-50 and Fig. 3-51.

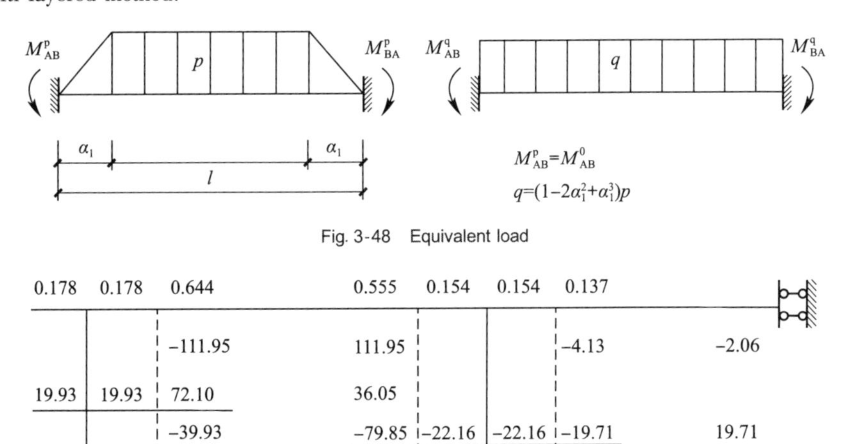

Fig. 3-48 Equivalent load

0.178	0.178	0.644	0.555	0.154	0.154	0.137	
		-111.95	111.95			-4.13	-2.06
19.93	19.93	72.10	36.05				
		-39.93	-79.85	-22.16	-22.16	-19.71	19.71
7.11	7.11	25.71	12.86				
			-7.14	-1.98	-1.98	-1.76	1.76
27.04	27.04	-54.07	73.87	-24.14	-24.14	-25.60	19.41

Fig. 3-49 Calculation process of the moment distribution method

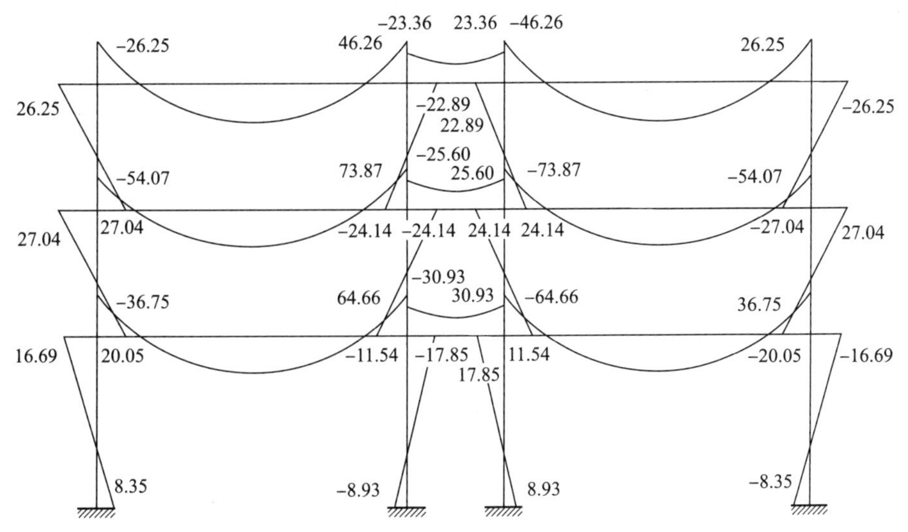

Fig. 3-50 Bending moment of the frame under dead loads (kN · m)

Since the loads and cross-section sizes of beams and columns of span AB and span CD are exactly the same, the internal forces of beams and columns of span CD are symmetrical to those of span AB. The shear forces and axial forces of each beam and each column of the frame can be obtained as shown in Fig. 3-52 to Fig. 3-55.

2. Calculation of internal force under wind loads

The structural calculation diagram under wind loads is shown in Fig. 3- 46. D-value method is adopted for internal force calculation of the frame under the horizontal load. The calculation of the inflection point heights of axis A column and axis B column is shown in Table 3-8 and Table 3-9.

The bending moment calculation of the frame under wind loads is shown in Table 3-10 and Table 3-11. And the bending moment of each beam and each column of the frame is shown in Fig. 3-56.

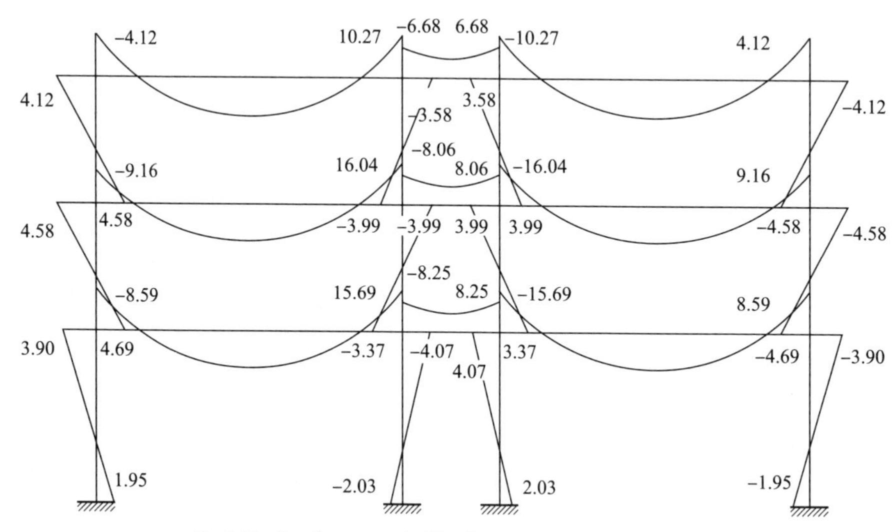

Fig. 3-51 Bending moment of the frame under live loads (kN · m)

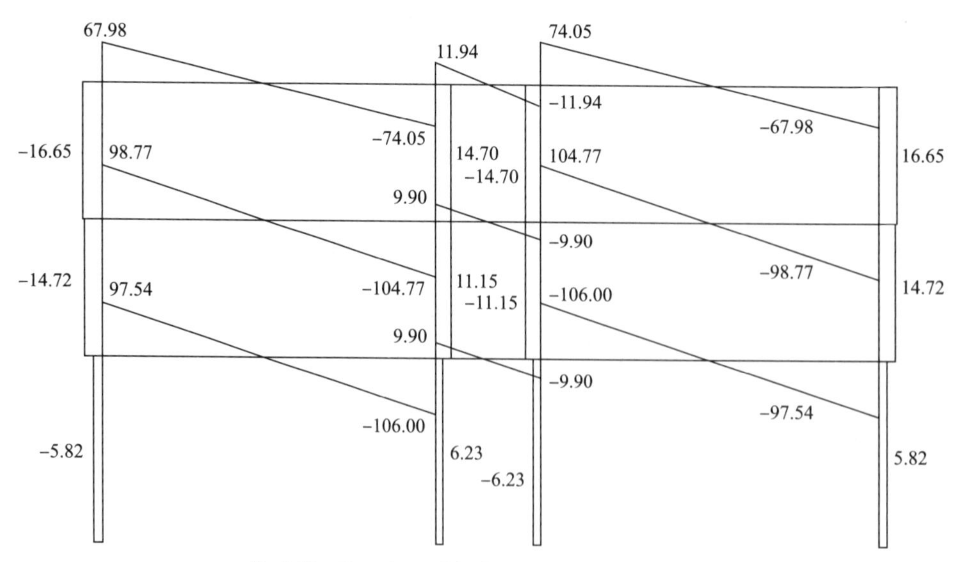

Fig. 3-52 Shear force of the frame under dead loads (kN)

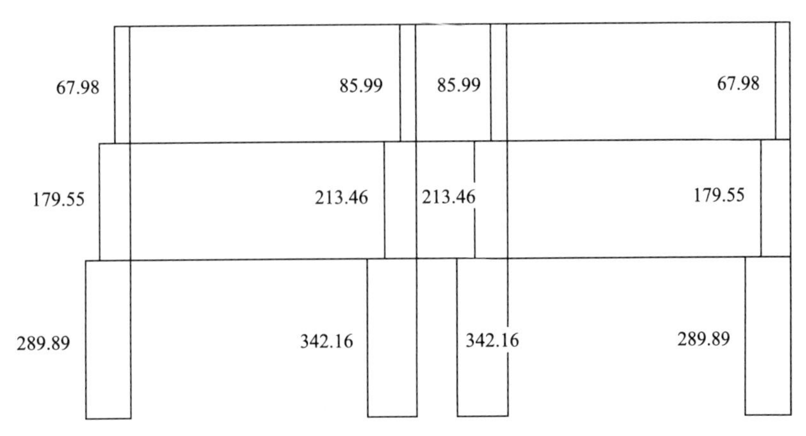

Fig. 3-53 Axial force of the frame under dead loads (kN)

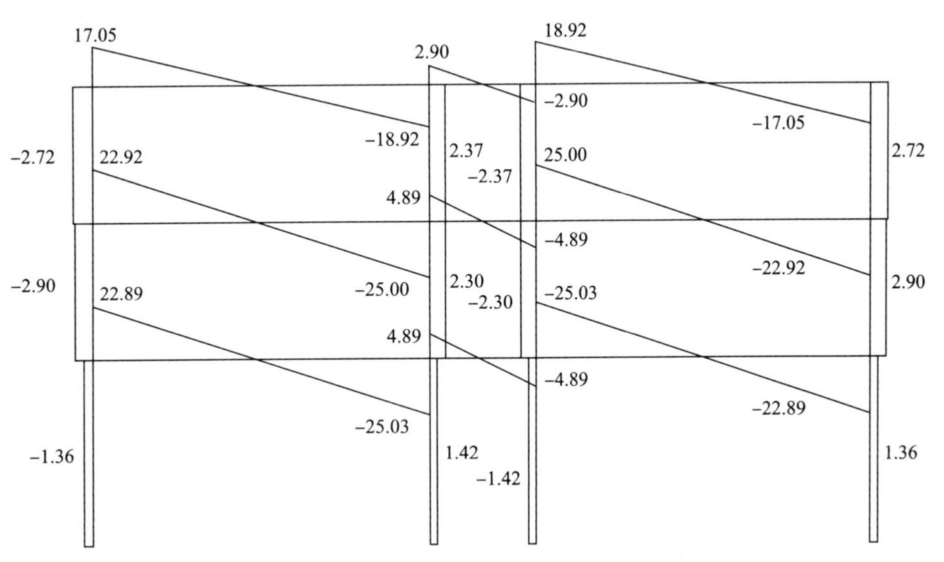

Fig. 3-54 Shear force of the frame under live loads (kN)

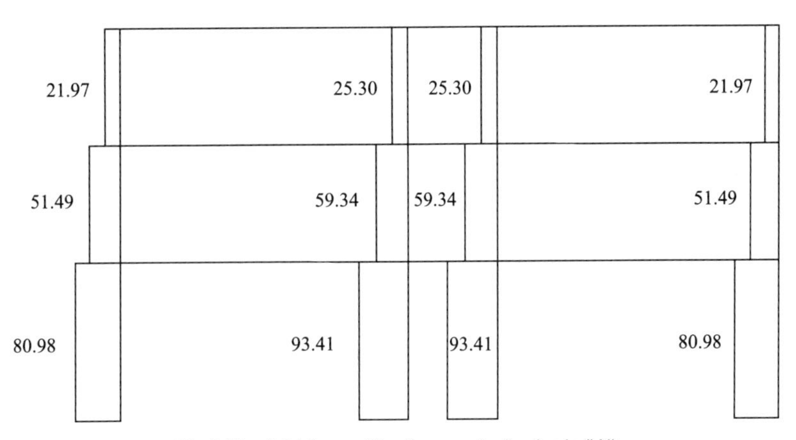

Fig. 3-55 Axial force of the frame under live loads (kN)

Inflection point height of axis *A* column **Table 3-8**

Floor	h(m)	K	y_0	y_1	y_2	y_3	y	yh(m)
3	3.2	1.22	0.36	0	0	0	0.36	1.15
2	3.2	1.22	0.45	0	0	0	0.45	1.44
1	4.3	1.48	0.58	0	0	0	0.58	2.49

Inflection point height of axis *B* column **Table 3-9**

Floor	h(m)	K	y_0	y_1	y_2	y_3	y	yh(m)
3	3.2	1.95	0.40	0	0	0	0.40	1.28
2	3.2	1.95	0.45	0	0	0	0.45	1.44
1	4.3	2.36	0.55	0	0	0	0.55	2.37

Calculation of the bending moment of axis *A* column or axis *D* column **Table 3-10**

Floor	h (m)	V_i (kN)	ΣD_{ij} (kN/m)	Axis A column or axis D column					
				D_i (kN/m)	V_{ij} (kN)	yh (m)	M_{iu}	M_{id}	M_b
3	3.2	7.98	57,796	13,500	1.86	1.15	3 .81	2.14	3.81
2	3.2	16.62	57,796	13,500	3.88	1.44	6.83	5.59	8.97
1	4.3	23.15	32,062	8112	5.86	2.49	10.61	14.59	16.20

Calculation of the bending moment of axis *B* column and axis *C* column

Table 3-11

Floor	h (m)	V_i (kN)	ΣD_{ij} (kN/m)	Axis B column or axis C column						
				D (kN/m)	V_{ij} (kN)	yh (m)	M_{iu}	M_{id}	M_{bl}	M_{br}
3	3.2	7.98	57,796	15,398	2.13	1.28	4.09	2.73	2.74	1.35
2	3.2	16.62	57,796	15,398	4.43	1.44	7.80	6.68	7.05	3.47
1	4.3	23.15	32,062	7919	5.72	2.37	11.04	13.56	11.87	5.85

The shear force calculation of the frame under wind loads is shown in Table 3-12 (Fig. 3-57).

The axial force calculation of the frame under wind loads is shown in Table 3-13 (Fig. 3-58).

The inter-storey displacement under the wind loads can be calculated by the following formula:

$$\Delta u_j = \frac{V_j}{\sum_{k=1}^{m} D'_{jk}} = \frac{V_j}{\sum_{k=1}^{m} \frac{12 i_{jk}}{h_j^2}} \qquad (3\text{-}38)$$

The displacement calculation is shown in Table 3-14.

It can be seen from the Table 3-14 that the maximum inter-storey displacement angle of the frame under wind loads is far less than 1/550 which meets the specification requirements.

3. Bending moment modulation of frame beam under vertical loads

The bending moment modulations of frame beam under dead loads and live loads are shown in Table 3-15 and Table 3-16, respectively.

4. Calculation of internal force at the support edge of frame beam

The internal force at the support edge of frame beam can be calculated according to section 3. 5. 1. The shear force and bending moment of control sections at beam ends are shown in Table 3-17—Table 3-21, respectively.

$$V_{\mathrm{b}}^{l}=V_{\mathrm{c}}^{l}-\frac{ql}{2},\quad V_{\mathrm{b}}^{\mathrm{r}}=V_{\mathrm{c}}^{\mathrm{r}}-\frac{ql}{2} \tag{3-39}$$

$$M_{\mathrm{b}}^{l}=M_{\mathrm{c}}^{l}-V_{\mathrm{c}}^{l}l/2$$

5. Calculation of bending moment and displacement considering *P*-*Δ* effect

The calculations of increasing coefficient of bending moment and displacement are shown in Table 3-22 and Table 3-23, respectively.

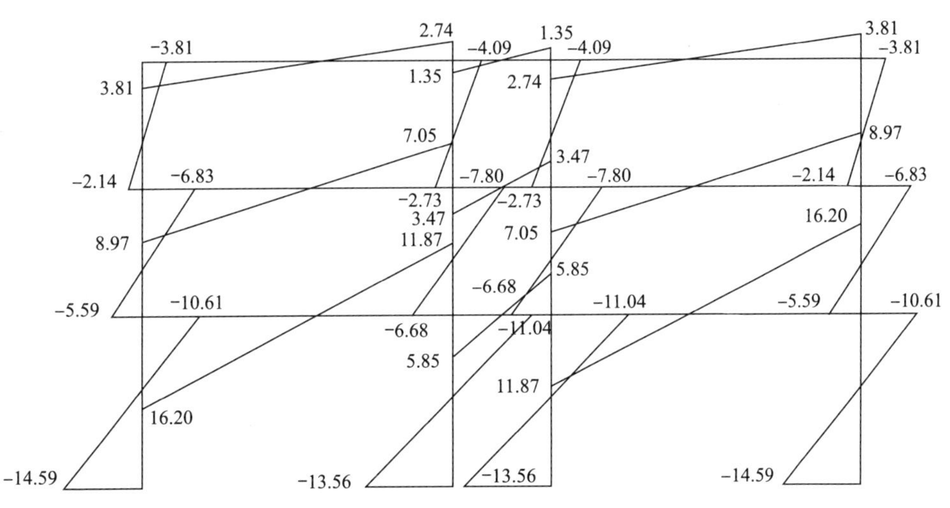

Fig. 3-56 Bending moment of the frame under wind loads (kN · m)

Calculation of shear force of frame under wind loads **Table 3-12**

Floor	Span *AB*	Span *BC*	Span *CD*	Axis *A*	Axis *B*	Axis *C*	Axis *D*
	V(kN)	*V*(kN)	*V*(kN)	*V*(kN)	*V*(kN)	*V*(kN)	*V*(kN)
3	-1.00	-1.05	-1.00	1.91	2.08	2.08	1.91
2	-2.40	-2.72	-2.40	3.97	4.34	4.34	3.97
1	-4.23	-4.57	-4.23	5.86	5.72	5.72	5.86

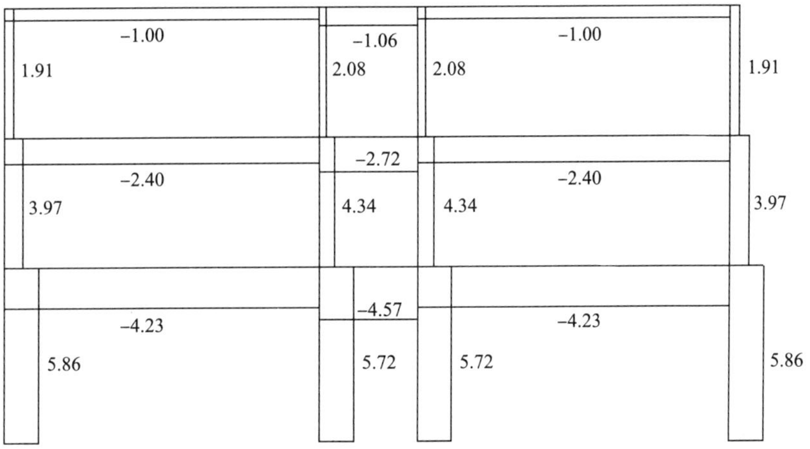

Fig. 3-57 Shear force of the frame under wind loads (kN · m)

Calculation of axial force of frame under wind loads **Table 3-13**

Floor	Axis A column	Axis B column	Axis C column	Axis D column
3	1.00	0.06	0.06	1.00
2	3.40	0.38	0.38	3.40
1	7.63	0.72	0.72	7.63

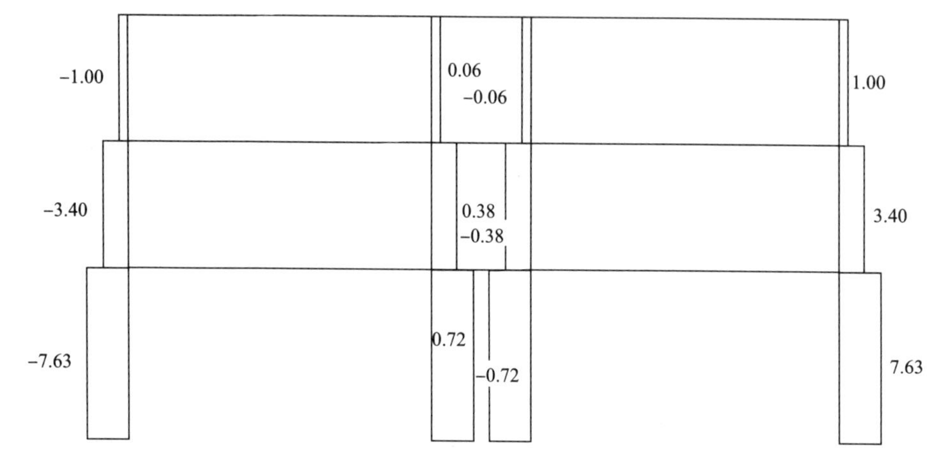

Fig. 3-58 Axial force of the frame under wind loads (kN)

Displacement calculation under wind loads **Table 3-14**

Floor	F_r(kN)	V_i(kN)	ΣD_{ij}(kN/m)	Δu(m)	$\Sigma \Delta u_i$(m)	$\Delta u_i/h_i$
3	7.98	7.98	57,796	0.00014	0.00115	0.00004
2	8.64	16.62	57,796	0.00029	0.00101	0.00009
1	6.53	23.15	32,062	0.00072	0.00072	0.00017

Bending moment modulation of frame beam under dead loads (kN · m) **Table 3-15**

Me						
Floor	*A-B*		*B-C*		*C-D*	
	L	R	L	R	L	R
3	-26.25	46.26	-23.36	23.36	-46.26	26.25
2	-54.07	73.87	-25.60	25.60	-73.87	54.07
1	-36.75	64.66	-30.13	30.13	-64.66	36.75
$Ma=0.8Me$						
Floor	*A-B*		*B-C*		*C-D*	
	L	R	L	R	L	R
3	-21.00	37.01	-18.69	18.69	-37.01	21.00
2	-43.26	59.10	-20.48	20.48	-59.10	43.26
1	-29.40	51.73	-24.10	24.10	-51.73	29.40

Continued

Floor	Mid-span bending moment					
	A-B		*B-C*		*C-D*	
	Me	*Ma*	*Me*	*Ma*	*Me*	*Ma*
3	82.32	87.78	-17.42	-13.12	82.32	87.78
2	115.65	126.13	-24.59	-18.56	115.65	126.13
1	117.21	127.36	-25.35	-18.42	117.21	127.36

Moment modulation of frame beam under live loads (kN · m) **Table 3-16**

Floor	*Me*					
	A-B		*B-C*		*C-D*	
	L	R	L	R	L	R
3	-4.12	10.27	-6.68	6.68	-10.27	4.12
2	-9.16	16.04	-8.06	8.06	-16.04	9.16
1	-8.59	15.69	-8.25	8.25	-15.69	8.59
	$Ma=0.8Me$					
Floor	*A-B*		*B-C*		*C-D*	
	L	R	L	R	L	R
3	-3.30	8.22	-5.34	5.34	-8.22	3.30
2	-7.33	12.83	-6.45	6.45	-12.83	7.33
1	-6.87	12.55	-6.60	6.60	-12.55	6.87
	Mid-span bending moment					
Floor	*A-B*		*B-C*		*C-D*	
	Me	*Ma*	*Me*	*Ma*	*Me*	*Ma*
3	22.47	23.92	-4.88	-3.51	22.47	23.92
2	26.95	29.34	-5.00	-3.39	26.95	29.34
1	27.42	29.80	-5.19	-3.54	27.42	29.80

The calculations of bending moment and displacement under wind loads considering P-Δ effect are shown in Table 3-24 and Table 3-25, respectively.

The maximum inter-storey displacement angle of the frame under wind load is far less than 1/550 which meets the specification requirements.

3.8.5 Internal force combination

Internal force combination Ⅰ: $S_d = 1.3S_{Gk} + 1.5S_{wk} + 1.5\times0.7\times S_{Qk}$

Internal force combination Ⅱ: $S_d = 1.3S_{Gk} + 1.5S_{wk} + 1.5\times0.6\times S_{Qk}$

The internal force combinations of beam and column are shown in Table 3-26—Table 3-28, respectively.

Shear force of control sections at beam ends under dead loads (kN) **Table 3-17**

Floor	Span *AB*				Span *BC*				Span *CD*			
	V_{cl}	V_{bl}	V_{cr}	V_{br}	V_{cl}	V_{bl}	V_{cr}	V_{br}	V_{cl}	V_{bl}	V_{cr}	V_{br}
3	67.98	66.64	-74.05	-72.71	11.94	11.34	-11.94	-11.34	74.05	72.71	-67.98	-66.64
2	98.77	96.84	-104.77	-102.84	9.90	9.41	-9.90	-9.41	104.8	102.84	-98.77	-96.84
1	97.54	95.61	-106.00	-104.07	9.90	9.41	-9.90	-9.41	106.0	104.1	-97.54	-95.61

Bending moment of control sections at beam ends under dead loads (kN · m) **Table 3-18**

Floor	Span *AB*				Span *BC*				Span *CD*			
	M_{cl}	M_{bl}	M_{cr}	M_{br}	M_{cl}	M_{bl}	M_{cr}	M_{br}	M_{cl}	M_{bl}	M_{cr}	M_{br}
3	-21.00	-16.75	37.01	32.38	-18.69	-17.94	18.69	17.94	-37.01	-32.38	21.00	16.75
2	-43.26	-37.09	59.10	52.55	-20.48	-19.86	20.48	19.86	-59.10	-52.55	43.26	37.09
1	-29.40	-23.30	51.73	45.11	-24.10	-23.48	24.10	23.48	-51.73	-45.11	29.40	23.30

Shear force of control sections at beam ends under live loads (kN) **Table 3-19**

Floor	Span *AB*				Span *BC*				Span *CD*			
	V_{cl}	V_{bl}	V_{cr}	V_{br}	V_{cl}	V_{bl}	V_{cr}	V_{br}	V_{cl}	V_{bl}	V_{cr}	V_{br}
3	21.97	21.56	-23.83	-23.42	1.47	1.29	-1.47	-1.29	23.83	23.42	-21.97	-21.56
2	29.52	28.98	-31.60	-31.06	2.44	2.15	-2.44	-2.15	31.60	31.06	-29.52	-28.98
1	29.49	28.95	-31.63	-31.09	2.44	2.15	-2.44	-2.15	31.63	31.09	-29.49	-28.95

Bending moment of control sections at beam ends under live loads (kN · m) **Table 3-20**

Floor	Span *AB*				Span *BC*				Span *CD*			
	M_{cl}	M_{bl}	M_{cr}	M_{br}	M_{cl}	M_{bl}	M_{cr}	M_{br}	M_{cl}	M_{bl}	M_{cr}	M_{br}
3	-3.30	-1.65	8.22	6.43	-5.34	-5.23	5.34	5.23	-8.22	-6.43	3.30	1.65

Continued

Floor	Span AB				Span BC				Span CD			
	M_{cl}	M_{bl}	M_{cr}	M_{br}	M_{cl}	M_{bl}	M_{cr}	M_{br}	M_{cl}	M_{bl}	M_{cr}	M_{br}
2	-7.33	-5.12	12.83	10.46	-6.45	-6.27	6.45	6.27	-12.8	-10.46	7.33	5.12
1	-6.87	-4.66	12.55	10.18	-6.60	-6.42	6.60	6.42	-12.6	-10.18	6.87	4.66

Bending moment of control sections at beam ends under wind loads (kN · m) **Table 3-21**

Floor	Span AB				Span BC				Span CD			
	M_{cl}	M_{bl}	M_{cr}	M_{br}	M_{cl}	M_{bl}	M_{cr}	M_{br}	M_{cl}	M_{bl}	M_{cr}	M_{br}
3	3.91	3.84	2.67	2.60	1.32	1.24	1.32	1.24	2.67	2.60	3.91	3.84
2	9.18	9.00	6.68	6.50	3.40	3.20	3.40	3.20	6.68	6.50	9.18	9.00
1	16.33	16.01	11.58	11.26	5.71	5.37	5.71	5.37	11.58	11.26	16.33	16.01

Calculation of increasing coefficient of bending moment **Table 3-22**

Floor	Column	Reduced linear stiffness ($\times 10^4$ kN · m)		K	a_c	D_h(kN)	ΣD_h(kN)
2nd or 3rd floor	Side column	$i_c=1.20$	$i_b=2.60$	2.16	0.52	23,400	102,600
	Middle column	$i_c=1.20$	$i_b=1.28$	3.23	0.62	27,900	
1st floor	Side column	$i_c=0.89$	$i_b=2.60$	2.92	0.70	17,386	72,524
	Middle column	$i_c=0.89$	$i_b=1.28$	4.36	0.76	18,876	

Floor	ΣD_h(kN)	N_k of column (kN)		ΣN_k(kN)	η_s(column)	H_s(beam)
		Dead load	Live load			
3	102,600	287.64	94.54	382.18	1.0037	1.0037
2	102,600	768.82	221.66	990.48	1.0097	1.0097
1	72,524	1250.00	348.78	1598.78	1.0225	1.0225

Calculation of increasing coefficient of displacement **Table 3-23**

Floor	Column	K	a_c	D_h(kN)	ΣD_h(kN)
2nd or 3rd floor	Side column	3.61	0.64	43,200	184,948
	Middle column	5.39	0.73	49,274	
1st floor	Side column	4.36	0.76	34,882	137,868
	Middle column	6.52	0.82	34,052	

Floor	ΣD_h(kN)	N_k of column (kN)		ΣN_k(kN)	η_s
		Dead load	Live load		
3	184,948	287.64	94.54	382.18	1.0021
2	184,948	768.82	221.66	990.48	1.0054
1	137,868	1250.00	348.78	1598.78	1.0117

Bending moment under wind loads considering P-Δ effect (kN · m) **Table 3-24**

Adjustment	Floor	Axis A			Axis B			
		M_{cu}	M_{cd}	M_{br}	M_{cu}	M_{cd}	M_{bl}	M_{br}
Before	3	3.81	2.14	3.81	4.09	2.73	2.74	1.35
After	3	3.82	2.15	3.82	4.10	2.74	2.75	1.35
Before	2	6.83	5.59	8.97	7.80	6.68	7.05	3.47
After	2	6.90	5.64	9.06	7.87	6.74	7.12	3.50
Before	1	10.61	14.59	16.20	11.04	13.56	11.87	5.85
After	1	10.85	15.92	16.56	11.29	13.87	12.14	5.98

Calculation of displacement under wind loads considering P-Δ effect (m)

Table 3-25

Floor	Δ_{ui}	Δ_{ui} after adjustment	$\Sigma\Delta_{ui}$
3	0.00014	0.00014	0.00116

Continued

Floor	Δ_{ui}	Δ_{ui} after adjustment	$\Sigma\Delta_{ui}$
2	0.00029	0.00029	0.00102
1	0.00072	0.00073	0.00073

Internal force combination of beam **Table 3-26**

Location		M or N	S_{Gk}	S_{Qk}	S_{wk}	Combination I		Combination II	
						→	←	→	←
3rd floor	A	M	-16.75	-1.65	3.93	-17.61	-29.40	-17.37	-29.16
		N	66.64	21.56	-1.00	107.77	110.77	104.54	107.54
	B-left	M	32.38	6.43	-2.75	44.72	52.97	43.76	52.01
		N	-72.71	-23.42	-1.00	-120.61	-117.6	-117.1	-114.1
	B-right	M	-17.94	-5.23	1.35	-26.79	-30.84	-26.00	-30.05
		N	11.34	1.29	-1.05	14.52	17.67	14.33	17.48
	Mid-span	M_{AB}	87.78	-23.92	0.72	90.08	87.92	93.67	91.51
		M_{BC}	-13.12	3.51	0	-13.37	-13.37	-13.90	-13.90
2nd floor	A	M	-37.09	-5.12	9.30	-39.64	-67.54	-38.88	-66.78
		N	96.84	28.98	-2.40	152.72	159.92	148.37	155.57
	B-left	M	52.55	10.46	-7.12	68.62	89.98	67.05	88.41
		N	-102.84	-31.06	-2.40	-169.91	-162.7	-165.2	-158.0
	B-right	M	-19.86	-6.27	3.50	-27.15	-37.65	-26.21	-36.71
		N	9.41	2.15	-2.72	10.41	18.57	10.09	18.25
	Mid-span	M_{AB}	126.13	-29.34	1.07	134.77	131.56	139.17	135.96
		M_{BC}	-18.56	3.39	0	-20.57	-20.57	-21.08	-21.08
1st floor	A	M	-23.30	-4.66	16.84	-9.92	-60.44	-9.22	-59.74
		N	95.61	28.95	-4.23	148.35	161.04	144.00	156.69
	B-left	M	45.11	10.18	-12.14	51.12	87.54	49.60	86.02
		N	-104.07	-31.09	-4.23	-174.28	-161.5	-169.2	-156.9
	B-right	M	-23.48	-6.42	5.98	-28.30	-46.24	-27.33	-45.27
		N	9.41	2.15	-4.57	7.64	21.35	7.31	21.02
	Mid-span	M_{AB}	127.36	-29.80	2.44	137.94	130.62	142.41	135.09
		M_{BC}	-18.42	3.54	0	-20.23	-20.23	-20.76	-20.76

Note:

"→" means under right wind load; "←" means under left wind load. The unit of M is "kN · m", The unit of N is "kN".

Internal force combination of column at axis *A* **Table 3-27**

Location		*M* or *N*	S_{Gk}	S_{Qk}	S_{wk}	Combination Ⅰ		Combination Ⅱ	
						→	←	→	←
3rd floor	Column top	*M*	26.25	4.12	-3.82	32.72	44.18	32.10	43.56
		N	67.98	21.97	-1.91	108.58	114.31	105.28	111.01
	Column bottom	*M*	-27.04	-4.58	2.15	-36.74	-43.19	-36.05	-42.50
		N	80.78	21.97	-1.91	125.22	130.95	121.92	127.65
2nd floor	Column top	*M*	27.04	4.58	-6.90	29.61	50.31	28.92	49.62
		N	179.55	51.49	-3.97	281.52	293.43	273.80	285.71
	Column bottom	*M*	-20.05	-4.69	5.64	-22.53	-39.45	-21.83	-38.75
		N	192.35	51.49	-3.97	298.16	310.07	290.44	302.35
1st floor	Column top	*M*	16.69	3.90	-10.85	9.52	42.07	8.93	41.48
		N	289.89	80.98	-5.86	453.10	470.68	440.95	458.53
	Column bottom	*M*	-8.35	-1.95	15.92	10.98	-36.78	11.27	-36.49
		N	302.69	80.98	-5.86	469.74	487.32	457.59	475.17

Note:

"→" means under right wind load; "←" means under left wind load. The unit of *M* is "kN · m", The unit of *N* is "kN".

Internal force combination of column at axis *B* **Table 3-28**

Location		*M* or *N*	S_{Gk}	S_{Qk}	S_{wk}	Combination Ⅰ		Combination Ⅱ	
						→	←	→	←
3rd floor	Column top	*M*	-22.89	-3.58	-4.10	-39.67	-27.37	-39.13	-26.83
		N	85.99	25.3	-2.08	135.23	141.47	131.44	137.68
	Column bottom	*M*	24.14	3.99	2.74	39.68	31.46	39.08	30.86
		N	98.79	25.3	-2.08	151.87	158.11	148.08	154.32
2nd floor	Column top	*M*	-24.14	-3.99	-7.87	-47.38	-23.77	-46.78	-23.17
		N	213.46	59.34	-4.34	333.30	346.32	324.39	337.41
	Column bottom	*M*	11.54	3.37	6.74	28.65	8.43	28.15	7.93
		N	226.26	59.34	-4.34	349.94	362.96	341.03	354.05
1st floor	Column top	*M*	-17.85	-4.07	-11.29	-44.41	-10.54	-43.80	-9.93
		N	342.16	93.41	-5.72	534.31	551.47	520.30	537.46
	Column bottom	*M*	8.93	2.03	13.87	34.55	-7.06	34.24	-7.37
		N	354.96	93.41	-5.72	550.95	568.11	536.94	554.10

Note:

"→" means under right wind load; "←" means under left wind load. The unit of *M* is "kN · m", The unit of *N* is"kN".

3.8.6 Reinforcement design of members

According to the internal force combination results, the most unfavorable internal force of each cross-section can be selected for section reinforcement calculation. It must be pointed out that the internal force obtained from the internal force combination maybe not the most unfavorable internal force. For example, in the case of large eccentric compression, it may be more dangerous when N is small but not the minimum and M is larger. Therefore, if it is necessary, the most unfavorable group of internal forces should be recombined and calculated according to the eccentric compression of the column section for reinforcement. Finally, the reinforcement of the control sections can be determined considering the structural requirements, and then the structural construction drawings are made. The above concrete contents are omitted here.

Exercises

3.1 How to choose the cross-section size of beam and column of frame structure?

3.2 How to determine the calculation diagrams of frame structure under vertical loads and horizontal loads?

3.3 What are the basic assumptions of multi-layered method?

3.4 Please describe the calculation procedure of multi-layered method to calculate the internal force of frame structure under vertical loads.

3.5 Why are the linear stiffness and moment transfer coefficients of the columns (except the bottom layer column) decreased into 0.9 and 1/3 in multi-layered method?

3.6 What are the characteristics of the shear force and bending moment of frame structure under horizontal loads?

3.7 Please list the calculation procedure of inflection method to calculate the internal force of frame structure under horizontal loads.

3.8 How to calculate the lateral stiffness of column in the D-value method?

3.9 Please list the calculation procedure of D-value method to calculate the internal force and lateral displacement of frame structure under horizontal loads.

3.10 How to consider second-order effect for the frame structure?

3.11 What is the allowable value of elastic inter-storey displacement angle? Why is the displacement needed to satisfy $[\theta_e]$?

3.12 How to determine the control sections of beam and column of frame structures?

3.13 Why is it necessary to adjust the bending moment of beam end under the vertical load? How to adjust?

3.14 Please calculate the internal forces of the frame structure (Fig. 3-59) under vertical loads and draw the internal forces(M, V, N).

3.15 Please calculate the internal force and lateral displacement of the frame structure(Fig. 3-60) under horizontal loads and draw the internal force (M, V, N). The linear stiffness of beams and columns are list in the figure $i = 2{,}600$ kN · m.

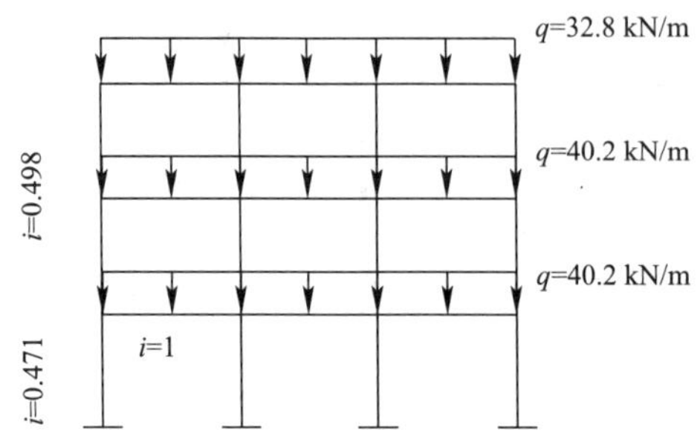

Fig. 3-59 Calculation diagram of exercise 3. 14

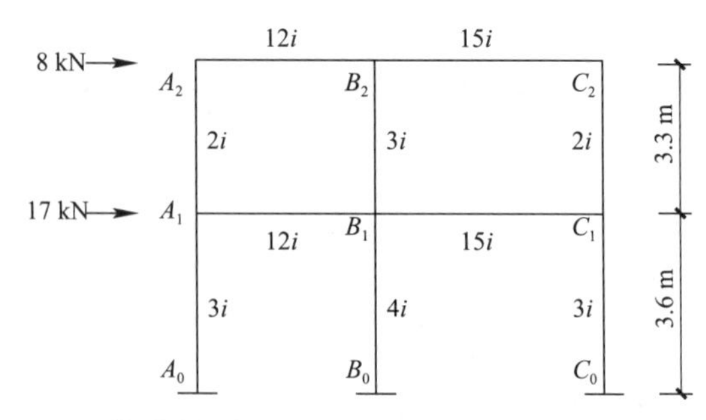

Fig. 3-60 Calculation diagram of exercise 3. 15

Chapter 4
Masonry Structure

Prologue

Main points

1. Overview masonry materials, including brick and mortar.
2. Resistance calculation of masonry walls and columns under compressive load.
3. Design method for walls and columns of masonry building.
4. Introduction of the lintel, ring beam and structural column.

Learning requirements

1. Know the mechanical characteristics of the brick and mortar, and the basic principles of their arrangement and selection.
2. Understand the mechanical mechanism of masonry under pressure and tension.
3. Master the calculation method for resistance of masonry members under global and concentrated compressive load.
4. Master the design method for masonry building, including stability checking and resistance checking.
5. Know the significance and design of the lintel, ring beam and structural column.

4.1 Introduction

Masonry is the assemblage of masonry units (i.e., brick, concrete block, stone, etc.) laid in a specified pattern and joined together with mortar. Masonry can be either reinforced or unreinforced, but this chapter is limited to unreinforced brick masonry. The design and construction of unreinforced brick masonry is governed by ***Code for Design of Masonry Structures*** (GB 50003—2011). The latest version of GB 50003 was issued in 2011.

4.1.1 Historical background

The first masonry material to be used was probably stone. In the ancient Near East, evolution of housing was from huts, to apsidal houses (Fig. 4-1a), and finally to rectangular houses (Fig. 4-1b). The earliest examples of the first permanent houses can be found near Lake Hullen, Israel (9000 BC—8000 BC), where dry-stone huts, circular and semisubterranean, were found. In addition, the Egyptian Pyramids (2800 BC—2000 BC), the Colosseum in Rome (70 AD—90 AD), the Great Wall of China (700 BC—1600 AD) and the Taj Mahal (1630 AD—1650 AD) are some of the most significant masonry structures, scattered around the world.

The Great Wall of China (万里长城) is a series of fortifications that were built across the historical northern borders of ancient Chinese States and Imperial China as protection against various nomadic groups from the Eurasian Steppe, as shown in Fig. 4-2. Several walls were built from as early as the 7th century BC, with selective stretches later joined together by Qin Shi Huang (220 BC—206 BC), the first emperor of China. Later on, many successive dynasties built and maintained multiple stretches of border walls. Collectively, they stretch from Liaodong in the east to Lop Lake in the west, from the present-day Sino-Russian border in the north to Tao River (Taohe) in the south; along an arc that roughly delineates the edge of the Mongolian steppe; spanning over 20,000 km in total. Today, the defensive system of the Great Wall is generally recognized as one of the most impressive architectural feats in history. The Great Wall is mainly composed of tens of thousands of stones. This great project also shows the wisdom and industrious spirit of the Chinese people.

(a) Beehive houses from a village in Cyprus(5650 BC)

(b) Rectangular dwellings from a village in Iraq

Fig. 4-1 Examples of prehistoric architecture of masonry in the ancient Near East

Fig. 4-2 The Great Wall of China

With the industrial revolution, brick production developed, and by the mid-nineteenth century most of the buildings were constructed of ceramic brick masonry. A clear example of this is the Monadnock Building in Chicago, United States of America, as shown in Fig. 4-3, which with 16 floors has walls that reach 1.82 m thick. With the constant evolution of construction techniques, building codes, automatic calculation programs and experimental testing, masonry structures tend to exhibit better behaviors, allowing the construction of walls with a thinner base, enabling its construction in areas of greater seismicity. In Europe, the building solutions using unreinforced structural masonry represent about 15% to more than 50% of the new housing construction, taking as reference countries with low seismicity (e.g., Germany, Netherlands, or Norway) but also countries with high seismicity (e.g., Italy). In China, the masonry structure became an attractive and efficient solution from a perspective of cost-benefit analysis for buildings in regions of low to high seismicity, e.g., hotels, houses (Fig. 4-4), apartments, office buildings, schools, commercial buildings or warehouses. At present, the output of masonry units (i.e., bricks and blocks) is more than that of other countries in the world combined.

4.1.2 Advantages and disadvantages

The basic advantage of masonry construction is that it is possible to use the same element to perform a variety of functions. Specifically, masonry may simultaneously provide structure, subdivision of space, thermal and acoustic insulation as well as fire and weather protection. As a material, it is relatively cheap but durable and produces external wall with very acceptable appearance. Masonry construction is flexible in terms of building layout and can be constructed without very large capital expenditure on the part of the builder. In most cases, the construction of masonry structure can be finished by workers and without expensive and complex equipment, which is very important for economically underdeveloped countries with backward industries.

As the old saying goes, every coin has two sides. Masonry structure also has its disadvantages. Masonry construction requires an excellent amount of time. The design of the structure is limited by the type of masonry. Particularly, a large amount of manpower could also be essential. Due to heavy weight, masonry structure requires a large foundation. Also, cracks and settlement may occur. The integrity of masonry structure is poor, which results in serious damage under the earthquake. In addition, masonry has low strength, especially low tensile, shear and bending strength. Table 4-1 lists the comparison of brick masonry with concrete and steel. It can be seen that the strength-weight ratio and tension-compression strength ratio of brick masonry are far less than that of concrete and steel.

Fig. 4-3 Monadnock building

Fig. 4-4 Masonry houses in China

The strength comparison among brick masonry, concrete and steel Table 4-1

Material	Brick masonry (MU10, M5)	Concrete (C20)	Steel (Q235)
Compress strength-weight ratio	$1.50/19 \approx 1/12.7$	$9.6/24 \approx 1/2.5$	$235/8.5 \approx 1/0.33$
Tensile strength-weight ratio	$0.13/1.50 \approx 1/11.5$	$1.1/9.6 \approx 1/8.7$	$235/235 \approx 1/1$

4.1.3 Frontiers

Although masonry is a building material with great success in modern building construction worldwide, there are still many challenges in material, design theory and construction technology, preventing it from becoming the most popular and powerful building material.

1. Masonry materials

The most fatal limitation of masonry is its strength. Hence, improving the strength of masonry at the lowest cost, especially the tensile strength and shear strength, has always been an important task for researchers in the community of civil engineering. Besides, the use of industrial wastes such as fly ash and steel slag as raw materials to make masonry is an important part of the development of low-carbon and green buildings. This is also in line with the sustainable development concept that China has always advocated.

2. Design theory

Since the 1980s, a considerable amount of research and practical experience has led to the improvement and refinement of the design code of masonry structure. In the latest version of the Chinese ***Code for the Design of Masonry***

Structures (GB 50003—2011), the structural design of masonry building is approaching a level similar to that applying to steel and concrete. However, it is still a great challenge to fully understand the nonlinear behavior of masonry under earthquake action and carry out reasonable seismic design.

3. Construction technology

At present, most masonry structures are still built manually. The performance of masonry structure, which is largely determined by the capability and experience of workers, has great uncertainty. At the same time, the labor costs can also be high because of the high labor intensity of artificial masonry construction and the long construction time. Research on intelligent masonry construction equipment and prefabricated construction methods with high efficiency is currently a hot spot in the field of masonry structure.

4.2 Brick and mortar

4.2.1 Introduction

Masonry is a well proven building material possessing excellent properties in terms of appearance, durability and cost in comparison with alternatives. However, the quality of the masonry in a building depends on the materials used, and hence all masonry materials must conform to certain minimum standards. The most common components of masonry are brick and mortar. The object of this section is to describe the properties of the two materials making up the masonry.

4.2.2 Brick

Brick is the most widely used masonry unit. According to different production methods and materials, bricks are divided into fired common bricks, fired perforated bricks, autoclaved fly-ash bricks, autoclaved sand-lime bricks, et al. Among them, the fired common brick is a widely used one and very suitable for general buildings.

Fired common bricks are solid bricks made of fired coal gangue, shale, fly ash or clay. It is worth noting that, because the collection of clay will destroy farmland, China and many other countries have banned the use of clay to produce fired common bricks. Industrial wastes such as coal gangue and fly ash are highly recommended. Fired common bricks must be free from deep and extensive cracks, from damage to edges and corners and also from expansive particles of lime. These must conform to the national standard GB 50003—2011. The dimension of fired common bricks is 240 mm × 115 mm × 53 mm, as shown in Fig. 4-5.

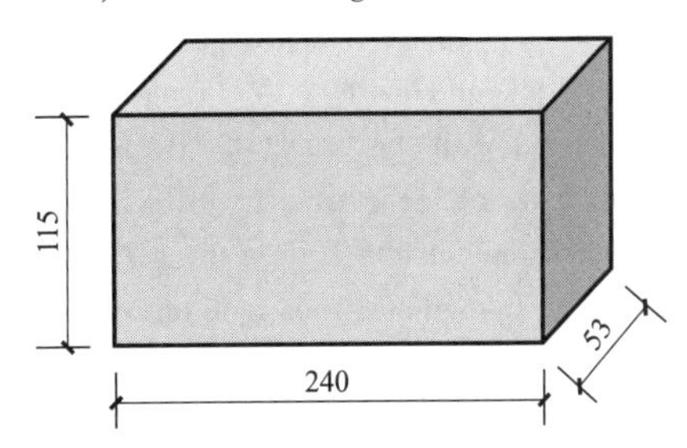

Fig. 4-5 Dimension of the fired common brick (mm)

From the structural point of view, the compressive strength of the unit is the controlling factor. Fired common bricks with various strengths are available to suit a wide range of architectural and engineering requirements. Table 4-2 gives a classification of fired common bricks according to the compressive strength. In the table, MU indicates that the type of the masonry unit is the fired common brick; and the number (i.e., 30, 25…) refers to the average compressive strength of the fired common brick with the unit of MPa.

Classification of fired common bricks according to the compressive strength (MPa)

Table 4-2

Class	MU30	MU25	MU20	MU15	MU10
Average compressive strength	30	25	20	15	10

4.2.3 Mortar

Mortar is a mixture of inorganic binders, aggregates and water, together with additions and admixtures if required. Mortar holds masonry units together to form a composite structural material and compensates for their dimensional tolerances. As such, mortar is a factor in the compressive, shear and flexural strengths of the masonry assemblage. Functions and requirements of mortar are: ① Development of early strength; ② Workability, i.e., ability to spread easily; ③ Water retentivity, i.e., the ability of mortar to retain water against the suction of brick. If water is not retained and is extracted quickly by a high-absorptive brick, there will be insufficient water left in the mortar joint for hydration of the cement, resulting in poor bond between brick and mortar; ④ Proper development of bond with the brick; ⑤ Resistance to cracking and rain penetration; ⑥ Resistance to frost and chemical attack, e.g. by soluble sulphate; ⑦ Immediate and long-term appearance.

Depending upon the materials used for mortar mixture preparation, the mortar could be classified as pure cement mortar, mixed mortar and no cement mortar. Pure cement mortar is a type of mortar where only cement is used as binding material. Mixed mortar is a type of mortar where cement, lime, gypsum, etc. are used as binding material. And no cement mortar is a type of mortar where no cement is used as binding material. The compressive strength of mortar used for masonry structure includes M15, M10, M7.5, M5.0, M2.5 and M0, in which M represents the mortar and the number (i.e., 15 and 10) is the standard value of compressive strength of the mortar cube with the unit of "MPa". M0 means that the mortar has not yet cured in the construction stage, so its compressive strength is 0.

4.2.4 Arrangement

As mentioned above, masonry is a heterogeneous material that consists of units and mortar joints, as shown in Fig. 4-6. There are a huge number of arrangements of units to form masonry walls and columns. Just for fired common brick masonry, some examples of usual arrangements are shown in Fig. 4-7. The most critical principle is that vertical mortar joints cannot be aligned up and down. Otherwise, it will be very unfavorable to the bearing capacity of masonry components.

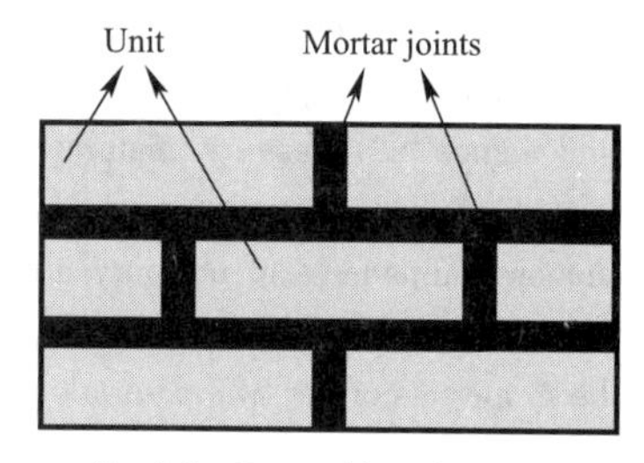

Fig. 4-6 Composition of masonry

The combination of blocks in different directions can form different component dimensions. Limited by the fixed size of fired common bricks, the general thicknesses of fired common brick masonry walls are 120 mm, 240 mm, 370 mm and 490 mm, whose cross sections are shown in Fig. 4-8; and the general dimensions of fired common brick masonry columns are the combinations of 120 mm, 240 mm, 370 mm and 490 mm.

4.2.5 Choice of bricks and mortar

Fired common bricks contain countless pores. In the very humid environment, all these pores may be filled with water. As a result, cracks may appear inside the brick due to large tensile stress in the cold winter, since water expands one-tenth on freezing. Therefore, fired common bricks must be carefully selected to avoid damage due to frost. In the humid environment, the strength of fired common bricks should not be less than MU15; and in the very humid condition and in the saturated condition, the strength of fired common bricks should not be less than MU20.

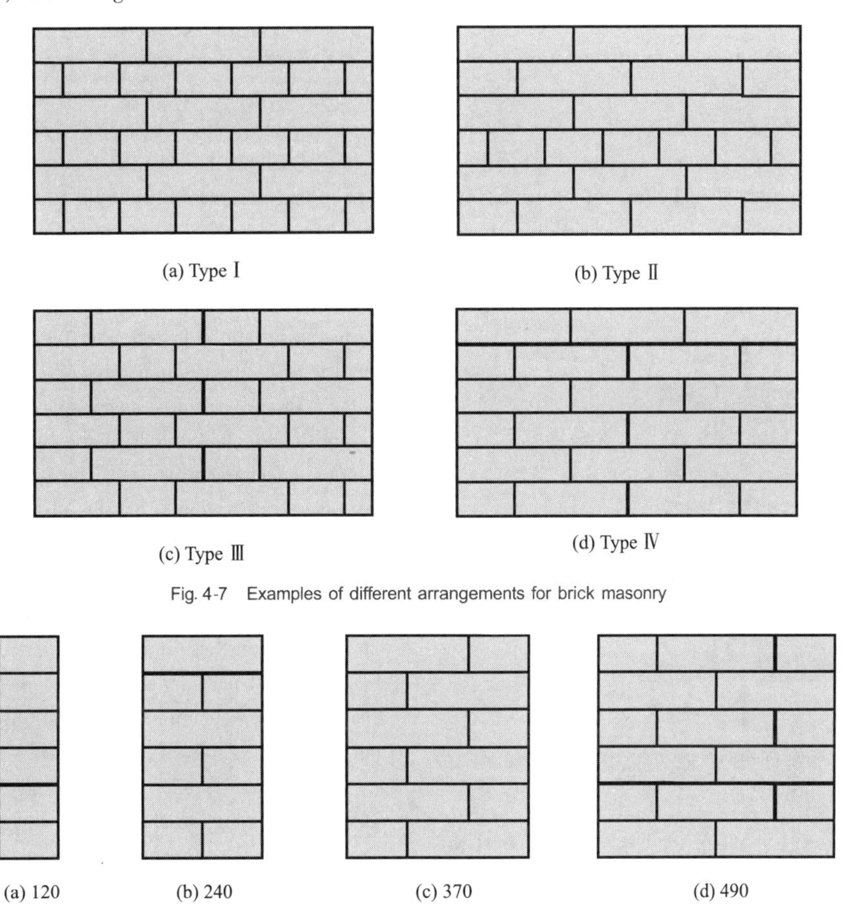

Fig. 4-7 Examples of different arrangements for brick masonry

Fig 4-8 General dimensions of fired common brick masonry components (mm)

4.3 Masonry properties

4.3.1 Introduction

The design of masonry structure requires a clear understanding of the behavior of the composite unit-mortar material under various stresses. Primarily, masonry walls and columns are vertical loadbearing components in which resistance to compressive stress is the predominating factor in design. However, walls and columns are frequently required to resist horizontal shear forces or lateral pressure from wind and therefore the strength of masonry in tension must also be considered. Currently, values for the design strength of masonry have been derived from tests on small specimens.

4.3.2 Compressive strength

A small masonry specimen, as shown in Fig. 4-9, is adopted to test the compressive strength of the fired common brick masonry. It is assumed that there is no deformation in the two steel plates. Therefore, the stress in the masonry specimen is uniform if the internal inconsistency of the masonry is ignored. Fig 4-10 shows the crack distribution in the masonry specimen. From beginning to failure, three phases could be used to describe the characteristics of crack appearance and development.

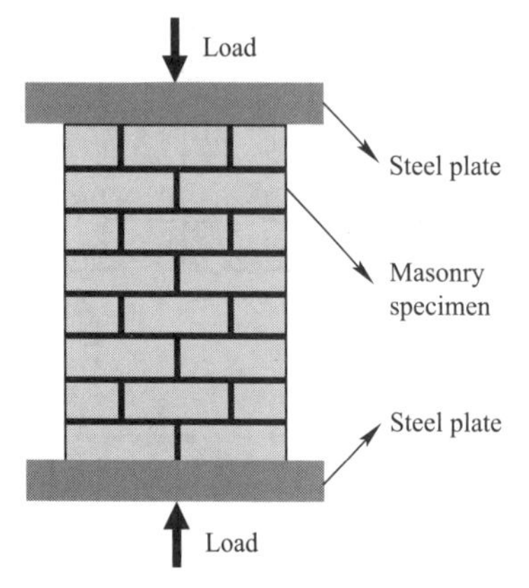

Fig. 4-9 Masonry specimen used for testing the compressive strength

Phase Ⅰ: When load increases from 0 to (50%—70%) P_u, in which P_u is the ultimate strength of masonry, cracks appear in different bricks. In this phase, cracks do not develop if load does not increase.

Phase Ⅱ: When load increases to (80%—90%) P_u, cracks develop and connect together. In this phase, cracks continue to develop even load does not increase.

Phase Ⅲ: When load increases to P_u, the masonry specimen is divided into several slender columns by connected cracks. The specimen fails due to crushing of bricks and instability of these slender columns.

A number of important points have been derived from compression tests on the small masonry specimen.

1. Strictly speaking, because the mortar bed under the brick is un-uniform, the brick is not evenly supported on the mortar bed. As a result, the brick in the masonry is not only compressed but is subjected to the combined action of pressure, bending moment and shear.

2. The Poisson's ratio of mortar is greater than that of brick. Under the same vertical compressive stress, the transverse deformation of mortar is greater than that of brick. If the slippage of the interface between the mortar and the brick is not considered, the transverse deformation of mortar is restrained by brick, resulting in additional compressive stress in mortar. According to the principle of force and reaction, additional tensile stress is generated in the brick.

3. In the masonry specimens, under the pressure of the upper half of the masonry, the deformation characteristics of the lower half of the masonry are similar to the elastic foundation, resulting in a bending deformation.

4. Vertical mortar joints are uneven. Stress concentration occurs in the mortar joints.

Because of these reasons mentioned above, the strength of masonry in compression is much smaller than the compressive strength of the unit and mortar. Table 4-3 lists the compress strength of fired common brick masonry used for design. According to the table, the compress strength of fired common brick masonry used for design is only 2.67 MPa when the fired common brick with 20 MPa and the mortar with 10 MPa are adopted.

It is apparent that the masonry strength increases with the increase of brick strength and mortar strength. The shape of the brick also influences the strength of masonry built from it. The higher the height of the brick, the higher its bending, shear and tensile strength, and the strength of the masonry increases accordingly. The smoother the surface of the brick, the smaller its bending and shearing effects, and the higher the strength of the masonry. The longer the length of the brick, the greater its bending and shearing effects, and the lower the strength of the masonry.

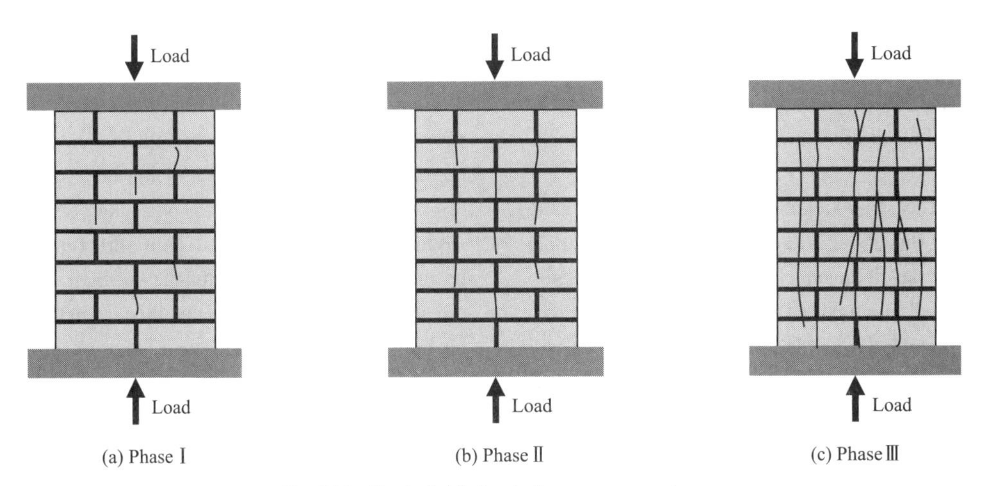

Fig. 4-10 Crack distribution in the masonry specimen

The compress strength of fired common brick masonry used for structural design(MPa) **Table 4-3**

Strength of fired common brick	Mortar strength					Mortar strength
	M15	M10	M7.5	M5.0	M2.5	0
MU30	3.94	3.27	2.93	2.59	2.26	1.15
MU25	3.60	2.98	2.68	2.37	2.06	1.05
MU20	3.22	2.67	2.39	2.12	1.84	0.94
MU15	2.79	2.31	2.07	1.83	1.60	0.82
MU10	—	1.89	1.69	1.50	1.30	0.67

Masonry has a very long tradition of building by craftsmen. Therefore, another important point is the construction quality, which affects the uniformity of mortar. It in turn affects the bending and shear action of brick. In the masonry design code, the construction quality is Grade A, B and C from good to bad, determined by the thickness, the uniformity, the proportion, the mixture method (i.e., by craftsmen or by machines) of the mortar joint. The design strength of the masonry is adjusted according to its construction quality.

4.3.3 Tensile strength

1. Direct tensile strength

As shown in Fig. 4-11, direct tensile stresses in masonry can arise as a result of in-plane loading effects. These may be caused by wind, by eccentric gravity loads, by thermal or moisture movements or by foundation movement. The tensile resistance of masonry, particularly across bed joints, depends on adhesion between units and mortar and therefore is not generally relied upon in structural design. The mechanism of unit-mortar adhesion is not fully understood. It should only be used in design with great caution. The tensile strength of fired common brick masonry used for design is listed in Table 4-4.

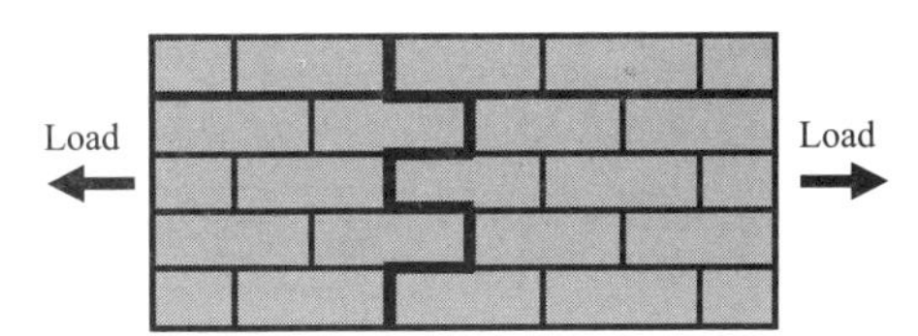

Fig. 4-11 Direct tension case

2. Flexural tensile strength

Masonry panels used essentially as cladding for buildings have to withstand lateral wind pressure and suction. The stability is derived from the self-weight of a wall, but generally this is insufficient to provide the necessary resistance to wind forces, and therefore reliance has to be placed on the flexural tensile strength of the masonry.

The same factors as influence direct tensile bond, discussed in the preceding section, apply to the development of flexural tensile strength. If a wall is supported only at its base, as shown in Table 4-4 Case Ⅲ, its lateral resistance will depend on the flexural tensile strength developed across the joint bed. If it is supported also on its vertical edges, as shown in Table 4-4 Case Ⅱ, lateral resistance will also depend on the flexural strength of the brickwork in the direction at right angles to the joint bed. The strength in this direction is typically about three times as great as across the joint bed. If the brick-mortar adhesion is good, the bending strength parallel to the bed joint direction will be limited by the flexural tensile strength of the units. If the adhesion is poor, this strength will be limited mainly by the shear strength of the unit-mortar interface in the bed joints. The flexural strength of fired common brick masonry used for design is also listed in Table 4-4.

Tensile strength of fired common brick masonry (MPa) **Table 4-4**

Class	Legend	Mortar strength			
		≥M10	M7.5	M5.0	M2.5
Direct tension	Case Ⅰ	0.19	0.16	0.13	0.09
Flexural tension	Case Ⅱ	0.33	0.29	0.23	0.17
	Case Ⅲ	0.17	0.14	0.11	0.08

4.4 Load resistance of masonry components

4.4.1 Introduction

This section deals with the bearing capacity of masonry components which are subjected to vertical loads. In practice, the design of load-bearing components reduces to the determination of the strength of the masonry and the dimension of the component required to support the design loads. Once the required strength is calculated, suitable types of masonry/mortar combinations can be selected from tables.

Like reinforced concrete components, the primary objective in designing load-bearing masonry components is to ensure an adequate margin of safety against the attainment of the ultimate limit state. In general terms, this is achieved by ensuringthat:

Design load ≤ Resistance

In which, the term on the left-hand side is determined from the applied loading and the term on the right is related to the dimension of the component and the strength of the masonry. The design load, which could be pressure, tension, bending moment, etc., is computed according to the ***Load Code for the Design of Building Structures*** (GB 50009—2012). The latest version was published in 2012. The calculation of the resistance of the masonry component will be introduced in the following several sections.

4.4.2 Modification factors for strength

Because these design strengths of fired common brick masonry listed in Table 4-3 and 4-4 are estimated by experiments using small specimens, they are not applicable to all design cases. For some special cases, the following adjustment coefficients γ_a should be adopted.

1) For unreinforced masonry components, when the area of the cross section A is less than 0.3 m^2, the adjustment coefficient is $A+0.7$, in which the unit of the area A is "m^2".

2) When the masonry is built using the pure cement mortar and the strength of the pure cement mortar is less than M5.0, the adjustment coefficient is 0.9 for the compression resistance and 0.8 for the tensile resistance.

3) When checking the resistance of these components during construction, the adjustment coefficient is 1.1.

4) When the construction quality is Grade B, the adjustment coefficient is unnecessary. But when the construction quality is Grade A and C, the adjustment coefficients 1.07 and 0.89 should be adopted, respectively.

It should be noted that when several special cases mentioned above occur at the same time, all these adjustment coefficients should be adopted and multiplied.

[Example 4-1]

Question: There is an unreinforced masonry column. Its dimension is 370 mm×490 mm. The fired common brick MU15 and the pure cement mortar M2.5 are adopted. The construction quality is Grade C. What is the strength of the fired common brick masonry when checking the resistance of the column?

[Solution]

According to Table 4-3, the strength value is 1.60 MPa.

The area A of the column is $0.37\times0.49=0.1813\ m^2<0.3\ m^2$.

The first adjustment coefficient is $A+0.3=0.1813+0.7=0.8813$.

The pure cement mortar whose strength is less than M5.0 isused, the second adjustment coefficient is 0.9.

The construction quality is Grade C. The third adjustment coefficient is 0.89.

The final strength of the fired common brick masonry is $1.60\times0.8813\times0.9\times0.89=1.13$ MPa.

4.4.3 Resistance under global compressive load

Columns and walls in masonry building are usually subjected to vertical loads arising from the self-weight of the masonry and the adjacent supported floors. In this case, the global compressive load is applied to columns and walls. As a result, compressive stress is un-uniformly or uniformly distributed across the entire cross section of the column and wall, as shown in Fig. 4-12.

The type of failure which would occur depends on the slenderness ratio, i.e., the ratio of the effective height to the effective thickness. For short stocky columns, where the slenderness ratio is low, failure would result from compression of the material, whereas for long thin columns and higher values of slenderness ratio, failure would occur from lateral instability. The slenderness ratio is computed by:

$$\beta=\gamma_{\beta}\frac{H_0}{h} \tag{4-1}$$

Where,

β —— the slenderness ratio;

γ_{β} —— the adjusted coefficient;

H_0 —— the effective height of the column or wall;

h —— the thickness of the column or wall.

When $\beta\leqslant3$, the column or wall is regarded as a short component; otherwise, the column or wall is regarded as a slender component.

(a) Uniform stress

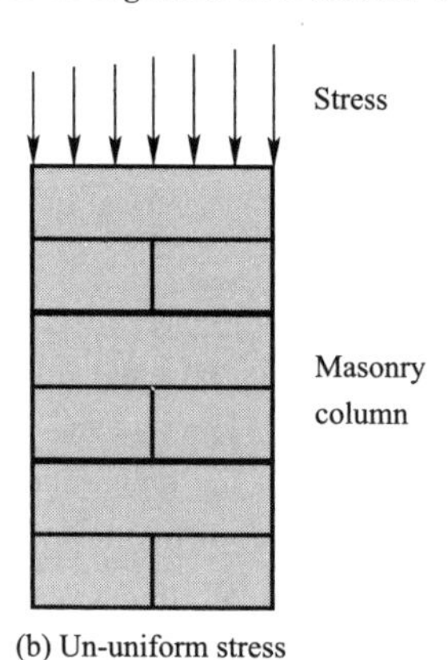

(b) Un-uniform stress

Fig. 4-12 Stress distribution under the global compressive load

The adjusted coefficient γ_{β} is equal to 1.0 when the fired common brick is adopted.

The effective height H_0 relates to the degree of restraint imposed by the floors and beams which frame into the wall or columns, which will be introduced in section 4.5.

For single leaf columns or walls, h is taken as the actual thickness. h is equal to the length of the short side under axial load; and h is equal to the length of the side along the eccentricity under eccentric load. For walls with piers, as shown in Fig. 4-13, h is taken as the equivalent thickness h_T, which is equal to $3.5i$. Here, i is the radius of gyration of the cross section.

The resistance of columns or walls under global compressive load is given by:

$$N_{u,1}=\varphi fA \tag{4-2}$$

Where,

$N_{u,1}$ —— the ultimate resistance of columns or walls under global compressive load;

φ —— the designing coefficient;

f —— the compressive strength of masonry in consideration with the strength adjustment coefficient;

A —— the area of the cross section.

The designing coefficient φ is calculated as

$$\varphi=\frac{1}{1+12\left(\frac{e}{h}\right)^2} \qquad \beta\leqslant 3$$

$$\varphi=\frac{1}{1+12\left[\frac{e}{h}+\sqrt{\frac{1}{12}\left(\frac{1}{\varphi_0}-1\right)}\right]^2} \qquad \beta>3$$

(4-3)

Where,

e —— the eccentricity of the compressive load;

φ_0—— the stability coefficient of columns or walls under axial load.

The stability coefficient φ_0 is calculated as:

$$\varphi_0=\frac{1}{1+\alpha\beta^2} \qquad (4\text{-}4)$$

Where,

α —— the strength coefficient associated with the compressive strength of mortar. α is taken as 0.0015 when the compressive strength of mortar is greater than or equal to M5.0; and α is taken as 0.002 and 0.009 when the compressive strength of mortar is equal to M2.5 and M0, respectively.

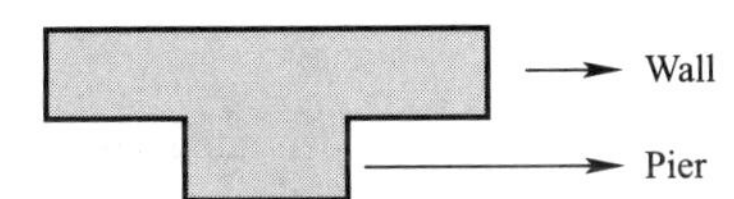

Fig. 4-13 Sketch of the cross section of a wall with a pier

[Example 4-2]

Question: A masonry column with section size of 370 mm×490 mm is built with the fired common brick MU15 and mixed mortar M5.0. The construction quality is Grade B. The effective height of the column is 3.2 m. The global compressive load applied on the column bottom is 205 kN. There is no eccentricity. Judge whether the masonry column is safe from the compressive bearing capacity point of view.

[Solution]

Calculate the slenderness ratio

$$\beta=\gamma_\beta\frac{H_0}{h}=1.0\times\frac{3.2}{0.37}=8.65$$

Calculate the designing coefficient

$$\varphi=\varphi_0=\frac{1}{1+\alpha\beta^2}=\frac{1}{1+0.0015\times 8.65}=0.99$$

According to Table 4-3, the compressive strength f is equal to 1.83 MPa.

Calculate the cross-section area of the column

$$A=0.37\times 0.49=0.18\ \text{m}^2$$

Because the cross-section area is less than 0.3 m^2, the adjustment coefficient is:

$$\gamma_a=0.7+0.18=0.88$$

Calculate the resistance of the column

$$N_{u,1}=\varphi f A=0.99\times 0.88\times 1.83\times 0.18\times 10^3$$
$$=286.97\ \text{kN}$$

The resistance is greater than the global compressive load, so the masonry column is safe.

[Example 4-3]

Question: A masonry walls with piers is made of the fired common brick MU15 and pure cement mortar M2.5. The dimension of the masonry wall is shown in Fig. 4-14. The effective height of the column is 5.0 m. The construction quality is Grade B. Calculate the resistance under the global compressive load acting on the point A.

[Solution]

Calculate the area

$$A=1\times 0.24+0.24\times 0.25=0.3\ \text{m}^2$$

Calculate the moment of inertia

$$y_1=\frac{1\times 0.24\times 0.12+0.24\times 0.25\times 0.365}{0.3}$$
$$=0.169\ \text{m}$$

$$y_2=0.49-0.169=0.321\ \text{m}$$

$$I=\frac{1}{3}\times 1\times 0.169^3+\frac{1}{3}\times(1-0.24)\times(0.24-0.169)^3+\frac{1}{3}\times 0.24\times 0.321^3=0.0043\ \text{m}^4$$

Calculate the radius of gyration of the cross section

$$i=\sqrt{\frac{I}{A}}=\sqrt{\frac{0.0043}{0.3}}=0.12\ \text{m}$$

Calculate the equivalent thickness

$$h_T=3.5i=3.5\times 0.12=0.42\ \text{m}$$

Calculate the eccentricity

$$e=y_2-0.15=0.321-0.15=0.171\ \text{m}$$

$$\frac{e}{y_2}=\frac{0.171}{0.321}=0.53$$

Calculate the slenderness ratio

$$\beta=\gamma_\beta \frac{H_0}{h_T}=1.0\times\frac{5.0}{0.42}=11.9$$

Calculate the stability coefficient

$$\varphi_0=\frac{1}{1+\alpha\beta^2}=\frac{1}{1+0.0015\times11.9^2}=0.825$$

Calculate the designing coefficient

$$\varphi=\frac{1}{1+12\left[\frac{e}{h}+\sqrt{\frac{1}{12}\left(\frac{1}{\varphi_0}-1\right)}\right]^2}=\frac{1}{1+12\left[\frac{0.171}{0.42}+\sqrt{\frac{1}{12}\left(\frac{1}{0.825}-1\right)}\right]^2}=0.22$$

According to Table 4-3, the compressive strength f is equal to 1.60 MPa.

Because the pure cement mortar whose strength is less than M5.0 is adopted, the adjustment coefficient is 0.9.

Calculate the resistance

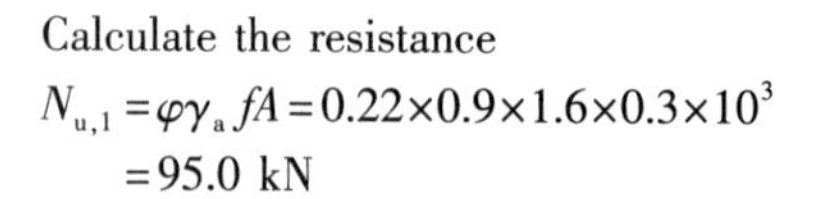

$$N_{u,1}=\varphi\gamma_a fA=0.22\times0.9\times1.6\times0.3\times10^3=95.0 \text{ kN}$$

4.4.4 Resistance under concentrated compressive load

In some cases, the stress only distributes in the partial cross section of the masonry component, for examples, the spread masonry foundation or walls bearing concrete columns, as shown in Fig. 4-15; the wall supporting the concrete beam or slab, as shown in Fig. 4-16. The compressive load transferred the masonry component is termed as the concentrated compressive load.

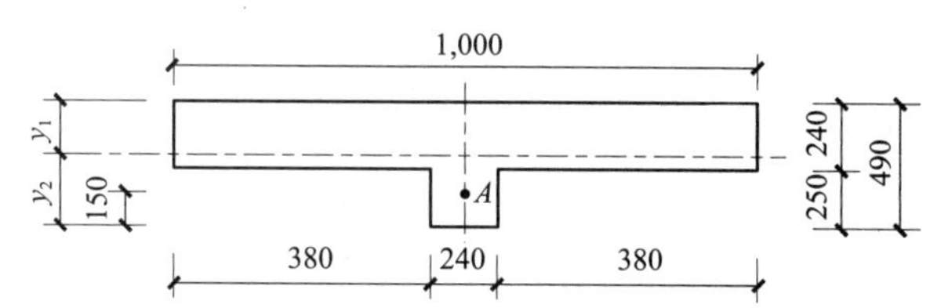

Fig. 4-14 Dimension of the masonry wall with a pier (mm)

In Fig. 4-15, if the eccentricity of the vertical load on top of the column is ignored, the concentrated load is assumed to be uniformly distributed over the bearing area. And the rest area of the masonry foundation or wall is not subjected to compressive stress. This case is termed as uniformly concentrated compression.

In Fig. 4-16, the concrete beam would bend due to the vertical load. As a result, the stress transferred to the masonry from the concrete beam is not uniformly distributed over the bearing area. Similarly, the rest area of the masonry wall is not subjected to compressive stress. This case is termed as un-uniformly concentrated compression.

For uniformly concentrated compression, the resistance of the fired common brick masonry component is:

$$N_{u,2}=\gamma f A_l \tag{4-5}$$

Where,

$N_{u,2}$—— the ultimate resistance of columns or walls under concentrated compressive load;

γ—— the enhancement coefficient;

f—— the compressive strength of masonry in consideration with the strength adjustment coefficient;

A_l—— the concentrated load bearing area.

The adjustment coefficient is equal to 1 even if the area of the concentrated load bearing area A_l is less than 0.3 m^2.

The enhancement coefficient γ is equal to:

$$\gamma=1+0.35\sqrt{\frac{A_0}{A_l}-1} \tag{4-6}$$

Where,

A_0—— the spread area, which is determined by Fig. 4-17. The limitation of the enhancement coefficient γ for case Ⅰ, Ⅱ, Ⅲ, Ⅳ in Fig. 4-17 is 2.5, 2.0, 1.5, 1.25, respectively.

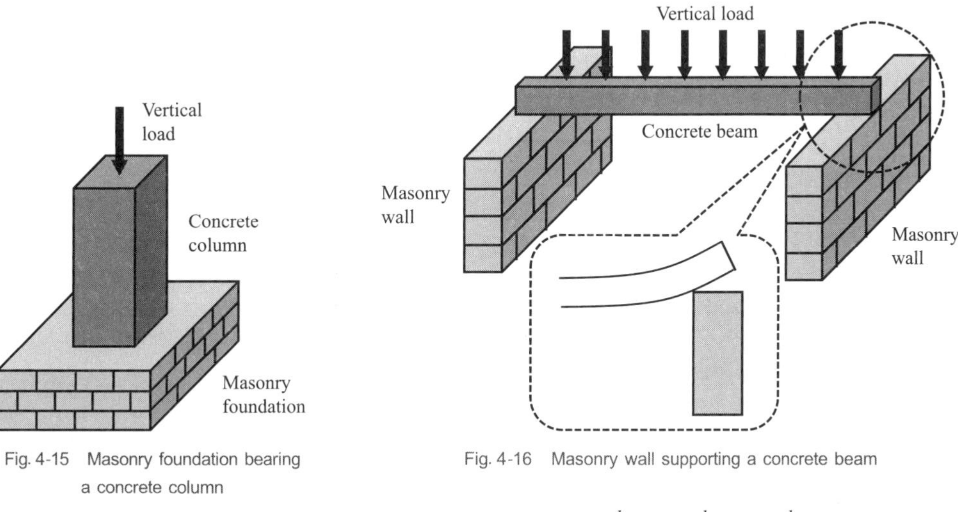

Fig. 4-15 Masonry foundation bearing a concrete column

Fig. 4-16 Masonry wall supporting a concrete beam

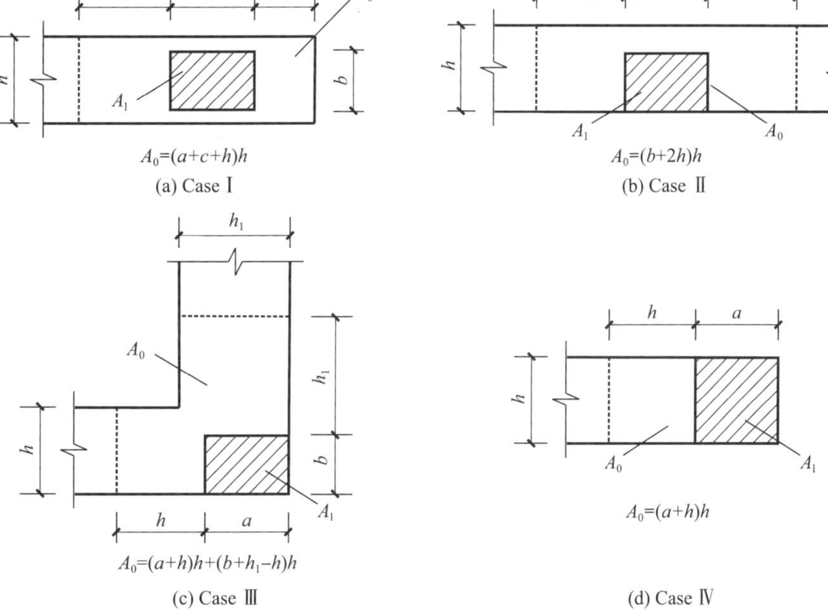

Fig. 4-17 Spread area in different cases

For the non-uniformly concentrated compression, the typical case is the masonry wall supporting the concrete beam, as shown in Fig. 4-18. The basic principle of design can be expressed as:

$$\psi N_0+N_1 \leqslant \eta\gamma f A_1 \tag{4-7}$$

Where

ψ —— the reduction coefficient;

N_0—— the vertical load transferred from the upper wall;

N_1—— the concentrated load transferred from the concrete beam;

η —— the shape coefficient;

A_1—— the area of the concentrated load bearing area.

The shape coefficient η is equal to 0.7 when the beam is perpendicular to the wall; while it is equal to 1.0 when the beam is parallel to the wall.

The reduction coefficient is calculated by:

$$\psi=1.5-0.5\frac{A_0}{A_1} \tag{4-8}$$

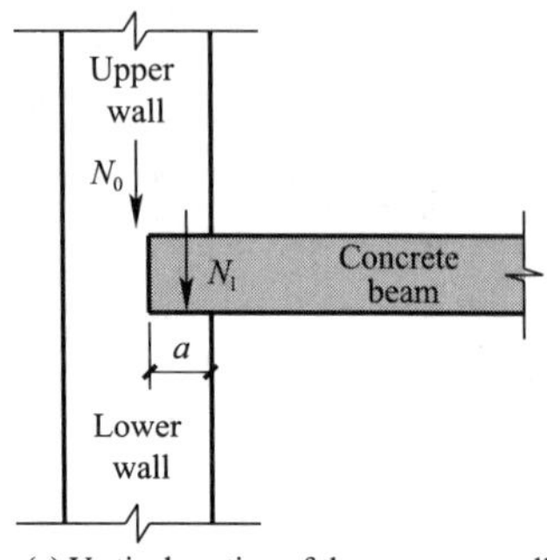

(a) Vertical section of the masonry wall

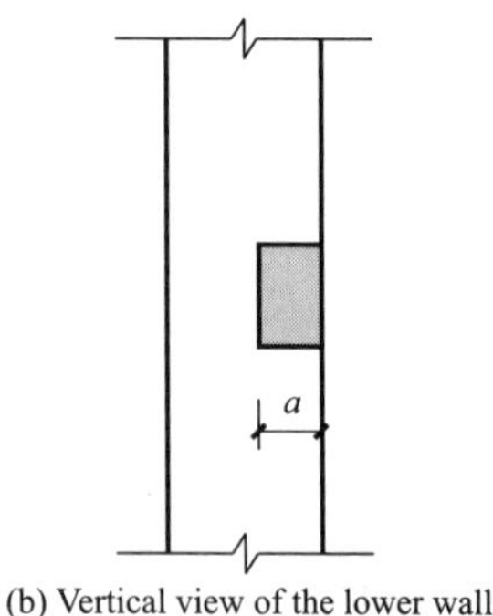

(b) Vertical view of the lower wall

Fig. 4-18 Load applied on masonry wall supporting the concrete beam

Where,

A_0—— the spread area computed by Fig. 4-17(b), A_l is the concentrated load bearing area. When A_0/A_l is greater than or equal to 3, ψ is equal to 0. The concentrated load bearing area, defined by:

$$A_l = a_0 b \tag{4-9}$$

Where,

a_0—— the equivalent support length;

b —— the width of the concrete beam.

The equivalent support length is equal to:

$$a_0 = 10\sqrt{\frac{h_c}{f}} \leqslant a \tag{4-10}$$

Where,

h_c—— the height of the concrete beam;

a —— the real support length shown in Fig. 4-18(b).

【Example 4-4】

Question: A masonry wall supports a concrete column, as shown in Fig. 4-19. The size of the concrete column is 200 mm×200 mm. The masonry wall is made of the fired common brick MU10 and mixed mortar M2.5. The construction quality is Grade C. The axial load transferred from the concrete column to the masonry wall is 60 kN. Judge whether the masonry wall is safe from the compressive bearing capacity point of view.

【Solution】

Calculate the concentrated load bearing area

$A_l = 200 \times 200 = 40{,}000\ \text{mm}^2$

Calculate the spread area

$A_0 = 370 \times (100 + 200 + 370) + 370 \times (200 + 150) = 377{,}400\ \text{mm}^2$

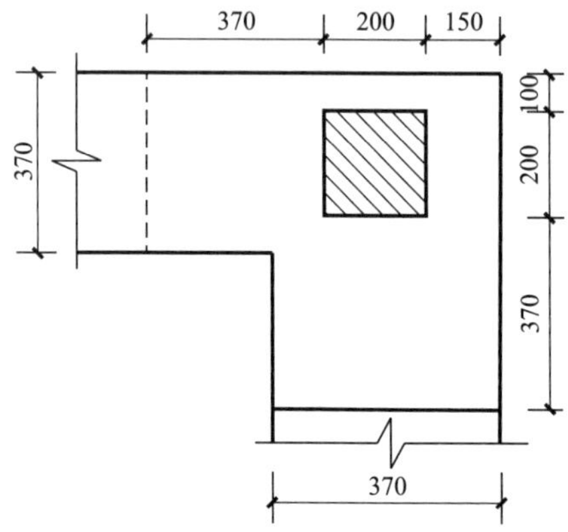

Fig. 4-19 Dimension of the masonry wall supporting a concrete column (mm)

Calculate the enhancement coefficient

$$\gamma = 1 + 0.35\sqrt{\frac{A_0}{A_l} - 1} = 1 + 0.35\sqrt{\frac{377{,}400}{40{,}000} - 1} = 2.02$$

Because the enhancement coefficient is greater than 1.5, the enhancement coefficient is equal to 1.5.

According to Table 4-3, the compressive strength f is equal to 1.30 MPa.

Because the construction quality is Grade C, the adjustment coefficient is 0.9.

Calculate the resistance

$N_{n,2} = \gamma\gamma_a f A_l = 1.5 \times 0.89 \times 1.30 \times 40{,}000 \times 10^{-3} = 69.4\ \text{kN}$

The resistance is greater than the axial load, so the masonry wall is safe.

【Example 4-5】

Question: A masonry wall supports a concrete beam, as shown in Fig. 4-20. The size of the beam $b \times h$ is 200 mm×500 mm. The real support length of the concrete beam is 240 mm. The vertical load transferred from the upper wall N_0 is 82 kN; and the concentrated load transferred from the concrete beam N_l is 102 kN. The masonry wall is made of the fired common brick MU10 and mixed mortar M2.5. Judge whether the masonry wall is safe from the compressive bearing capacity point of view.

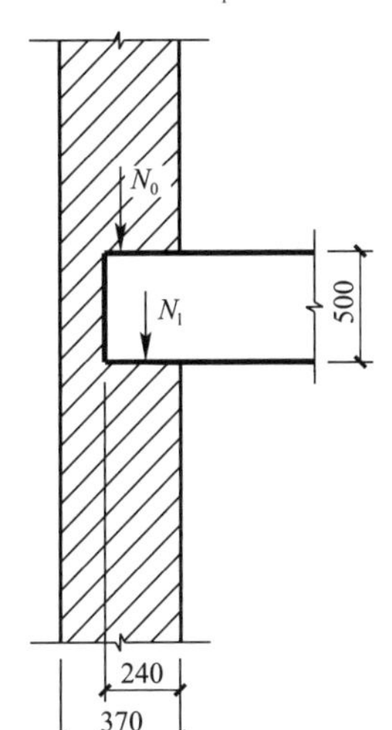

(a) Vertical section of the masonry wall

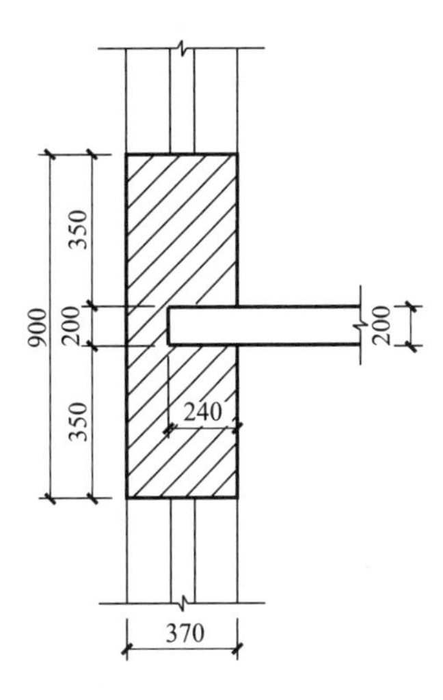

(b) Vertical view of the lower wall

Fig. 4-20 Dimension of the masonry wall supporting a concrete beam (mm)

[Solution]

Calculate the equivalent support length

$$a_0 = 10\sqrt{\frac{h_c}{f}} = 10\times\sqrt{\frac{500}{1.3}} = 196 \text{ mm}$$

Calculate the concentrated load bearing area

$$A_l = a_0 b = 196\times200 = 39{,}200 \text{ mm}^2 = 0.0392 \text{ m}^2$$

Calculate the spread area

$$b+2h = 200+2\times370>900 \text{ mm}$$

$$A_0 = (b+2h)\times h = 900\times370 = 333{,}000 \text{ mm}^2 = 0.333 \text{ m}^2$$

Because A_0/A_l is greater than 3, ψ is equal to 0.

Calculate the enhancement coefficient

$$\gamma = 1+0.35\sqrt{\frac{A_0}{A_b}-1} = 1+0.35\sqrt{8.5-1} = 1.959$$

According to Table 4-3, the compress strength f is equal to 1.30 MPa.

Calculate the resistance

$N_{n,2} = \eta\gamma A_l f = 0.7\times1.959\times0.0392\times1.3\times10^3 = 69.88 \text{ kN} < \psi N_0 + N_l = 0+102 = 102$ kN

The masonry wall is safe.

4.5 Design for masonry buildings

4.5.1 Introduction

In general, a masonry building includes vertical load-bearing components and horizontal load-bearing components. Specifically, the vertical load-bearing components are masonry walls, masonry columns and foundations; horizontal load-bearing members are roofs, slabs and beams. In order to improve the seismic resistance of masonry buildings and reduce the damage under earthquake, ring beams and structural columns are often used, as shown in Fig. 4-21.

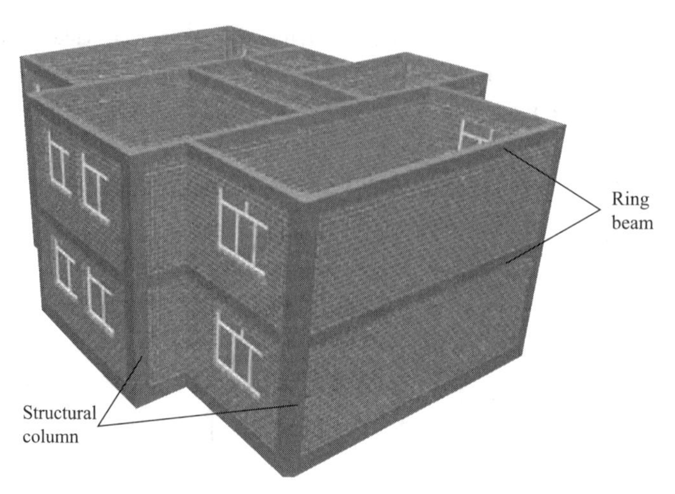

Fig. 4-21 Schematic diagram of ring beams and structural columns

Various structures, such as light steel structure, wood structure, precast hollow slab and reinforced concrete slab, can be used for the roof and floor of masonry building. In this section, the roof and floor are limited to the cast-in-situ reinforced concrete slab and the total length of the wall is limited to 32 m. These masonry buildings whose walls are longer than 32 m require special designs and beyond the scope of this book.

4.5.2 Effective height of masonry walls and columns

The effective height is related to the degree of lateral resistance to movement provided by supports. For columns in masonry buildings, the effective height is equal to their real heights. For walls, their effective height is equal to:

$$H_0=\begin{cases}1.0H & s>2H\\ 0.4s+0.2H & 2H\geqslant s>H\\ 0.6s & s\leqslant H\end{cases} \quad (4\text{-}11)$$

Where,

H —— the real height of the masonry wall;

s —— the distance between two transverse masonry walls, as shown in Fig. 4-22.

4.5.3 Stability of masonry walls and columns

In masonry building, walls and columns must not only meet the bearing capacity requirements, but also ensure their stabilities. The basic principle of checking the stability is:

$$\beta\leqslant[\beta] \quad (4\text{-}12)$$

Where,

β —— the slenderness ratio;

$[\beta]$ —— the ultimate slenderness ratio.

Here, the slenderness ratio is calculated by:

$$\beta=\frac{H_0}{h} \quad (4\text{-}13)$$

It should be noted that the calculation of the slenderness ratio in formula (4-12) is different from that in formula (4-1). The ultimate slenderness ratio is found in Table 4-5.

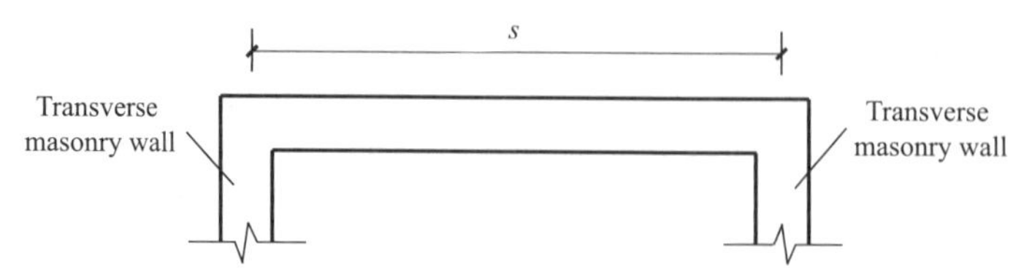

Fig. 4-22 Schematic diagram of transverse masonry walls

Ultimate slendernessratio of masonry walls and columns

Table 4-5

Mortar strength	Wall	Column
M2.5	22	15
M5.0	24	16
≥M7.5	26	17

1. Wall with window

When there are windows on the wall, the ultimate slenderness ratio should be corrected. Formula (4-12) is rewritten as:

$$\beta \leqslant \mu_1 \mu_2 [\beta] \tag{4-14}$$

Where,

μ_1 — the correcting coefficient for self-bearing walls;

μ_2 — the correcting coefficient considering the window.

When the thickness of the self-bearing wall is 240 mm, μ_1 is 1.2; when the thickness of the self-bearing wall is 90 mm, μ_1 is 1.5. If the thickness of the self-bearing wall is between 240 mm and 90 mm, the linear interpolation of 1.2 and 1.5 could be used.

The correcting coefficient considering the window is computed by:

$$\mu_2 = 1 - 0.4 \frac{b_s}{s} \tag{4-15}$$

Where,

b_s — the length of the window;

s — the distance between two adjacent walls, as shown in Fig. 4-23.

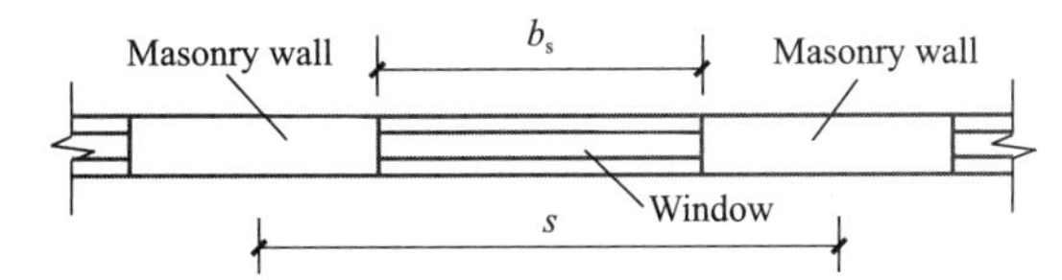

Fig. 4-23 Dimension of masonry wall with windows

2. Wall with piers

The checking of the stability of masonry walls with piers is divided into two steps. The first step is checking of the stability of the whole masonry wall, and the second step is checking of the stability of the masonry wall between two piers.

When checking the stability of the whole masonry wall, the whole masonry wall can be simplified as a short wall with a pier, as shown in Fig. 4-24. The equivalent thickness h_T is adopted to calculate the slenderness ratio. The length of the short wall b_f is obtained according to the following rules.

1) For multi-storey building, when there are windows, the length of the short wall b_f is equal to the length of the wall between two windows. When there is no window, the length of the short wall b_f is equal to one third of the wall height, which should be equal to or less than the distance between two piers.

2) For single-storey building, the length of the short wall b_f is equal to the summation of the width of the pier and two thirds of the wall height, which should be equal to or less than the distance between two piers and the length of the wall between two windows.

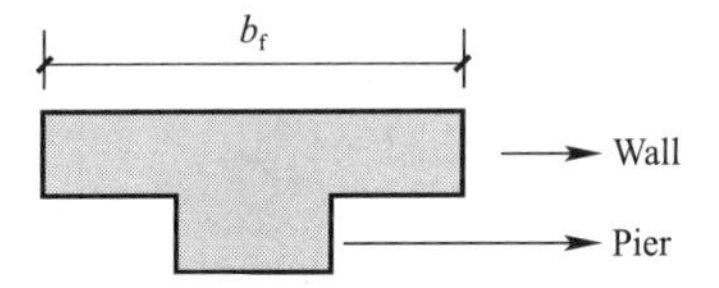

Fig. 4-24 A short wall with a pier

When checking of the stability of the masonry wall between two piers, formula. (4-14) could be used. s in formula. (4-15) is equal to the distance between two adjacent piers. The equivalent thickness is adopted to replace the wall thickness.

4.5.4 Design for masonry walls

The typical layout of a masonry building is shown in Fig. 4-25. Masonry walls are divided into exterior walls and interior walls. The stress distribution in the exterior wall has much difference with that in the interior wall, so they should be designed separately.

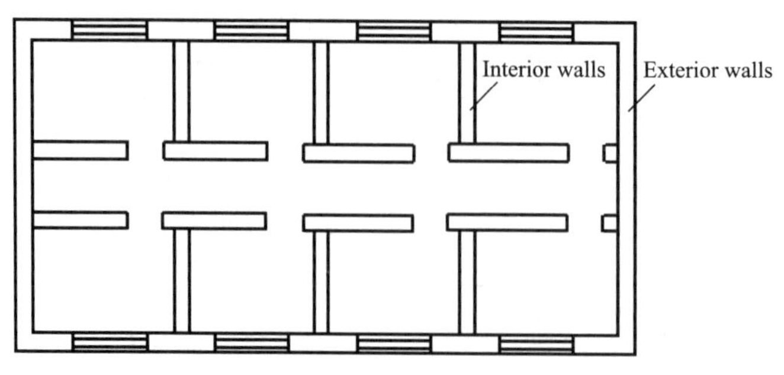

Fig. 4-25 Typical layout of a masonry building

1. Design for exterior wall

1) Computing model

As shown in Fig. 4-26, a part between two adjacent windows is taken as the calculation element. Because the beam and slab of the floor are embedded in the exterior wall, the cross section of the exterior wall is reduced dramatically. As a result, the constraint of the exterior wall on the rotation of the beam and slab can be ignored. Although the bottom end of the exterior wall is not weakened, the axial pressure is large and the bending moment is relatively small, so the bottom end can also be considered as a hinge. Therefore, under the vertical load, each storey of the exterior wall can be approximately regarded as a vertical beam with two simply supported ends, as shown in Fig. 4-26(b). Under the lateral load, such as wind, the exterior wall can be simplified as a vertical continuous beam, as shown in Fig. 4-26(c).

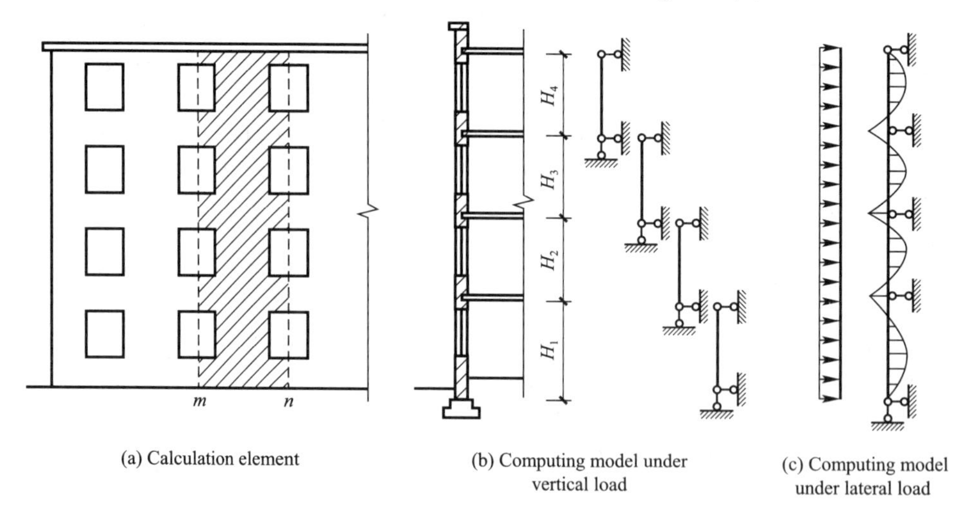

Fig. 4-26 Computing model of the exterior wall

2) Dangerous sections

There are generally four potential dangerous sections in each storey, which are the top of the wall, the upper end of the window, the lower end of the window, and the bottom the wall, as shown in Fig. 4-27. To ensure the wall safety, the resistance in section Ⅰ-Ⅰ and section Ⅳ-Ⅳ should be carefully checked. If the thickness of the exterior is identical from roof to foundation, only the first storey needs to be checked.

3) Load

The load acts on the cross section Ⅰ-Ⅰ under the vertical load is:

$$N_I = N_u + N_l \qquad M_I = N_l e_l \tag{4-16}$$

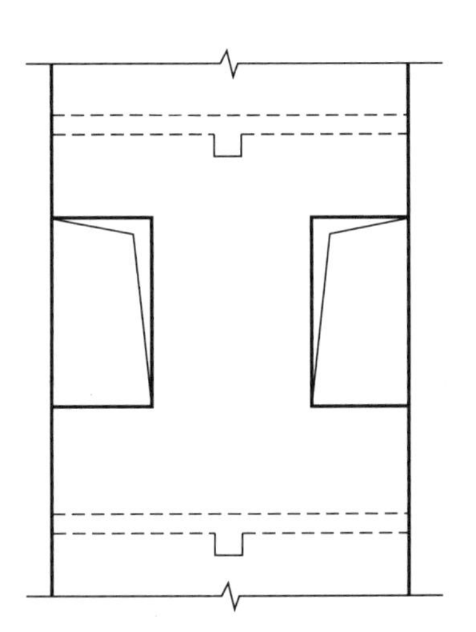

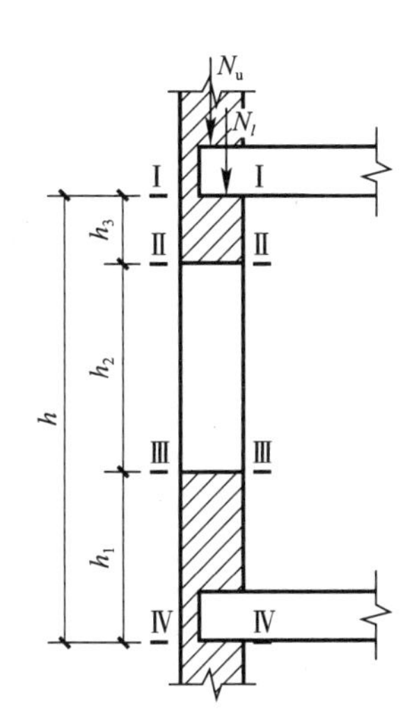

Fig. 4-27 Potential dangerous sections of the exterior wall

And the load acts on the cross section Ⅳ-Ⅳ under the vertical load is:

$$N_{\text{IV}} = N_u + N_l + G$$
$$M_{\text{IV}} = 0 \tag{4-17}$$

Where,

N_u—— the load transferred from the upper wall;

N_l—— the load transferred from the concrete beam or slab;

e_l—— the eccentricity of the load N_l, which is equal to $0.5h$-$0.4a_0$;

a_0—— the equivalent support length;

G —— the weight of the exterior wall from section I-I to section Ⅳ-Ⅳ.

The load acts on the exterior wall under the lateral load is:

$$M = \frac{1}{12} w H_i^2 \tag{4-18}$$

Where,

M —— the bending moment induced by lateral load;

w —— the distributed wind load acting on the exterior wall;

H_i—— the height of the i-th storey.

2. Design for interior wall

The lateral load has little influence on the interior wall. Therefore, only vertical load should be considered when checking the resistance of the interior wall.

1) Computing model

Generally, the interior wall with the width of 1 m is taken as the calculation element, as shown in Fig. 4-28. Each storey of the interior wall is also regarded as a vertical beam with two simply supported ends.

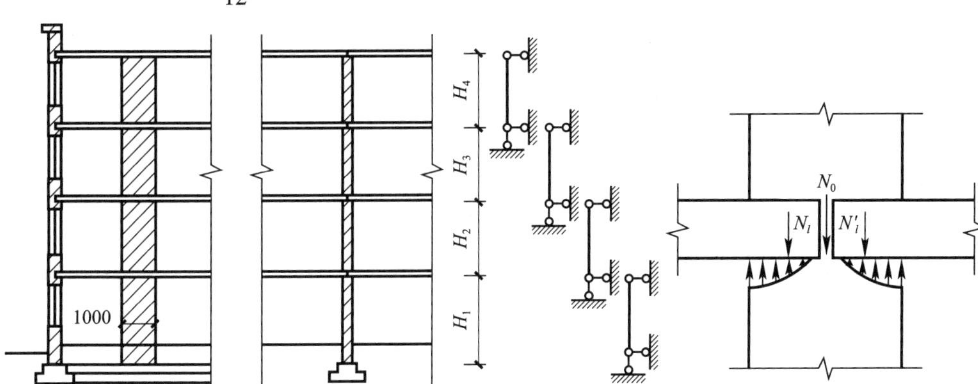

Fig. 4-28 Computing model of the interior wall (mm)

2) Dangerous sections

The potential dangerous sections in each storey are the same as that of the exterior wall.

3) Load

The load acts on the interior wall includes N_l transferred from the left slab or beam and N_l' transferred from the right slab or beam. The load action point of N_l and N_l' is $0.4a_0$ from the edge. In practice, bending moment caused by the difference of N_l and N_l' is usually small. The eccentricity can be ignored.

4.6 Design for lintels, ring beams and structural columns

4.6.1 Introduction

In addition to walls and columns, there are lintels, ring beams and structural columns in a masonry building. Although these components hardly bear vertical and lateral loads, they play a vital role in preventing uneven deformation, cracks, earthquake damage and so on. In particular, to improve the seismic resistance, the role of ring beams and structural columns may be greater than that of masonry walls and columns. In this section, the design for lintels, ring beams, and structural columns is introduced.

4.6.2 Lintel

The lintel is a member located on the top of the door and window. It bears the weight of the wall above the door and window, as well as the load transferred from the beam and slab. Four types of lintels, including the brick arch, the brick beam, the reinforced brick beam and the reinforced concrete beam, are usually employed, as shown in Fig. 4-29.

For generally masonry building, the brick beam can be adopted when the span is less than 1.2 m; the reinforced brick beam can be employed when the span is less than 1.5 m. For these masonry building with strong vibration or un-uniform settlement, the reinforced concrete beam should be used. Furthermore, the reinforced concrete beam should be used when the span is greater than 1.5 m. At present, the first three types of lintels are frequently replaced by the reinforced concrete beams.

4.6.3 Ring beam

In masonry building, the reinforced concrete beam which is placed continuously along the horizontal direction on the top of the masonry wall is called the ring beam. The ring beam can enhance the integrity and space stiffness of the masonry building. At the same time, the un-uniform settlement would be alleviated.

The set of the ring beam should obey the follow rules.

1) For single-storey masonry building made of fired common bricks, one ring beam should be set if the height of the building is 5-8 m. The number of ring beams should be increased when the building is higher than 8 m.

2) For multi-storey building, two ring beams should be set on the top and the bottom of the building when the number of storeys is 3-4. Ring beams should be set in each two storeys in addition to the ring beams setting on the top and the bottom of the building when the number of storeys is larger than 4.

The ring beam should be continuously located in a horizontal plane and forms a closed ring. When the ring beam is cut off by the door or window, the additional ring beam with the same section should be added to the top of the hole. The overlapping length of the additional ring beam shall not be less than twice of the vertical spacing between the original ring beam and the additional ring beam. Moreover, the minimal overlapping length of the additional ring beam is 1 m.

It is better that the width of the ring beam is the same as the thickness of the masonry wall. When the wall thickness is equal to or greater than 240 mm, the width of the ring beam should be equal to or greater than two thirds of the wall thickness. The height of the ring beam should be equal to or greater than 120 mm. The number and the diameter of reinforcements should not be less than 4 and 10 mm, respectively.

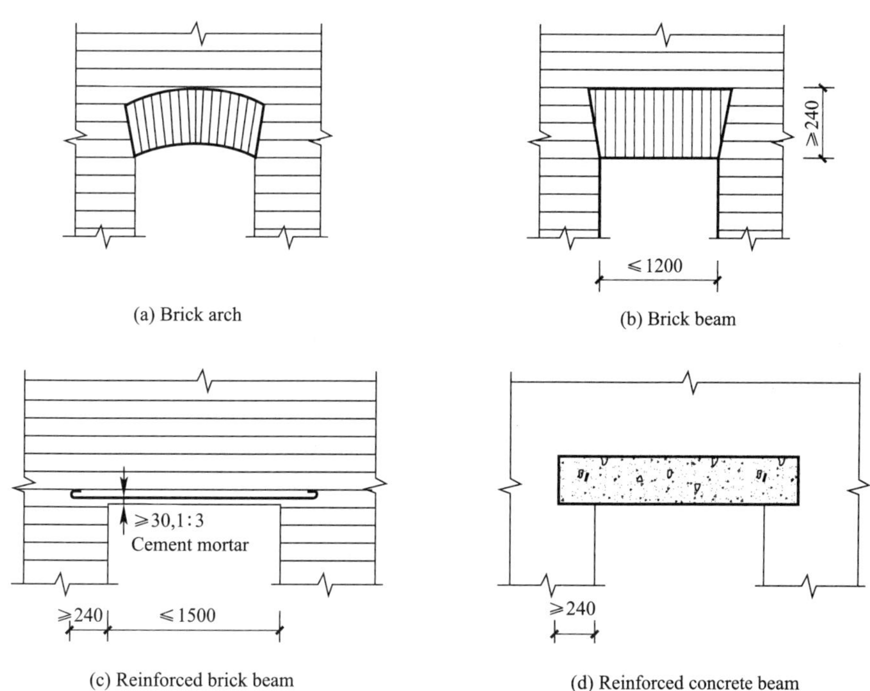

(a) Brick arch

(b) Brick beam

(c) Reinforced brick beam

(d) Reinforced concrete beam

Fig. 4-29 Different types of lintel (mm)

4.6.4 Structural column

Because of the poor integrity and seismic resistance of masonry building, seismic analysis shows that structural columns, which are made of reinforced concrete and set in appropriate locations of multi-storey masonry building, can work with ring beams to reduce the earthquake damages.

In real practice, the minimum dimension of the structural column is 240 mm×180 mm. The minimum reinforcements are 4Φ12 and the minimum spacing of stirrups is 250 mm. At the top and bottom of the structural column, the stirrup spacing should be properly reduced.

Exercises

4.1 A masonry wall supports a concrete column, as shown in Fig. 4-30. The size of the concrete column is 240 mm×240 mm. The masonry wall is made of the fired common brick MU10 and mixed mortar M2.5. The construction quality is Grade A. The axial load transferred from the concrete column

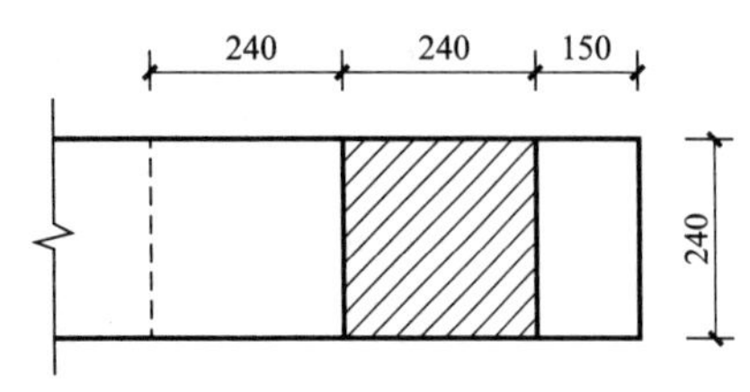

Fig. 4-30 Dimension of the masonry wall for exercise 4.1 (mm)

to the masonry wall is 50 kN. Judge whether the masonry wall is safe from the compressive bearing capacity point of view.

4.2 A masonry walls with a pier is made of the fired common brick MU10 and pure cement mortar M5.0. The dimension of the masonry wall is shown in Fig. 4-31. The effective height of the column is 6.5 m. The construction quality is Grade C. Calculate the resistance under the global compressive load acting on the points A, B and O, respectively.

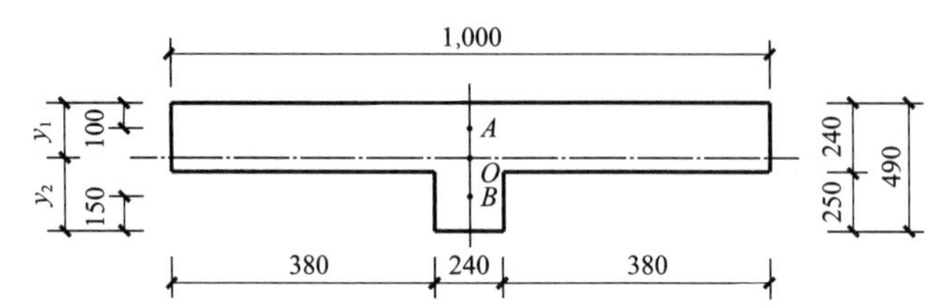

Fig. 4-31 Dimension of the masonry wall with a pier for exercise 4.2 (mm)

4.3 A masonry wall supports a concrete beam, as shown in Fig. 4-32. The size of the beam $b \times h$ is 250 mm×600 mm. The real support length of the concrete beam is 240 mm. The vertical load transferred from the upper wall N_0 is 265 kN; and the concentrated load transferred from the concrete beam N_l is 92 kN. The masonry wall is made of the fired common brick MU15 and mixed mortar M2.5. Judge whether the masonry wall is safe from the compressive bearing capacity point of view.

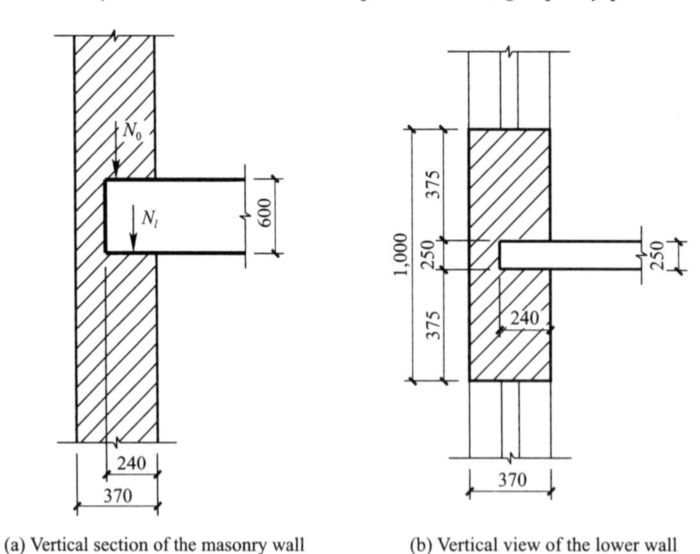

Fig. 4-32 Dimension of the masonry wall for exercise 4.3 (mm)

4.4 A masonry wall with piers is shown in Fig. 4-33. The masonry wall is made of the fired common brick MU10 and mixed mortar M5.0. The total length of the wall is 27 m. The distance between two adjacent piers is 4.5 m. A window with width of 2.0 m is located between each two adjacent piers. The height and thickness of the wall are 5.4 m and 240 mm, respectively. The size of the pier is 370 mm×250 mm. Check the stability of the masonry wall.

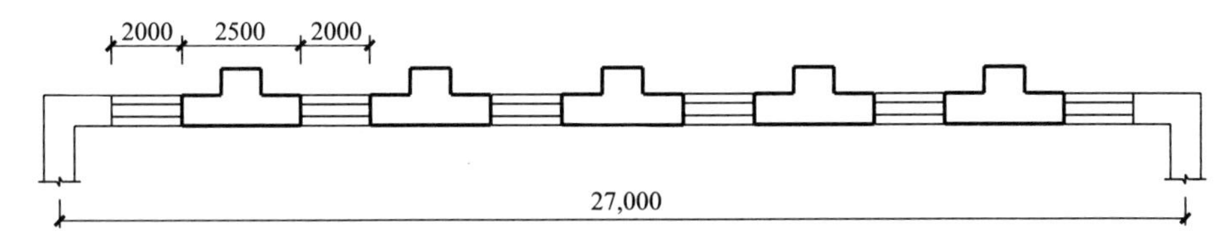

Fig. 4-33 Dimension of the masonry wall for exercise 4. 4 (mm)

Appendix 1
Basic parameter of electric bridge crane (5-50/5 t) Basic parameters and size series of general service electric bridge crane (ZQ1-62)

Appendix table

<table>
<tr><th rowspan="2">Lifting capacity Q(t)</th><th rowspan="2">Span L_k (m)</th><th colspan="4">Size</th><th colspan="4">Crane working level A4-A5</th></tr>
<tr><th>Width B (mm)</th><th>Track width K(mm)</th><th>H (mm)</th><th>B_1 (mm)</th><th>P_{max} (kN)</th><th>P_{min} (t)</th><th>m_1 (t)</th><th>m_2 (t)</th></tr>
<tr><td rowspan="5">5</td><td>16.5</td><td>4650</td><td>3500</td><td rowspan="5">1870</td><td rowspan="5">230</td><td>76</td><td>3.1</td><td>16.4</td><td rowspan="5">2.0 (Single braking) 2.1 (Double braking)</td></tr>
<tr><td>19.5</td><td rowspan="2">5150</td><td rowspan="2">4000</td><td>85</td><td>3.5</td><td>19.0</td></tr>
<tr><td>22.5</td><td>90</td><td>4.2</td><td>21.4</td></tr>
<tr><td>25.5</td><td rowspan="2">6400</td><td rowspan="2">5250</td><td>10</td><td>4.7</td><td>24.4</td></tr>
<tr><td>28.5</td><td>105</td><td>6.3</td><td>28.5</td></tr>
<tr><td rowspan="5">10</td><td>16.5</td><td>5550</td><td>4400</td><td rowspan="3">2140</td><td rowspan="5">230</td><td>115</td><td>2.5</td><td>18.0</td><td rowspan="5">3.8 (Single braking) 3.9 (Double braking)</td></tr>
<tr><td>19.5</td><td rowspan="2">5550</td><td rowspan="2">4400</td><td>120</td><td>3.2</td><td>20.3</td></tr>
<tr><td>22.5</td><td>125</td><td>4.7</td><td>22.4</td></tr>
<tr><td>25.5</td><td rowspan="2">6400</td><td rowspan="2">5250</td><td rowspan="2">2190</td><td>135</td><td>5.0</td><td>27.0</td></tr>
<tr><td>28.5</td><td>140</td><td>6.6</td><td>31.5</td></tr>
<tr><td rowspan="5">15</td><td>16.5</td><td>5650</td><td rowspan="3">4400</td><td>2050</td><td>230</td><td>165</td><td>3.4</td><td>24.1</td><td rowspan="5">5.3 (Single braking) 5.5 (Double braking)</td></tr>
<tr><td>19.5</td><td rowspan="2">5550</td><td rowspan="4">2140</td><td rowspan="4">260</td><td>170</td><td>4.8</td><td>25.5</td></tr>
<tr><td>22.5</td><td>185</td><td>5.8</td><td>31.6</td></tr>
<tr><td>25.5</td><td rowspan="2">6400</td><td rowspan="2">5250</td><td>195</td><td>6.0</td><td>38.0</td></tr>
<tr><td>28.5</td><td>210</td><td>6.8</td><td>40.0</td></tr>
<tr><td rowspan="5">15/3</td><td>16.5</td><td>5650</td><td rowspan="3">4400</td><td>2050</td><td>230</td><td>165</td><td>3.5</td><td>25.0</td><td rowspan="5">6.9 (Single braking) 7.4 (Double braking)</td></tr>
<tr><td>19.5</td><td rowspan="2">5550</td><td rowspan="4">2150</td><td rowspan="4">260</td><td>175</td><td>4.3</td><td>28.5</td></tr>
<tr><td>22.5</td><td>185</td><td>5.0</td><td>32.1</td></tr>
<tr><td>25.5</td><td rowspan="2">6400</td><td rowspan="2">5250</td><td>195</td><td>6.0</td><td>36.0</td></tr>
<tr><td>28.5</td><td>210</td><td>6.8</td><td>40.5</td></tr>
<tr><td rowspan="5">20/5</td><td>16.5</td><td>5650</td><td rowspan="3">4400</td><td>2200</td><td>230</td><td>195</td><td>3.0</td><td>25.0</td><td rowspan="5">7.5 (Single braking) 7.8(Double braking)</td></tr>
<tr><td>19.5</td><td rowspan="2">5550</td><td rowspan="4">2300</td><td rowspan="4">260</td><td>205</td><td>3.5</td><td>28.0</td></tr>
<tr><td>22.5</td><td>215</td><td>4.5</td><td>32.0</td></tr>
<tr><td>25.5</td><td rowspan="2">6400</td><td rowspan="2">5250</td><td>230</td><td>5.3</td><td>30.5</td></tr>
<tr><td>28.5</td><td>240</td><td>6.5</td><td>41.0</td></tr>
<tr><td rowspan="5">30/5</td><td>16.5</td><td>6050</td><td>4600</td><td rowspan="5">2600</td><td>260</td><td>270</td><td>5.0</td><td>34.0</td><td rowspan="5">11.7 (Single braking) 11.8 (Double braking)</td></tr>
<tr><td>19.5</td><td rowspan="2">6150</td><td rowspan="2">4800</td><td rowspan="4">300</td><td>280</td><td>6.5</td><td>36.5</td></tr>
<tr><td>22.5</td><td>290</td><td>7.0</td><td>42.0</td></tr>
<tr><td>25.5</td><td rowspan="2">6650</td><td rowspan="2">5250</td><td>310</td><td>7.8</td><td>47.5</td></tr>
<tr><td>28.5</td><td>320</td><td>8.8</td><td>51.5</td></tr>
</table>

Continued

<table>
<tr><th rowspan="2">Lifting capacity Q(t)</th><th rowspan="2">Span L_k (m)</th><th colspan="4">Size</th><th colspan="4">Crane working level A4-A5</th></tr>
<tr><th>Width B (mm)</th><th>Track width K(mm)</th><th>H (mm)</th><th>B_1 (mm)</th><th>P_{max} (kN)</th><th>P_{min} (t)</th><th>m_1 (t)</th><th>m_2 (t)</th></tr>
<tr><td rowspan="5">50/5</td><td>16.5</td><td rowspan="3">6350</td><td rowspan="3">4800</td><td>2700</td><td rowspan="5">300</td><td>395</td><td>7.5</td><td>44.0</td><td rowspan="5">14.0 (Single braking)
14.5(Double braking)</td></tr>
<tr><td>19.5</td><td rowspan="4">2750</td><td>415</td><td>7.5</td><td>48.0</td></tr>
<tr><td>22.5</td><td>425</td><td>8.5</td><td>52.0</td></tr>
<tr><td>25.5</td><td rowspan="2">6800</td><td rowspan="2">5250</td><td>445</td><td>8.5</td><td>56.0</td></tr>
<tr><td>28.5</td><td>460</td><td>9.5</td><td>61.0</td></tr>
</table>

Notes:

1. The dimensions and lifting capacity listed in the table are the maximum limits manufactured in this standard.

2. m_1 is the total mass of the crane which is calculated according to the mass of the trolley with double brakes and the closed control room; m_2 is the total mass of trolley.

3. This table does not include cranes with working levels of A6 and A7, which can be checked (ZQ1-62) when necessary.

4. The weight unit of this table is ton (t), which shall be converted into the legal gravity calculation unit kilonewton (kN), so the median value of the table shall be multiplied by 9.81. For simplification, the value in the table is approximately multiplied by 10.0.

5. Lifting capacity 50/5 t means the lifting capacity of the main hook is 50 t and that of the auxiliary hook is 5 t.

6. H is height above rail top; B_1 is the distance from track center to end; P_{max} is the maximum wheel pressure; and P_{min} is the minimum wheel pressure.

Appendix 2
Coefficients of reaction force and horizontal displacement of the single-step column top

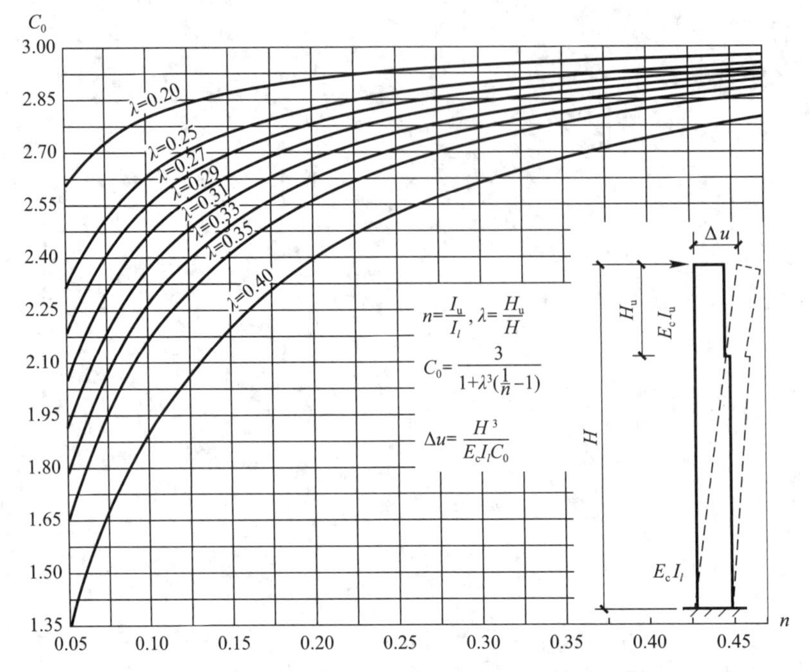

Appendix 2-1 C_0 of column top under unit concentrated horizontal force

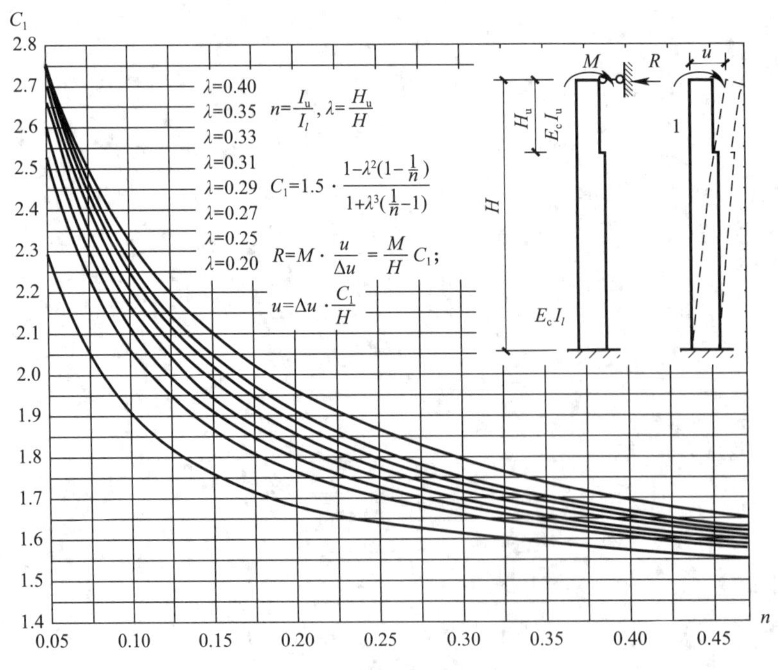

Appendix 2-2 C_1 of column top under moment

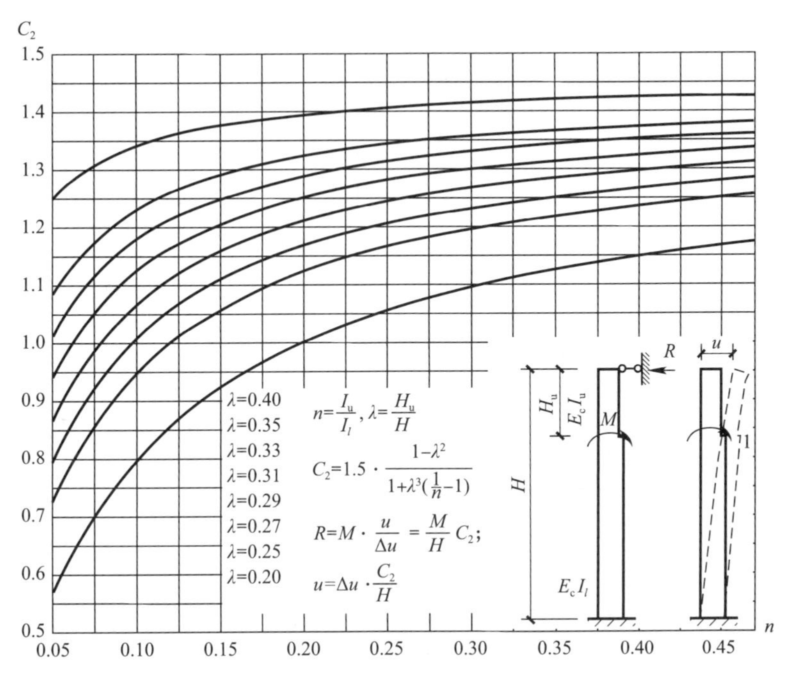

Appendix 2-3 C_2 of column corbel under moment

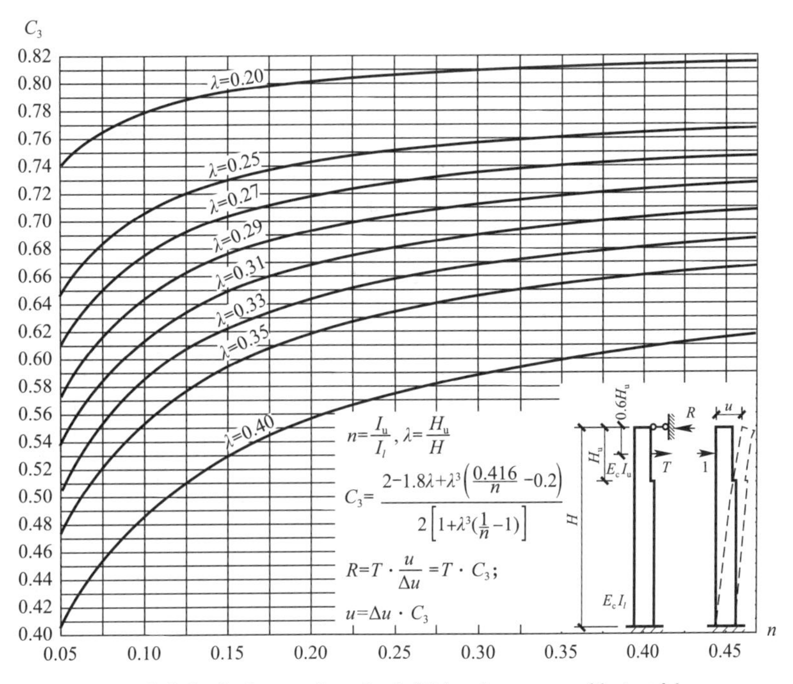

Appendix 2-4 C_3 of upper column ($y=0.6H_u$) under concentrated horizontal force

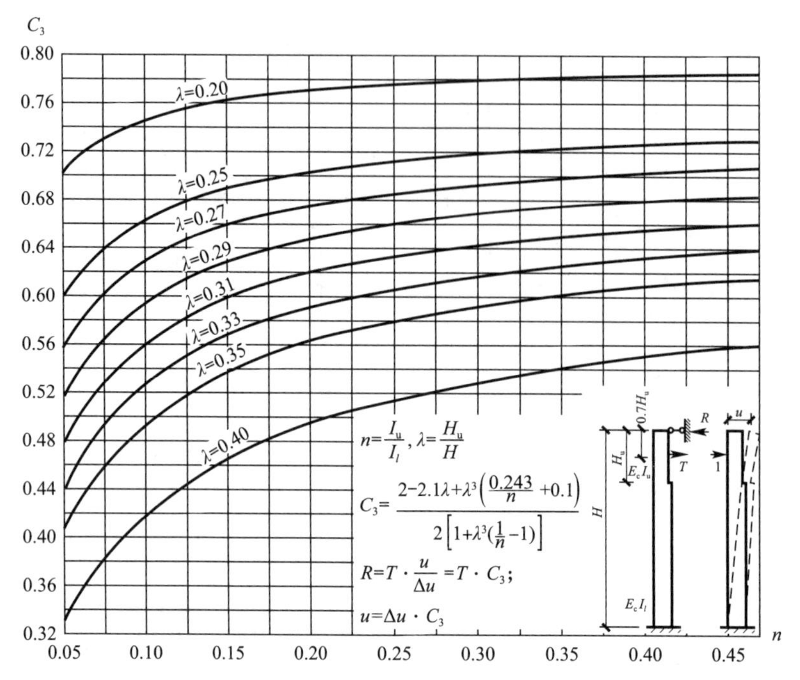

Appendix 2-5 C_3 of upper column ($y=0.7H_u$) under concentrated horizontal force

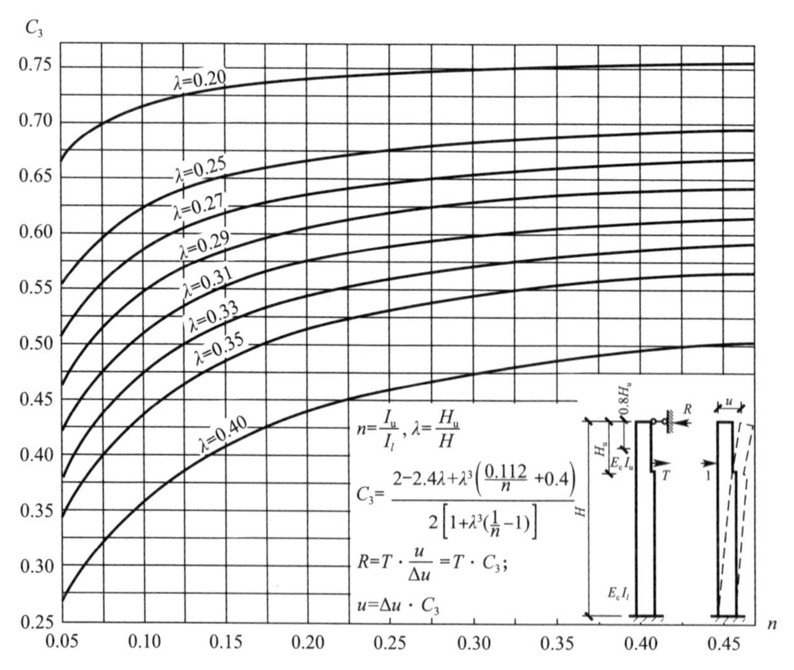

Appendix 2-6 C_3 of upper column ($y=0.8H_u$) under concentrated horizontal force

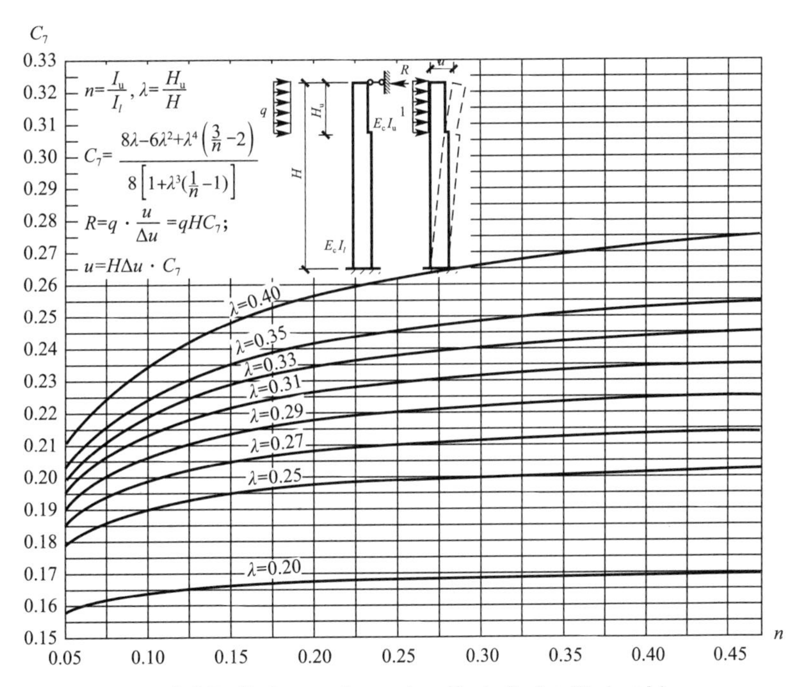

Appendix 2-7 C_7 of upper column under uniformly distributed horizontal force

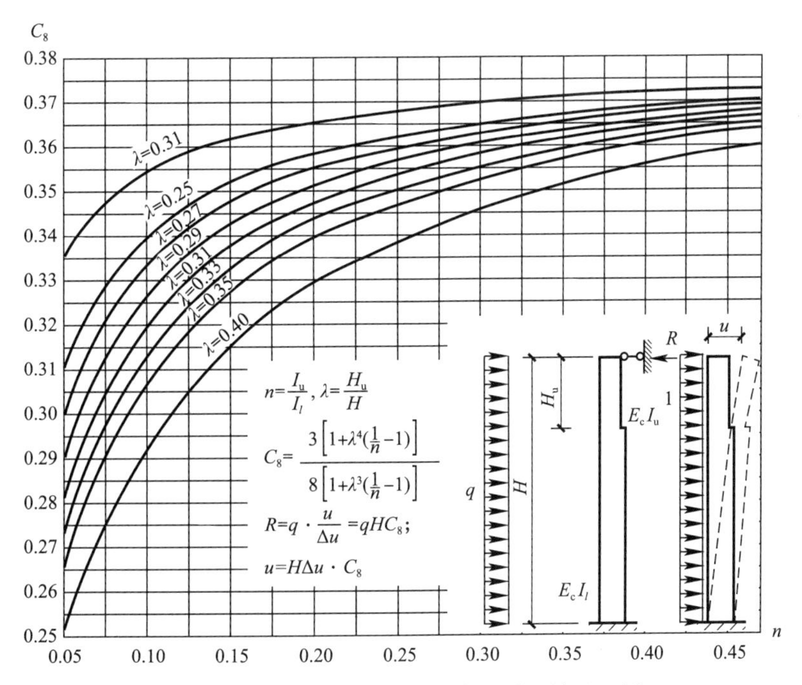

Appendix 2-8 C_8 of column under uniformly distributed horizontal force

Appendix 3
Ratio of standard inflection point height of regular frame under uniform horizontal load or inverted triangular distributed horizontal load

Ratio of standard inflection point height y_0 of regular frame under uniform horizontal load **Appendix table 3-1**

n	j \ K	0.1	0.2	0.3	0.4	0.5	0.6	0.7	0.8	0.9	1.0	2.0	3.0	4.0	5.0
1	1	0.80	0.75	0.70	0.65	0.65	0.60	0.60	0.60	0.60	0.55	0.55	0.55	0.55	0.55
2	2	0.45	0.40	0.35	0.35	0.35	0.35	0.40	0.40	0.40	0.40	0.45	0.45	0.45	0.45
	1	0.95	0.80	0.75	0.70	0.65	0.65	0.65	0.60	0.60	0.60	0.55	0.55	0.55	0.50
3	3	0.15	0.20	0.20	0.25	0.30	0.30	0.30	0.35	0.35	0.35	0.40	0.45	0.45	0.45
	2	0.55	0.50	0.45	0.45	0.45	0.45	0.45	0.45	0.45	0.45	0.45	0.50	0.50	0.50
	1	1.00	0.85	0.80	0.75	0.70	0.70	0.65	0.65	0.65	0.60	0.55	0.55	0.55	0.55
4	4	-0.05	0.05	0.15	0.20	0.25	0.30	0.30	0.35	0.35	0.35	0.40	0.45	0.45	0.45
	3	0.25	0.30	0.30	0.35	0.35	0.40	0.40	0.40	0.40	0.45	0.45	0.50	0.50	0.50
	2	0.65	0.55	0.50	0.50	0.45	0.45	0.45	0.45	0.45	0.45	0.50	0.50	0.50	0.50
	1	1.10	0.90	0.80	0.75	0.70	0.70	0.65	0.65	0.65	0.60	0.55	0.55	0.55	0.55
5	5	-0.20	0.00	0.15	0.20	0.25	0.30	0.30	0.30	0.35	0.35	0.40	0.45	0.45	0.45
	4	0.10	0.20	0.25	0.30	0.35	0.35	0.40	0.40	0.40	0.40	0.45	0.45	0.50	0.50
	3	0.40	0.40	0.40	0.40	0.40	0.45	0.45	0.45	0.45	0.45	0.50	0.50	0.50	0.50
	2	0.65	0.55	0.50	0.50	0.50	0.50	0.50	0.50	0.50	0.50	0.50	0.50	0.50	0.50
	1	1.20	0.95	0.80	0.75	0.75	0.70	0.70	0.65	0.65	0.65	0.55	0.55	0.55	0.55
6	6	-0.30	0.00	0.10	0.20	0.25	0.25	0.30	0.30	0.35	0.35	0.40	0.45	0.45	0.45
	5	0.00	0.20	0.25	0.30	0.35	0.35	0.40	0.40	0.40	0.40	0.45	0.45	0.50	0.50
	4	0.20	0.30	0.35	0.35	0.40	0.40	0.40	0.45	0.45	0.45	0.45	0.50	0.50	0.50
	3	0.40	0.40	0.40	0.45	0.45	0.45	0.45	0.45	0.45	0.45	0.50	0.50	0.50	0.50
	2	0.70	0.60	0.55	0.50	0.50	0.50	0.50	0.50	0.50	0.50	0.50	0.50	0.50	0.50
	1	1.20	0.95	0.85	0.80	0.75	0.70	0.70	0.65	0.65	0.65	0.55	0.55	0.55	0.55
7	7	-0.35	-0.05	0.10	0.20	0.20	0.25	0.30	0.30	0.35	0.35	0.40	0.45	0.45	0.45
	6	-0.10	0.15	0.25	0.30	0.35	0.35	0.35	0.40	0.40	0.40	0.45	0.45	0.50	0.50
	5	0.10	0.25	0.30	0.35	0.40	0.40	0.40	0.45	0.45	0.45	0.45	0.50	0.50	0.50
	4	0.30	0.35	0.40	0.40	0.40	0.45	0.45	0.45	0.45	0.45	0.50	0.50	0.50	0.50
	3	0.50	0.45	0.45	0.45	0.45	0.45	0.45	0.45	0.45	0.45	0.50	0.50	0.50	0.50
	2	0.75	0.60	0.55	0.50	0.50	0.50	0.50	0.50	0.50	0.50	0.50	0.50	0.50	0.50
	1	1.20	0.95	0.85	0.80	0.75	0.70	0.70	0.65	0.65	0.65	0.55	0.55	0.55	0.55
8	8	-0.35	-0.15	0.10	0.15	0.25	0.25	0.30	0.30	0.35	0.35	0.40	0.45	0.45	0.45
	7	-0.10	0.15	0.25	0.30	0.35	0.35	0.40	0.40	0.40	0.40	0.45	0.50	0.50	0.50
	6	0.05	0.25	0.30	0.35	0.40	0.40	0.40	0.45	0.45	0.45	0.45	0.50	0.50	0.50
	5	0.20	0.30	0.35	0.40	0.40	0.45	0.45	0.45	0.45	0.45	0.50	0.50	0.50	0.50

Continued

n	K / j	0.1	0.2	0.3	0.4	0.5	0.6	0.7	0.8	0.9	1.0	2.0	3.0	4.0	5.0
8	4	0.35	0.40	0.40	0.45	0.45	0.45	0.45	0.45	0.45	0.45	0.50	0.50	0.50	0.50
	3	0.50	0.45	0.45	0.45	0.45	0.45	0.45	0.45	0.50	0.50	0.50	0.50	0.50	0.50
	2	0.75	0.60	0.55	0.55	0.50	0.50	0.50	0.50	0.50	0.50	0.50	0.50	0.50	0.50
	1	1.20	1.00	0.85	0.80	0.75	0.70	0.70	0.65	0.65	0.65	0.55	0.55	0.55	0.55
9	9	−0.40	−0.05	0.10	0.20	0.25	0.25	0.30	0.30	0.35	0.35	0.45	0.45	0.45	0.45
	8	−0.15	0.15	0.25	0.30	0.35	0.35	0.35	0.40	0.40	0.40	0.45	0.45	0.50	0.50
	7	0.05	0.25	0.30	0.35	0.40	0.40	0.40	0.45	0.45	0.45	0.45	0.50	0.50	0.50
	6	0.15	0.30	0.35	0.40	0.40	0.45	0.45	0.45	0.45	0.45	0.50	0.50	0.50	0.50
	5	0.25	0.35	0.40	0.40	0.45	0.45	0.45	0.45	0.45	0.45	0.50	0.50	0.50	0.50
	4	0.40	0.40	0.40	0.45	0.45	0.45	0.45	0.45	0.45	0.45	0.50	0.50	0.50	0.50
	3	0.55	0.45	0.45	0.45	0.45	0.45	0.45	0.45	0.50	0.50	0.50	0.50	0.50	0.50
	2	0.80	0.65	0.55	0.55	0.50	0.50	0.50	0.50	0.50	0.50	0.50	0.50	0.50	0.50
	1	1.20	1.00	0.85	0.80	0.75	0.70	0.70	0.65	0.65	0.65	0.55	0.55	0.55	0.55
10	10	−0.40	−0.05	0.10	0.20	0.25	0.30	0.30	0.30	0.35	0.35	0.40	0.45	0.45	0.45
	9	−0.15	0.15	0.25	0.30	0.35	0.35	0.40	0.40	0.40	0.40	0.45	0.45	0.50	0.50
	8	0.00	0.25	0.30	0.35	0.40	0.40	0.40	0.45	0.45	0.45	0.45	0.50	0.50	0.50
	7	0.10	0.30	0.35	0.40	0.40	0.45	0.45	0.45	0.45	0.45	0.50	0.50	0.50	0.50
	6	0.20	0.35	0.40	0.40	0.45	0.45	0.45	0.45	0.45	0.45	0.50	0.50	0.50	0.50
	5	0.30	0.40	0.40	0.45	0.45	0.45	0.45	0.45	0.45	0.50	0.50	0.50	0.50	0.50
	4	0.40	0.40	0.45	0.45	0.45	0.45	0.45	0.45	0.45	0.50	0.50	0.50	0.50	0.50
	3	0.55	0.50	0.45	0.45	0.45	0.50	0.50	0.50	0.50	0.50	0.50	0.50	0.50	0.50
	2	0.80	0.65	0.55	0.55	0.55	0.50	0.50	0.50	0.50	0.50	0.50	0.50	0.50	0.50
	1	1.30	1.00	0.85	0.80	0.75	0.70	0.70	0.65	0.65	0.65	0.60	0.55	0.55	0.55
11	11	−0.40	0.05	0.10	0.20	0.25	0.30	0.30	0.30	0.35	0.35	0.40	0.45	0.45	0.45
	10	−0.15	0.15	0.25	0.30	0.35	0.35	0.40	0.40	0.40	0.40	0.45	0.45	0.50	0.50
	9	0.00	0.25	0.30	0.35	0.40	0.40	0.40	0.45	0.45	0.45	0.45	0.50	0.50	0.50
	8	0.10	0.30	0.35	0.40	0.40	0.45	0.45	0.45	0.45	0.45	0.50	0.50	0.50	0.50
	7	0.20	0.35	0.40	0.45	0.45	0.45	0.45	0.45	0.45	0.45	0.50	0.50	0.50	0.50
	6	0.25	0.35	0.40	0.45	0.45	0.45	0.45	0.45	0.45	0.45	0.50	0.50	0.50	0.50
	5	0.35	0.40	0.40	0.45	0.45	0.45	0.45	0.45	0.45	0.50	0.50	0.50	0.50	0.50
	4	0.40	0.45	0.45	0.45	0.45	0.45	0.45	0.50	0.50	0.50	0.50	0.50	0.50	0.50
	3	0.55	0.50	0.50	0.50	0.50	0.50	0.50	0.50	0.50	0.50	0.50	0.50	0.50	0.50
	2	0.80	0.65	0.60	0.55	0.55	0.50	0.50	0.50	0.50	0.50	0.50	0.50	0.50	0.50
	1	1.30	1.00	0.85	0.80	0.75	0.70	0.70	0.65	0.65	0.65	0.60	0.55	0.55	0.55

Continued

n	j \ K	0.1	0.2	0.3	0.4	0.5	0.6	0.7	0.8	0.9	1.0	2.0	3.0	4.0	5.0
≥12	1	-0.40	-0.05	0.10	0.20	0.25	0.30	0.30	0.30	0.35	0.35	0.40	0.45	0.45	0.45
	2	-0.15	0.15	0.25	0.30	0.35	0.35	0.40	0.40	0.40	0.40	0.45	0.45	0.50	0.50
	3	0.00	0.25	0.30	0.35	0.40	0.40	0.40	0.45	0.45	0.45	0.50	0.50	0.50	0.50
	4	0.10	0.30	0.35	0.40	0.40	0.45	0.45	0.45	0.45	0.45	0.50	0.50	0.50	0.50
	5	0.20	0.35	0.40	0.40	0.45	0.45	0.45	0.45	0.45	0.45	0.50	0.50	0.50	0.50
	6	0.25	0.35	0.40	0.45	0.45	0.45	0.45	0.45	0.45	0.45	0.50	0.50	0.50	0.50
	7	0.30	0.40	0.40	0.45	0.45	0.45	0.45	0.45	0.50	0.50	0.50	0.50	0.50	0.50
	8	0.35	0.40	0.45	0.45	0.45	0.45	0.45	0.50	0.50	0.50	0.50	0.50	0.50	0.50
	…	0.40	0.40	0.45	0.45	0.45	0.45	0.50	0.50	0.50	0.50	0.50	0.50	0.50	0.50
	4	0.45	0.45	0.45	0.45	0.50	0.50	0.50	0.50	0.50	0.50	0.50	0.50	0.50	0.50
	3	0.60	0.50	0.50	0.50	0.50	0.50	0.50	0.50	0.50	0.50	0.50	0.50	0.50	0.50
	2	0.80	0.65	0.60	0.55	0.55	0.50	0.50	0.50	0.50	0.50	0.50	0.50	0.50	0.50
	↑1	1.30	1.00	0.85	0.80	0.75	0.70	0.70	0.65	0.65	0.65	0.55	0.55	0.55	0.55

Note:

i_1 i_2

i

i_3 i_4

$$K=\frac{i_1+i_2+i_3+i_4}{2i}$$

Ratio of standard inflection point height y_0 of regular frame under inverted triangular distributed horizontal load Appendix table 3-2

n	j \ K	0.1	0.2	0.3	0.4	0.5	0.6	0.7	0.8	0.9	1.0	2.0	3.0	4.0	5.0
1	1	0.80	0.75	0.70	0.65	0.65	0.60	0.60	0.60	0.60	0.55	0.55	0.55	0.55	0.55
2	2	0.50	0.45	0.40	0.40	0.40	0.40	0.40	0.40	0.40	0.45	0.45	0.45	0.45	0.50
	1	1.00	0.85	0.75	0.70	0.70	0.65	0.65	0.65	0.60	0.60	0.55	0.55	0.55	0.55
3	3	0.25	0.25	0.25	0.30	0.30	0.35	0.35	0.35	0.40	0.40	0.45	0.45	0.45	0.50
	2	0.60	0.50	0.50	0.50	0.50	0.45	0.45	0.45	0.45	0.45	0.50	0.50	0.50	0.50
	1	1.15	0.90	0.80	0.75	0.75	0.70	0.70	0.65	0.65	0.65	0.60	0.55	0.55	0.55
4	4	0.10	0.15	0.20	0.25	0.30	0.30	0.35	0.35	0.35	0.40	0.45	0.45	0.45	0.45
	3	0.35	0.35	0.35	0.40	0.40	0.40	0.40	0.45	0.45	0.45	0.45	0.50	0.50	0.50
	2	0.70	0.60	0.55	0.50	0.50	0.50	0.50	0.50	0.50	0.50	0.50	0.50	0.50	0.50
	1	1.20	0.95	0.85	0.80	0.75	0.70	0.70	0.70	0.65	0.65	0.55	0.55	0.55	0.55

Continued

n	j \ K	0.1	0.2	0.3	0.4	0.5	0.6	0.7	0.8	0.9	1.0	2.0	3.0	4.0	5.0
5	5	-0.05	0.10	0.20	0.25	0.30	0.30	0.35	0.35	0.35	0.35	0.40	0.45	0.45	0.45
	4	0.20	0.25	0.35	0.35	0.40	0.40	0.40	0.40	0.40	0.45	0.45	0.50	0.50	0.50
	3	0.45	0.40	0.45	0.45	0.45	0.45	0.45	0.45	0.45	0.45	0.50	0.50	0.50	0.50
	2	0.75	0.60	0.55	0.55	0.50	0.50	0.50	0.50	0.50	0.50	0.50	0.50	0.50	0.50
	1	1.30	1.00	0.85	0.80	0.75	0.70	0.70	0.65	0.65	0.65	0.65	0.55	0.55	0.55
6	6	-0.15	0.05	0.15	0.20	0.25	0.30	0.30	0.35	0.35	0.35	0.40	0.45	0.45	0.45
	5	0.10	0.25	0.30	0.35	0.35	0.40	0.40	0.40	0.45	0.45	0.45	0.50	0.50	0.50
	4	0.30	0.35	0.40	0.40	0.45	0.45	0.45	0.45	0.45	0.45	0.50	0.50	0.50	0.50
	3	0.50	0.45	0.45	0.45	0.45	0.45	0.45	0.45	0.45	0.50	0.50	0.50	0.50	0.50
	2	0.80	0.65	0.55	0.55	0.55	0.55	0.50	0.50	0.50	0.50	0.50	0.50	0.50	0.50
	1	1.30	1.00	0.85	0.80	0.75	0.70	0.70	0.65	0.65	0.65	0.60	0.55	0.55	0.55
7	7	-0.20	0.05	0.15	0.20	0.25	0.30	0.30	0.35	0.35	0.35	0.45	0.45	0.45	0.45
	6	0.05	0.20	0.30	0.35	0.35	0.40	0.40	0.40	0.40	0.45	0.45	0.50	0.50	0.50
	5	0.20	0.30	0.35	0.40	0.40	0.45	0.45	0.45	0.45	0.45	0.50	0.50	0.50	0.50
	4	0.35	0.40	0.40	0.45	0.45	0.45	0.45	0.45	0.45	0.45	0.50	0.50	0.50	0.50
	3	0.55	0.50	0.50	0.50	0.50	0.50	0.50	0.50	0.50	0.50	0.50	0.50	0.50	0.50
	2	0.80	0.65	0.60	0.55	0.55	0.55	0.50	0.50	0.50	0.50	0.50	0.50	0.50	0.50
	1	1.30	1.00	0.90	0.80	0.75	0.70	0.70	0.70	0.65	0.65	0.60	0.55	0.55	0.55
8	8	-0.20	0.05	0.15	0.20	0.25	0.30	0.30	0.35	0.35	0.35	0.45	0.45	0.45	0.45
	7	0.00	0.20	0.30	0.35	0.35	0.40	0.40	0.40	0.40	0.45	0.45	0.50	0.50	0.50
	6	0.15	0.30	0.35	0.40	0.40	0.45	0.45	0.45	0.45	0.45	0.50	0.50	0.50	0.50
	5	0.30	0.45	0.40	0.45	0.45	0.45	0.45	0.45	0.45	0.45	0.50	0.50	0.50	0.50
	4	0.40	0.45	0.45	0.45	0.45	0.45	0.45	0.50	0.50	0.50	0.50	0.50	0.50	0.50
	3	0.60	0.50	0.50	0.50	0.50	0.50	0.50	0.50	0.50	0.50	0.50	0.50	0.50	0.50
	2	0.85	0.65	0.60	0.55	0.55	0.55	0.50	0.50	0.50	0.50	0.50	0.50	0.50	0.50
	1	1.30	1.00	0.90	0.80	0.75	0.70	0.70	0.70	0.65	0.65	0.60	0.55	0.55	0.55
9	9	-0.25	0.00	0.15	0.20	0.25	0.30	0.30	0.35	0.35	0.40	0.45	0.45	0.45	0.45
	8	-0.00	0.20	0.30	0.35	0.35	0.40	0.40	0.40	0.40	0.45	0.45	0.50	0.50	0.50
	7	0.15	0.30	0.35	0.40	0.40	0.45	0.45	0.45	0.45	0.45	0.50	0.50	0.50	0.50
	6	0.25	0.35	0.40	0.40	0.45	0.45	0.45	0.45	0.45	0.50	0.50	0.50	0.50	0.50
	5	0.35	0.40	0.45	0.45	0.45	0.45	0.45	0.45	0.50	0.50	0.50	0.50	0.50	0.50
	4	0.45	0.45	0.45	0.45	0.45	0.50	0.50	0.50	0.50	0.50	0.50	0.50	0.50	0.50
	3	0.60	0.50	0.50	0.50	0.50	0.50	0.50	0.50	0.50	0.50	0.50	0.50	0.50	0.50
	2	0.85	0.65	0.60	0.55	0.55	0.55	0.55	0.50	0.50	0.50	0.50	0.50	0.50	0.50
	1	1.35	1.00	0.90	0.80	0.75	0.75	0.70	0.70	0.65	0.65	0.60	0.55	0.55	0.55

Continued

n	j \ K	0.1	0.2	0.3	0.4	0.5	0.6	0.7	0.8	0.9	1.0	2.0	3.0	4.0	5.0
10	10	-0.25	0.00	0.15	0.20	0.25	0.30	0.30	0.35	0.35	0.40	0.45	0.45	0.45	0.45
	9	-0.05	0.20	0.30	0.35	0.35	0.40	0.40	0.40	0.40	0.45	0.45	0.50	0.50	0.50
	8	0.10	0.30	0.35	0.40	0.40	0.40	0.45	0.45	0.45	0.45	0.50	0.50	0.50	0.50
	7	0.20	0.35	0.40	0.40	0.45	0.45	0.45	0.45	0.45	0.50	0.50	0.50	0.50	0.50
	6	0.30	0.40	0.40	0.45	0.45	0.45	0.45	0.45	0.45	0.50	0.50	0.50	0.50	0.50
	5	0.40	0.45	0.45	0.45	0.45	0.45	0.45	0.50	0.50	0.50	0.50	0.50	0.50	0.50
	4	0.50	0.45	0.45	0.45	0.50	0.50	0.50	0.50	0.50	0.50	0.50	0.50	0.50	0.50
	3	0.60	0.55	0.50	0.50	0.50	0.50	0.50	0.50	0.50	0.50	0.50	0.50	0.50	0.50
	2	0.85	0.65	0.60	0.55	0.55	0.55	0.55	0.50	0.50	0.50	0.50	0.50	0.50	0.50
	1	1.35	1.00	0.90	0.80	0.75	0.75	0.70	0.70	0.65	0.65	0.60	0.55	0.55	0.55
11	11	-0.25	0.00	0.15	0.20	0.25	0.30	0.30	0.30	0.35	0.35	0.45	0.45	0.45	0.45
	10	-0.05	0.20	0.25	0.30	0.35	0.40	0.40	0.40	0.40	0.45	0.45	0.50	0.50	0.50
	9	0.10	0.30	0.35	0.40	0.40	0.40	0.45	0.45	0.45	0.45	0.50	0.50	0.50	0.50
	8	0.20	0.35	0.40	0.40	0.45	0.45	0.45	0.45	0.45	0.45	0.50	0.50	0.50	0.50
	7	0.25	0.40	0.40	0.45	0.45	0.45	0.45	0.45	0.45	0.50	0.50	0.50	0.50	0.50
	6	0.35	0.40	0.45	0.45	0.45	0.45	0.45	0.50	0.50	0.50	0.50	0.50	0.50	0.50
	5	0.40	0.45	0.45	0.45	0.45	0.50	0.50	0.50	0.50	0.50	0.50	0.50	0.50	0.50
	4	0.50	0.50	0.50	0.50	0.50	0.50	0.50	0.50	0.50	0.50	0.50	0.50	0.50	0.50
	3	0.65	0.55	0.50	0.50	0.50	0.50	0.50	0.50	0.50	0.50	0.50	0.50	0.50	0.50
	2	0.85	0.65	0.60	0.55	0.55	0.55	0.55	0.50	0.50	0.50	0.50	0.50	0.50	0.50
	1	1.35	1.05	0.90	0.80	0.75	0.75	0.70	0.70	0.65	0.65	0.60	0.55	0.55	0.55
≥12	↓1	-0.30	0.00	0.15	0.20	0.25	0.30	0.30	0.30	0.35	0.35	0.40	0.45	0.45	0.45
	2	-0.10	0.20	0.25	0.30	0.35	0.40	0.40	0.40	0.40	0.40	0.45	0.45	0.45	0.50
	3	0.05	0.25	0.35	0.40	0.40	0.40	0.45	0.45	0.45	0.45	0.45	0.50	0.50	0.50
	4	0.15	0.30	0.40	0.40	0.45	0.45	0.45	0.45	0.45	0.45	0.45	0.50	0.50	0.50
	5	0.25	0.35	0.50	0.45	0.45	0.45	0.45	0.45	0.45	0.45	0.50	0.50	0.50	0.50
	6	0.30	0.40	0.50	0.45	0.45	0.45	0.45	0.50	0.50	0.50	0.50	0.50	0.50	0.50
	7	0.35	0.40	0.55	0.45	0.45	0.45	0.50	0.50	0.50	0.50	0.50	0.50	0.50	0.50
	8	0.35	0.45	0.55	0.45	0.50	0.50	0.50	0.50	0.50	0.50	0.50	0.50	0.50	0.50
	…	0.45	0.45	0.55	0.45	0.50	0.50	0.50	0.50	0.50	0.50	0.50	0.50	0.50	0.50
	4	0.55	0.50	0.50	0.50	0.50	0.50	0.50	0.50	0.50	0.50	0.50	0.50	0.50	0.50
	3	0.65	0.55	0.50	0.50	0.50	0.50	0.50	0.50	0.50	0.50	0.50	0.50	0.50	0.50
	2	0.70	0.70	0.60	0.55	0.55	0.55	0.55	0.50	0.50	0.50	0.50	0.50	0.50	0.50
	↑1	1.35	1.05	0.90	0.80	0.75	0.70	0.70	0.70	0.65	0.65	0.60	0.55	0.55	0.55

Modified value y_1 of the ratio of linear stiffness of upper beams of that of lower beams on y_0 **Appendix table 3-3**

I \ K	0.1	0.2	0.3	0.4	0.5	0.6	0.7	0.8	0.9	1.0	2.0	3.0	4.0	5.0
0.4	0.55	0.40	0.30	0.25	0.20	0.20	0.20	0.15	0.15	0.15	0.05	0.05	0.05	0.05
0.5	0.45	0.30	0.20	0.20	0.15	0.15	0.15	0.10	0.10	0.10	0.05	0.05	0.05	0.05
0.6	0.30	0.20	0.15	0.15	0.10	0.10	0.10	0.10	0.05	0.05	0.05	0.05	0	0
0.7	0.20	0.15	0.10	0.10	0.10	0.10	0.05	0.05	0.05	0.05	0.05	0	0	0
0.8	0.15	0.10	0.05	0.05	0.05	0.05	0.05	0.05	0.05	0	0	0	0	0
0.9	0.05	0.05	0.05	0.05	0	0	0	0	0	0	0	0	0	0

Note:

i_1 i_2 i i_3 i_4

$I=\dfrac{i_1+i_2}{i_3+i_4}$ when $i_1+i_2>i_3+i_4$, $I=\dfrac{i_3+i_4}{i_1+i_2}$, and add negative sign "–" for y_1

$$K=\frac{i_1+i_2+i_3+i_4}{2i}$$

Modified values y_2 and y_3 of upper floor height changes and lower floor height changes on y_0 **Appendix table 3-4**

a_2	a_3 \ K	0.1	0.2	0.3	0.4	0.5	0.6	0.7	0.8	0.9	1.0	2.0	3.0	4.0	5.0
2.0		0.25	0.15	0.15	0.10	0.10	0.10	0.10	0.10	0.05	0.05	0.05	0.05	0.0	0.0
1.8		0.20	0.15	0.10	0.10	0.10	0.05	0.05	0.05	0.05	0.05	0.05	0.0	0.0	0.0
1.6	0.4	0.15	0.10	0.10	0.05	0.05	0.05	0.05	0.05	0.05	0.05	0.0	0.0	0.0	0.0
1.4	0.6	0.10	0.05	0.05	0.05	0.05	0.05	0.05	0.05	0.05	0.0	0.0	0.0	0.0	0.0
1.2	0.8	0.05	0.05	0.05	0.0	0.0	0.0	0.0	0.0	0.0	0.0	0.0	0.0	0.0	0.0
1.0	1.0	0.0	0.0	0.0	0.0	0.0	0.0	0.0	0.0	0.0	0.0	0.0	0.0	0.0	0.0
0.8	1.2	–0.05	–0.05	–0.05	0.0	0.0	0.0	0.0	0.0	0.0	0.0	0.0	0.0	0.0	0.0
0.6	1.4	–0.10	–0.05	–0.05	–0.05	–0.05	–0.05	–0.05	–0.05	0.05	0.0	0.0	0.0	0.0	0.0
0.4	1.6	–0.15	–0.10	–0.10	–0.05	–0.05	–0.05	–0.05	–0.05	–0.05	–0.05	0.0	0.0	0.0	0.0
	1.8	–0.20	–0.15	–0.10	–0.10	–0.10	–0.05	–0.05	–0.05	–0.05	–0.05	–0.05	0.0	0.0	0.0
	2.0	–0.25	–0.15	–0.15	–0.10	–0.10	–0.10	–0.10	–0.10	–0.05	–0.05	–0.05	–0.05	0.0	0.0

Note:

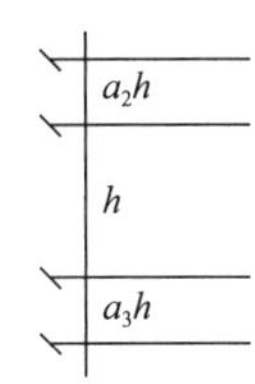

y_2—— it is obtained by K and a_2. It is positive when the upper floor height is higher.

y_3—— it is obtained by K and a_3.

Reference

[1] Ministry of Housing and Urban-Rural Development of the People's Republic of China. Code for Design of Concrete Structures (GB 50010—2010) [S]. Beijing: China Architecture and Building Press, 2019.

[2] Ministry of Housing and Urban-Rural Development of the People's Republic of China. Code for Design of Masonry Structures (GB 50003—2011) [S]. Beijing: China Planning Press, 2012.

[3] Lourenco P B. Masonry structures, overview [J]. Encyclopedia of Earthquake Engineering, 2014, DOI 10.1007/978-3-642-36197-5-111-1.

[4] Bill Mosley, John Bungey, Ray Hulse. Reinforced Concrete Design to Eurocode 2 [S], 7th edition. Palgrave Macmillan Publishing, 2012.

[5] John Morton. Design of masonry structures to Eurocode 6 [S]. ICE Publishing, 2012.

[6] McKenzie W M C. Design of structural masonry [M]. Palgrave Publishing, 2001.